Handbook of Textile Processors Series

The Substrates — Fibres, Yarn and Fabrics

Handbook of Textile Processors Series

The Substrates — Fibres, Yarn and Fabrics

Mathews Kolanjikombil

WOODHEAD PUBLISHING INDIA PVT LTD

New Delhi

Published by Woodhead Publishing India Pvt. Ltd.
Woodhead Publishing India Pvt. Ltd.,
303, Vardaan House, 7/28, Ansari Road,
Daryaganj, New Delhi - 110002, India
www.woodheadpublishingindia.com

First published 2018, Woodhead Publishing India Pvt. Ltd.
© Woodhead Publishing India Pvt. Ltd., 2018
Reprint, 2020

Woodhead Publishing India Pvt. Ltd. ISBN: 978-93-85059-37-7
Woodhead Publishing India Pvt. Ltd. e-ISBN: 978-93-85059-90-2

Typeset by Allen Smalley, Chennai

Printed and bound in India by Replika Press Pvt. Ltd.

Dedicated to my wife

Contents

Preface

It was my long time intention to write handbooks for shop floor textile processor. During my career as a textile processor in day to day work, I had come across on many occasions, the need for ready reference books for immediate help, rather than going through highly technical books which is more theoretical and not of much help practically on the production floor. A shop floor technician needs facts and figures, recipes as guideline, precautionary measures to be taken during the process, etc., for his immediate use and help in achieving results and targets in short time. All these information given in short, may be without much detailed theory as his time on the shop floor is limited.

I have been collecting information and recipes throughout my career, most of it practically tried, and I could put it on paper after my retirement from active involvement in production work. After roughly making manuscript, I came to the conclusion that it may not be able to fix it in one book. Hence, with the consultation with the publisher, I have decided to publish it as a series of 10 books – The Substrate – Fibre, Yarn and Fabrics, Preparation of the substrate, Fabric Dyeing Vols. I and II, Yarn Dyeing, Printing, Finishing, Laboratory Testing Methods, Tables and Useful Information and Important Machineries.

The book is a compilation of information.

The present book is the first in this series, which is about the different substrates used in the day to day processing. This book gives a general description of the substrate which includes the chemical structure, manufacturing methods, fundamental properties, physical and chemical characteristics which are very important in designing processing sequences, recipes, processing parameters and their major uses.

Even though the book is written with production personnel in mind, students of textile chemistry and engineering can find this book useful in their academics.

Any suggestions for improvement of the book, including any inclusions or omissions to be done in the future editions, are welcome.

K. Mathews

1.1   General

The history of textile fibre is as old as civilization, as we get reference of the usage of natural fibres thousand of years ago. During ancient times, the natural fibres like cotton, silk, wool and flax were used.

Later due to improvement in science and technology, synthetic fibres have been developed, which leads to engineer fibres as per requirements of the end use.

A fibre or textile fibre can be defined as the unit of mass which is capable of being spun into a yarn or made into a fabric by bonding or interlacing in a variety of methods including weaving, knitting, braiding, felting, twisting or webbing and which is the basic structural element for textile products. It is the smallest textile component, which may be man-made or natural. They have length at least 100 times to that of their diameter or width.

1.1.1   Classification of fibres

Fibres can be classified as follows:

1. Natural fibres

   (a) Animal fibres

   (b) Vegetable fibres

   (c) Mineral fibres

2. Man-made fibres

   (a) Organic fibres: regenerated fibres and synthetic fibres

   (b) Inorganic fibres

Fibres can also be classified based on type (natural and manufactured), length (short staple, long staple and continuous), size (ultrafine, fine, regular and coarse), etc.

Figure 1.1 summarizes the classification and examples.

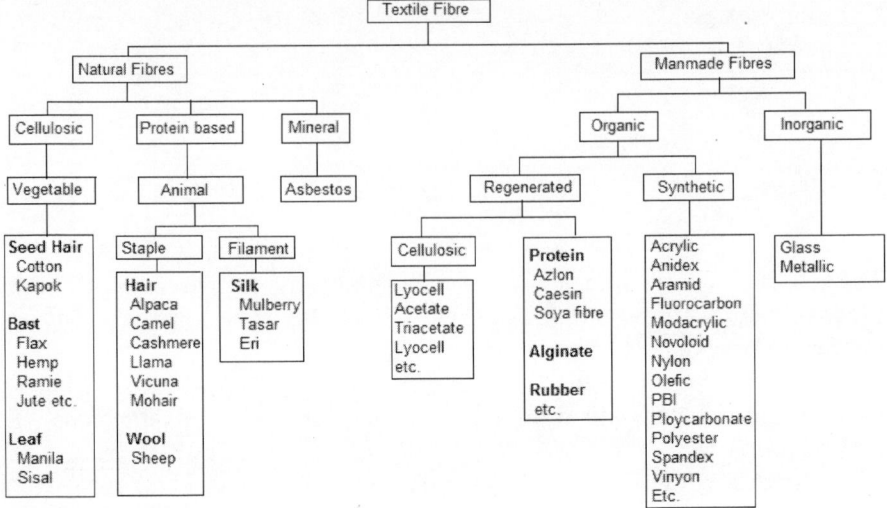

Note: PBI, polybenzimidazole.

Figure 1.1. Classification and examples of fibres

1.2   Natural fibres

Any hair like raw material directly obtainable from an animal, vegetable or mineral source that can be convertible after spinning into yarns and then into fabric. Natural fibres divided into three categories based on the source of material:

1. Plant
2. Animal
3. Mineral

1.2.1   Vegetable fibres

They can be further classified as:

(a) Seed fibre occurring on the seed, e.g., cotton, milk weed, kapok, cattail.

(b) Bast fibres (phloem fibre), e.g., flax, ramie, hemp, jute, sunn, kenaf, urena.

(c) Tendon fibre from stem or leaves, e.g., abaca, pineapple, manila hemp, sisal hemp, palm, New Zealand flax, yucca, palma istle, etc.

(d) Fibre occurring around the trunk, e.g., hemp palm.

(e) Fibre of fruit/nut shells (coconut fibre – Coir).

Cotton and linen are the most important among them (Table 1.1).

Table 1.1. Vegetable fibres of commercial interest

Commercial name	Source	Botanical name	Grown area
Bast or soft fibres			
China jute	Abutilon	*Abutilon theophrasti*	China
Flax		*Linum usitatissimum*	North and south temperate zones
Hemp		*Cannabis sativa*	All temperate zones
Jute		*Corchorus capsularis*	India
Kenaf		*Hibiscus cannabinus*	India, Iran, CIS, South America
Ramie		*Boehmeria nivea*	China, Japan, United States
Roselle		*Hibiscus sabdariffa*	Brazil, Indonesia (Java)
Sunn		*Crotalaria juncea*	India
Urena		*Urena lobata*	Zaire, Brazil
Abaca		*Musa textilis*	Borneo, Philippines, Sumatra
Cantala		*Agave cantala*	Philippines, Indonesia
Caroa		*Neoglaziovia variegata*	Brazil
Henequen		*Agave fourcroydes*	Australia, Cuba, Mexico
Istle		*Agave* (various species)	Mexico
Mauritius		*Furcraea gigantea*	Brazil, Mauritius, Venezuela, Tropics
Phormium		*Phormium tenax*	Argentina, Chile, New Zealand, India, West indies
Pineapple	Pina	*Ananas comasus*	Hawaii, Philippines, Indonesia, India, West Indies
Sansevieria		*Sansevieria* (entire genus)	Africa, Asia, South America
Sisal		*Agave sisalana*	Haiti, Java, Mexico, South America

Commercial name	Source	Botanical name	Grown area
Seed hair fibres			
Coir	Coconut husk	*Cocos nucifera*	Tropics, India, Mexico
Kapok		*Ceiba pentandra*	Tropics
Cotton		*Gossypium* sp. Cotton.	United States, Asia, Africa
Milkweed		*Chorisia* sp.	North America
Other fibres			
Broom root	Roots	*Muhlenbergia macroura*	Mexico
Broom corn	Flower head	*Sorghum vulgare technicum*	United States
Crin vegetal	Palm leaf	*Chamaerops humilis*	North Africa
Palmyra palm	Palm leaf stem	*Borassus flabellifer*	India
Pissava	Palm leaf base	*Attalea funifera*	Brazil
Raffia	Palm leaf	*Raphia raffia*	East Africa

1.2.2   Fibres of animal origin (natural protein fibres)

1. Animal hairs, e.g., wool (sheep), specialty hairs like alpaca, camel, cashmere, guanaco, llama, mohair (angora goat), vicuna, fur fibres like mink, muskrat, angora rabbit, etc.
2. Animal secretions, e.g., silk fibre like cultivated, dupioni, tussah (wild) and spider silk.

1.3   Man-made fibres

1.3.1   Regenerated natural fibres

1. Rayon: (a) Cuprammonium Fibres, (b) Viscose Rayon, like regular and high tenacity, high wet modulus, hollow fibres.
2. Acetate: (a) Secondary Acetate, (b) Tri-Acetate.

3. Protein: (a) Casein, (b) Zein, (c) Peanut, (d) Soyabean.

4. Miscellaneous: (a) Alginate, (b) Rubber.

1.3.2   Synthetic fibres

1. Condensate polymer fibres: (a) Nylon 6.6, Nylon 6, Nylon Type 11, 6, 10, Nromatic Type (Quina), Bicomponent Nylon, (b) Aramid like Kevlar and Nomex, (c) Polyester.

2. Addition polymer: (a) Anidex, (b) Acrylics, (c) Modacrylic, (d) Novoloid, (e) Nytril, (f) Olefin Fibres like Polyethylene and Polypropylene, (g) Saran, (h) Vinyl, (i) Vinyon.

3. Elastomers: (a) Spandex, (b) Rubber, (c) Lastrile.

4. Man-made mineral: (a) Glass, (b) Metallic.

5. Others: (a) Alginate, (b) Inorganic like Avceram, Fibrefrax, Thornel, (c) Organic like Polybenzimidazole, Teflon.

2
Cotton

2.1   General

Cotton fibres are seed hairs from plants of the order Malvales, family Malvaceae, tribe Gossypieae and genus *Gossypium*. Cotton is a soft fibre that grows around the seeds of the cotton plant and is the purest form of cellulose available in nature. After flowering, an elongated seed pod or boll is formed on the cotton plant in which the cotton fibres grow and each fibre is a single elongated cell that is flat twisted and ribbon like with a wide inner hollow (lumen).

A cotton seed pod may contain about 30 seeds and each seed contains around 2000–7000 seed hairs (fibres). The function of the hair/fibre is the disbursement of the seed, by helping it to float in the air. The outer surface is covered with a protective wax-like coating which gives fibre an adhesive quality, which is about 10% of the weight of the raw fibre consists of waxes, protein, pectate and minerals, otherwise the hair contains majority cellulose.

The following species are grown commercially:

- *Gossypium arboreum* and *Gossypium herbaceum*, known collectively as 'Desi' cottons, tree cotton, are native to India and Pakistan. These rough cottons are the shortest staple cottons cultivated (ranging from 3/8 to 3/4 in. [9.5–19 mm]) and are coarse (micronaire value >6.0) compared with the American Upland varieties. *G. herbaceum*, the original cotton of India, averages 1.2–1.8 m (3.9–5.9 ft) in height. The fibre is greyish white and grows from a seed encased in grey fuzz fibres (Fig. 2.1).

- *Gossypium barbadense*, originally of early South American origin, has the longest staple length and is commonly referred to as 'extra long staple' cotton. It is known as American Pima, Creole, Egyptian, or sea island cotton, native to tropical South America. The fibre is long and fine with a staple length usually greater than 1 3/8 in. (35 mm) and a micronaire value of below 4.0.

- *Gossypium hirsutum* is a upland cotton, native to Central America, Mexico, the Caribbean and Southern Florida, most commonly grown species in the world. The lengths, or staple lengths, of the upland cotton fibre vary from 7/8 to 1 1/2 in. (22–36 mm), and the micronaire value (an indicator of fibre fineness and maturity but not necessarily a reliable measure of either) ranges from 3.8 to 5.0. *G. hirsutum* is a shrubby plant that reaches a maximum height of 1.8 m (5.9 ft) and is used in apparel, home furnishings and industrial products.

- *Gossypium peruvianum* and *Gossypium purpurascade* are not of much commercial importance (Fig. 2.2).

The cotton plant is a tree or a shrub that grows naturally as a perennial, but for commercial purposes, it is grown as an annual crop. Botanically, cotton bolls are fruits.

Cotton is a warm-weather plant, cultivated in both hemispheres, mostly in North and South America, Asia, Africa and India (in tropical latitudes). Mostly it is cultivated in the northern hemisphere. It is primarily grown between 37° N and 32° S but can be grown as far north as 43° N latitude in Central Asia and 45° N in mainland China. Cotton is cultivated in North and South America, the Middle East, Africa, India, China and Australia.

Recently cotton production is being shifted to more environmentally friendly techniques such as organically produced cotton. Cotton grown without the use of any synthetically compounded chemicals (i.e., pesticides, fertilizers, defoliants, etc.) is considered as 'organic' cotton. It is produced under a system of production and processing that seeks to maintain soil fertility and the ecological environment of the crop. To be sold as organic, it must be certified. Certified organic cotton was introduced in 1989–1990 and over 20 countries have tried to produce organic cotton. Since 2001, Turkey has been the largest producer of organic cotton. The top organic cotton growing countries are: Turkey, India, China, Syria, Peru, the United States, Uganda, Tanzania, Israel and Pakistan.

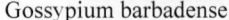

Gossypium barbadense   Gossypium herbaceum

Figure 2.1. Cotton of commercial importance

Cotton fibre can be classified in to three categories depending on the fibre length

1. *Long staple* – Plants with a staple length of between 30 and 65 mm. Staple cottons, well-known types such as Egyptian and Sea Island, are included in this group. Long staple cotton represents 3–5% of the world's production and are used for delicate fabrics with specific weight of less than 100 g/m² in the production of high-quality shirts and blouses, the best quality bed linen, underwear, etc.

2. *Medium staple* – Fibres with a staple length of between 20 and 30 mm. These medium length fibres are the most common form of cotton. Medium staple cotton (about 85% of the world production and are used in for the production of bed linen, table cloths, good quality denim (jeans) cloth as well as underwear. It is grown in Central *Asia*

Gossypium hirsutum

Gossypium arboreum

Gossypium peruvianium

Gossypium purpurascens

Figure 2.2. Common cotton varieties

(Uzbekistan, Tajikistan, Kazakhstan and Turkmenistan), West Africa (Chad, Mali, Ivory Coast and Burkina Faso), Europe (Greece and Spain), the Middle East (Turkey, Syria), the United States, Brazil and Pakistan.

3. *Short staple* – Fibres with a staple length of less than 20 mm from coarse, lower grade cotton, which is grown in Central Asia (Uzbekistan, Tajikistan, Kazakhstan and Turkmenistan), the United States and India. Short staple cotton is used especially for the production thick fabrics with specific weight greater than 250 g/m², for example, for denim, drill, flannel for work clothes, upholstery and carpets.

It has 8% moisture regain. (Raw conditioned 8.5%, saturation 20–25%, mercerized cotton 8.5–10.3%.) The fibre contains 90% cellulose and it is arranged in a way that gives cotton unique properties of strength, durability and absorbency. It is fresh, crisp, comfortable, absorbent and flexible, and it has no pilling problems and has good resistance to alkalis.

Harvesting: Harvesting is one of the final and most important steps in the production of a cotton crop, as the crop must be harvested before the inclement weather can damage the quality and reduce the yield (Table 2.1). Because of economic factors, virtually the entire crop (>99%) in the United States and Australia is harvested mechanically (Fig. 2.3). In rest of the world (~75%), hand harvesting of cotton, one boll at a time, is still quite prevalent, particularly in the less developed countries and in countries where the labour is cheaper.

Table 2.1. Harvesting time of different geographical regions and countries of the world

Geographical region	Country	Harvesting season
North and Central America	USA	July–January
	Mexico	June–January
	Guatemala	November–March
South America	Brazil	August–January or February–May
	Argentina	February–June
	Paraguay	February–June
	Columbia	July–September or December–March
	Peru	February–October

Geographical region	Country	Harvesting season
	Venezuela	February–May
Europe	Greece	September–November
	Spain	September–November
	Uzbekistan	September–November
	China	September–November
	India	July–January or December–May
Asia and Oceania	Pakistan	September–February
	Turkey	September–December
	Australia	April–June
	Iran	October–December
	Syria	September–November
	Egypt	September–October
	Sudan	January–April or September–May
Africa	S. Africa	April–May
	Ivory Coast	October–January
	Tanzania	May–July
	Nigeria	December–February

Picking: The process of plucking cotton bolls from cotton plant is called picking.

Hand harvesting   Mechanical harvesting

Figure 2.3. Types of harvesting

Figure 2.4. Spotting

Spotting: After seed cotton is collected immature bolls are discarded (Fig. 2.4).

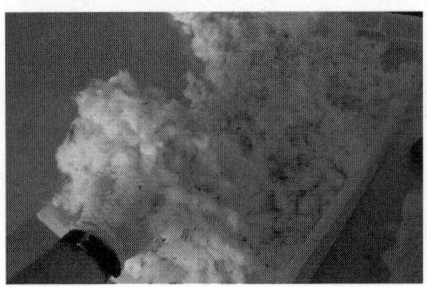 

Figure 2.5. Ginning

Ginning: The process of separating lint from the seed is called ginning. It is performed most often by mechanical means (Fig. 2.5). Ginning operations, which are considered a part of the harvest, are normally considered to include conditioning (to adjust moisture content), seed–fibre separation, cleaning (to remove plant trash) and packaging (Fig. 2.6).

Baling: After ginning, staple fibres are compacted by mechanical means. This is called Baling.

Spinning: It is the process of making yarns from unbundled fibres.

The process of spinning consists of the following steps:

- Upon arrival at the spinning mill, cotton bales are sampled according to lint quality and origin to ensure yarn homogeneity.

- They are then opened to make the lint fluffy by passage though bale openers.

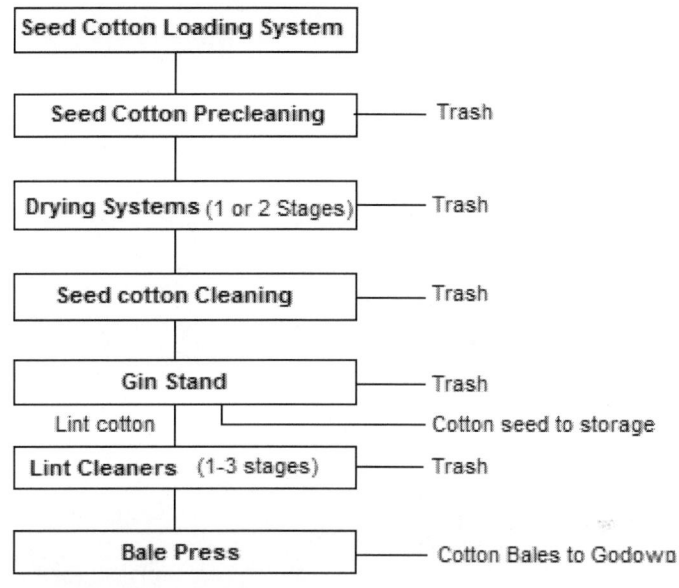

```
┌─────────────────────────────┐
│  Seed Cotton Loading System  │
└─────────────────────────────┘
                │
┌─────────────────────────────┐
│   Seed Cotton Precleaning    │──── Trash
└─────────────────────────────┘
                │
┌─────────────────────────────┐
│ Drying Systems (1 or 2 Stages)│──── Trash
└─────────────────────────────┘
                │
┌─────────────────────────────┐
│    Seed cotton Cleaning      │──── Trash
└─────────────────────────────┘
                │
┌─────────────────────────────┐
│         Gin Stand            │──── Trash
└─────────────────────────────┘
  Lint cotton                  ──── Cotton seed to storage
┌─────────────────────────────┐
│ Lint Cleaners  (1-3 stages)  │──── Trash
└─────────────────────────────┘
                │
┌─────────────────────────────┐
│         Bale Press           │──── Cotton Bales to Godown
└─────────────────────────────┘
```

Flow Chart of Ginning

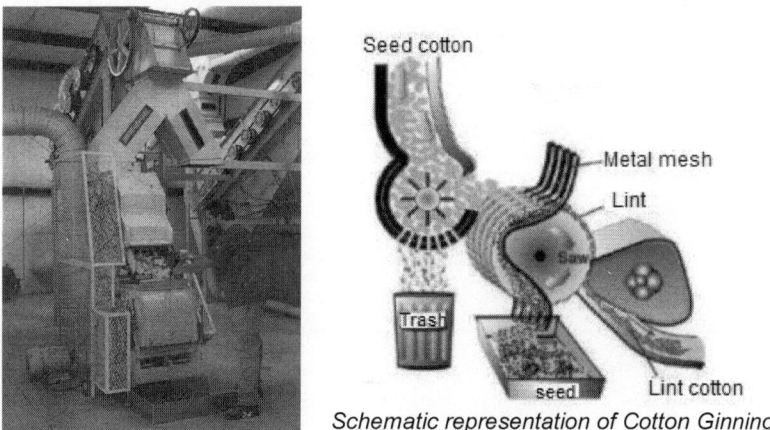

Schematic representation of Cotton Ginning

Figure 2.6. Ginning - the complete process

- *Cleaning*: Bale fibres are usually fed to air-jet (vortex) cleaners to remove extraneous matter from cotton lint.
- Carding separates fibres from each other, straightens fibres, aligns and condenses them into a single continuous strand and removes impurities (Fig. 2.7).

Carding

Combing

Figure 2.7. The process of carding and combing

- A sliver of approximately 1-m width is then obtained which is fed to several rubber rollers rotating at increasingly higher speed.
- *Bleaching*: It done using either hypochlorite or peroxide.
- Several slivers are drawn and twisted together to form the final yarn.

Cotton has poor wrinkle resistance, shrinkage, poor acid resistance, less abrasion resistance, susceptible to damage by moths and mildew, needs lots of maintenance and stains are difficult to remove. Its fibre length ranges from 1/2 to 2 in. It has 10% increase in strength when wet. It has a density of 1.54–1.56 g/cm^3. It has a twisted tube shape.

2.2   Structure of cotton fibre

As mentioned earlier, cotton fibre consists of approximately 95% cellulose. The structure of cotton cellulose is a linear polymer of β-D-glucopyranose. The noncellulosic constituents of the fibre are located principally in the cuticle, in the primary cell wall and in the lumen. Cotton fibres that have a high ratio of surface area to linear density generally exhibit a relatively higher noncellulosic content (Table 2.2).

Table 2.2. Typical composition of cotton fibres

Constituent	Composition (% by dry weight)	
	Typical %	Average %
Cellulose	95	88.0–96.0
Protein (% Nx6.25)	1.3	1.1–1.9
Pectic substances	0.9	0.7–1.2
Ash	1.2	0.7–1.6
Wax	0.6	0.4–1.0
Total sugars	0.3	0.1–1.0
Organic acids	0.8	0.5–1.0
Pigment trace	Trace	Trace
Others	1.4	1.1–1.5

The noncellulosic constituents include proteins, amino acids, other nitrogen containing compounds, wax, pectic substances, organic acids, sugars, inorganic salts and a very small amount of pigments. After cellulosic component, the major constituent is nitrogenous compounds which are normally expressed as protein. Most of the nitrogenous material occurs in the lumen of the fibre, most likely as protoplasmic residue and a small percentage from the primary wall. Underlying the waxy cuticle is the primary cell wall, which is composed of two distinct layers. The outermost layer is comprised primarily of pectin substances (usually designated as pectin) in the form of free pectic acid (linear polymer of (1→4)-D-galacturonic acid). The innermost layer is comprised of hemicelluloses, primarily in the form of xyloglucan and cellulose. Soluble sugars (about 0.1–1.0% of fibre dry weight) found on cotton originate from two sources: metabolic residues (plant sugars) located within the dried lumen and the outer fibre surface and insect sugars (insect 'honeydew' excretion) found on the outer surface of the fibre. Cotton wax (about

0.4–1.0% of fibre dry weight) comprises the cuticle on the outer surface of the fibre. The natural wax content serves as a protective barrier both to water penetration and to microbial degradation of the underlying polysaccharides. The wax serves as a lubricant that is essential for proper spinning of cotton fibre into yarn. The quantity of wax increases with the surface area of the cotton, and the finer cottons tend to have a larger percentage of wax. The wax is a mixture of high molecular weight, primarily long-chain saturated fatty acids and alcohols (with even numbers of carbon atoms, C_{28} to C_{34}), resins, saturated and unsaturated hydrocarbons, sterols and sterol glucosides, including montanyl triacontanoate (10–15%), montanol (25%), 1-triacontanol (18%) and β-sitosterol (10%) (Fig. 2.8).

Organic acids (0.5–1.0% of fibre dry weight) in the raw fibre, exclusive of pectic acid, are primarily 1-malic acid (up to 0.5%) and citric acid (up to 0.07%), are present in the lumen as metabolic residues and are removed during the normal scouring and bleaching due to their high water solubility. The inorganic salts (phosphates, carbonates and oxides) and salts of organic acids present in the raw fibre are reported as percent ash (about 1.2% of fibre dry weight) and expressed as the oxides of the elements present (excluding chlorine, which is expressed as such). The amounts of these cations present on the

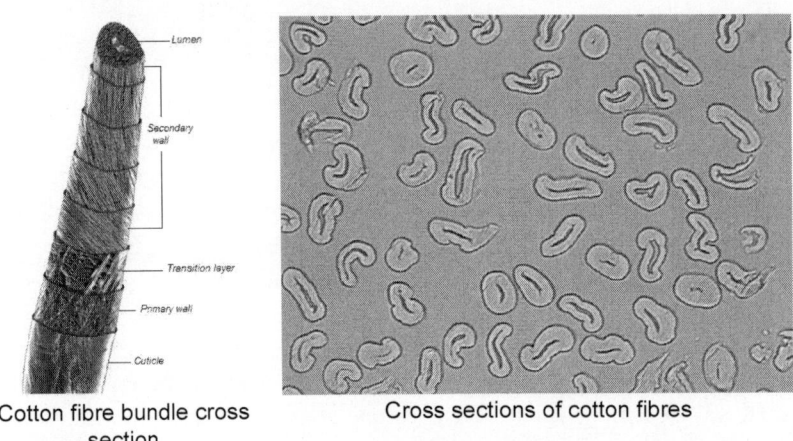

Cotton fibre bundle cross section

Cross sections of cotton fibres

Figure 2.8. Structure of cotton fibre

cotton fibre vary considerably because of maturity differences, environmental factors (e.g., rainfall) and agricultural practices, as well as the field and the handling procedures that affect deposition of material (plant parts and soil) on the fibre. During the production of cotton, the plant absorbs potassium and other metals as normal nutrients.

2.2.1   Metal content in cotton

Table 2.3. Metal content in cotton

Metal	Content in ppm
Potassium	2000–6500
Magnesium	400–1200
Calcium	400–1200
Sodium	100–300
Iron	30–90
Manganese	1–10
Copper	1–10
Zinc	1–10
Lead	n.d.
Cadmium	n.d.
Arsenic	Trace (<1)b

Metals are incorporated from the soil into plants as natural constituents. In addition to metals absorbed by plant tissue, soil and plant parts may be deposited directly onto the lint especially during harvesting. Ca, P, S, K and Fe are plant part elements and Mg, Al, Si, Fe, Cr, Se, Hg, Ni, Cu, K and Ca are soil elements. The metals are removed for the most part by proper scouring and bleaching processes that are used to prepare the fibre and fabric for dyeing and finishing (Table 2.3).

The mature cotton fibre has a noncellulosic covering called the cuticle that contains waxes, pectins and proteins. This cuticle is intermingled with the primary wall. The structure of the primary wall, which changes substantially during fibre development, is not well understood. It is responsible for maintaining the integrity of the fibre and may account for much of the strength of the cotton fibre. Most of the cuticle is dissolved and removed by industrial scouring of fabric, but it has important functions during spinning of the fibres into yarn and during weaving the yarn into fabric. One reason for scouring is that the waxes block access to the interior of the cotton fibre for molecules such as dyes.

Understanding of fibre structure requires knowledge of the structure of the cellulose molecule, the structure and perfection of its crystalline arrays, the packing of these arrays (elementary fibrils) into microfibrils, and then the arrangement of these microfibrils in the primary and secondary cell walls. The noncrystalline material is also important. Once the cellular fluid dries after the cotton boll opens, which is called lumen, it collapses leaving fibre cross sections with irregular, kidney bean shapes, which contain biological material. That material constitutes a small percentage of the total dry weight.

2.2.2   Chemical structure of cotton

As explained earlier cellulose constitutes around 95% of cotton fibre. Hydrolysis and oxidation studies point to the fact that the cellulose in cotton fibre is a 1-4-linked linear polymer of β-D-glucopyranose. These monomers are linked together by elimination of one molecule of water between the hydroxyl groups attached to the number 1 carbon atom of one glucose molecule and the number 4 carbon atom of another (Fig. 2.9).

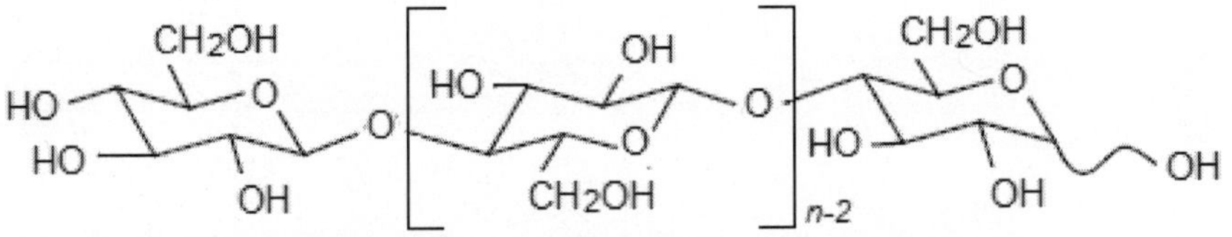

Figure 2.9. Chemical structure of cotton

The molecular cellulose chains have varying lengths. Measurements of the chain length require that cotton be in solution. Solvents for this purpose include cuprammonium hydroxide solution, phosphoric acid, nitric acid, quaternary ammonium bases, cadmium ethylenediamine hydroxide, cupriethylenediamine hydroxide, N,N-dimethylacetamide and lithium chloride (DMAC LiCl), and 1,3-dimethyl-2-imidazolidinone (DMI) and lithium chloride. DMAC LiCl, when used in conjunction with gel permeation chromatography, provides both the weight (M_w) and number average (M_n) molecular weight of cellulose in a nondegrading solvent without derivatization. In undegraded cotton fibre, the molecular chain length (degree of polymerization [DP]) may be higher than 20,000 monomeric D-glucopyranosyl units. This corresponds to a molecular weight of 3,240,000 Da. Each cellulose chain has a reducing end (O1-H) and a nonreducing end (O4-H). Reducing ends are especially reactive, but they are present in such small amounts in cellulose that they are often ignored.

Fibre strength cotton is influenced by the structural organization of the cellulose chains. Molecular weight of a polymer is one of the most important influences on its physical properties, and the determination of molecular weight distribution is critical for predicting performance of a polymer. For polymers, higher molecular weight and narrower molecular weight distribution are positively correlated with increased strength. Unfortunately, polymer characterization techniques generally depend upon dissolving of the polymers. Attempts to identify the true molecular weight of native cellulose have been limited mostly because cellulose is difficult to dissolve. During development, the composition of the cell wall of the cotton fibre changes continuously, ending with the cessation of the fibre's metabolic activity (Fig. 2.10).

2.2.3   Molecular physical structure

The physical and to some extent chemical characteristics of cotton fibre are influenced by the molecular and supramolecular arrangement of the cellulose molecule chains. They associate each other by forming intermolecular hydrogen bonds and hydrophobic bonds. They coalesce to form microfibrils also called crystallites. There are several different forms or polymorphs (cellulose I–IV and X with recent subclasses Iα and Iβ), depending on the source and treatment. Native cotton is cellulose I. In cotton, the microfibrils can organize into macrofibrils 60–300 nm wide. The macrofibrils are organized into fibres. Cotton fibres have a complex, reversing, helical arrangement of macrofibrils. There are both different unit cells and different packing arrangements in the unit cell. Mercerization with caustic soda or ammonia can convert cellulose I partially to cellulose II and III.

The cotton fibre is a porous, hydrophilic material that accounts for the comfort of cotton clothing. The moisture absorbed from the atmosphere and held under ambient conditions is expressed either as moisture content (amount of moisture as the percentage over the oven-dried weight) or more commonly as moisture regain (amount of moisture as a percentage of the oven-dry sample). Under ordinary atmospheric conditions, moisture regain is 7–11%. Upon immersion in liquid water, the cotton fibre swells and its internal pores fill with water. Pure cotton holds a substantial percentage of its dry weight in water under conditions of centrifugation. The values are ~30% for water of imbibition and 50% for the water retention. Centrifugation conditions are less severe in the latter case. Pores accessible to water molecules are not necessarily accessible to chemical agents. Many uses of cotton, e.g., easy care fabric, depend on chemical modification to impart the desired properties.

Preparation processes of cotton materials slightly increase the accessibility to internal volume, liquid ammonia treatment of the scoured–bleached cotton decreases it slightly, caustic mercerization substantially enhances, and crosslinking to impart durable press properties reduces the accessibility to internal pore volume substantially.

2.2.4   Chemical reactions

Even though 2-OH, 3-OH and 6-OH groups of the cellulose structure (see Fig. 2.9), but the regular occurrences of intermolecular and intramolecular hydrogen bonds in the crystalline regions of cotton cellulose render the involved hydroxyl groups unavailable to chemical agents under mild reaction conditions. Thus chemical agents that have access to the interior pores of the cotton fibre thus find potential reactive sites unavailable for reaction. The order of decreasing availability of hydroxyl groups in cotton is 2-OH > 6-OH > 3-OH. These are the main positions where reactions of dyes, crosslinking, reactive dyes (esterification), etc., take place.

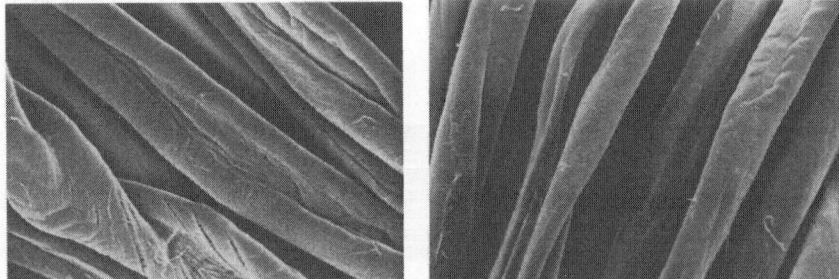

Figure 2.10. Scanning electron microscopic view of cotton unmercerized (LHS) and mercerized (RHS) (×1000)

Cellulose is readily attacked by oxidizing agents, such as hypochlorites, chlorous, chloric and perchloric acids, peroxides, dichromates, permanganates, periodic acid, periodate salts and nitrogen tetroxide.

2.2.5   Action of heat

Heat can dehydrate and decompose cotton. Heating cotton cellulose up to 120°C drives off moisture without any loss of strength. Heating to a higher 150°C has been shown to reduce solution viscosity, indicative of lowered molecular weight and tensile strength. Between 200°C and 300°C, volatile products and liquid pyrolyzate, mainly 1,6-anhydro-β-D-glucopyranose, commonly known as levoglucosan, evolve. At 450°C, only char remains.

2.2.6   Cotton fibre and yarn quality correlation

Cotton from various origin will have different characteristics, which may or may not be able to be generalized and is very confusing for a cotton buyer or a yarn manufacturer. Essential characteristics of cotton quality and characteristics of yarn quality are given from detailed experimental investigations. Some of the important conclusions which help to find correlation between yarn quality and cotton fibre quality are given below:

- *Staple length*: If the length of fibre is longer, it can be spun into finer counts of yarn which can fetch higher prices. It also gives stronger yarn.
- *Strength*: Stronger fibres give stronger yarns. Further, processing speeds can be higher so that higher productivity can be achieved with less end breakages.
- *Fibre fineness*: Finer fibres produce finer count of yarn and it also helps to produce stronger yarns.
- *Fibre maturity*: Mature fibres give better evenness of yarn. There will be less end breakages. Better dyes' absorbency is additional benefit.
- *Uniformity ratio*: If the ratio is higher, yarn is more even and there is reduced end breakages.
- *Elongation*: A better value of elongation will help to reduce end breakages in spinning and hence higher productivity with low wastage of raw material.
- *Nonlint content*: Low percentage of trash will reduce the process waste in blow room and cards. There will be less chances of yarn defects.
- *Sugar content*: Higher sugar content will create stickiness of fibre and create processing problem of licking in the machines.
- *Moisture content*: If moisture content is more than standard value of 8.5%, there will be more invisible loss. If moisture is less than 8.5%, then there will be tendency for brittleness of fibre resulting in frequent yarn breakages.
- *Feel*: If the feel of the cotton is smooth, it will be produce more smooth yarn which has potential for weaving better fabric.
- *Class*: Cotton having better grade in classing will produce less process waste and yarn will have better appearance.
- *Grey value*: Reading of calorimeter is higher, which means it can reflect light better and yarn will give better appearance.
- *Yellowness*: When value of yellowness is more, the grade becomes lower and lower grades produce weaker and inferior yarns.

- *Neppiness*: Neppiness may be due to entanglement of fibres in ginning process or immature fibres. Entangled fibres can be sorted out by careful processing but neppiness due to immature fibre will stay on in the end product and cause the level of yarn defects to go higher.

3

Bast Fibres

Bast fibre or skin fibre is a fibre collected from the phloem (the bast surrounding the stem of a certain plant).

The bast fibres have often higher tensile strength than other kinds and are ropes, yarn, paper, composites and burlap. A special property of bast fibres is that the fibre at that point represents a weak point. They are obtained by the process called retting. Bast fibres can be of two types:

Low lignin content – linen or flax (raw and bleached) and ramie

High lignin content – jute, hemp

3.1   Flax or linen

Flax is considered to be the oldest fibre in the Western world and CIS (formerly USSR) grows most of the flax fibre. Linen has been gradually loosing its position as an apparel fabric since 1950s, but the emergence of linen fibre as a component of blends and the importance as a cool fibre has stimulated considerable interest in the fashion world now (Fig. 3.1).

Linen or flax fibres are extracted from the stem or bast of flax plant (*Linum usitatissimum* L.). Fibres held together by gummy substance called as *pectin*. Flax fibre basically composed of cellulose. Flax is difficult to grow because of the soil preparation required before sowing, and the heavy applications of artificial fertilizers required. After a slow initial growth, the plants grow as rapidly as 1 inch/day for 30–40 days. Blossoms then begin to develop and stem growth ceases.

Flax is normally a 3-month crop, although this growing time varies with climatic and other growing conditions. It is attacked by several fungal and viral diseases, usually kept under control by chemical treatments of the seed or by cultivation of resistant varieties. Harvesting is usually carried out by pulling the plant from the ground. Pulling is considered superior to cutting since flax deteriorates at the cut. Yields of fibre per acre vary from 200 to 360 kg. Major flax-producing countries are Belgium, Scotland, France, Russia and Germany. Countrai flax produces the finest and strongest yarn from Belgium. Flax plant is also cultivated for the flax seed (oleaginous flax linseed) and cultivated in counties like Canada, Argentina, the United States, Egypt, India, etc. There are different varieties of flax used for fibre production and oil production.

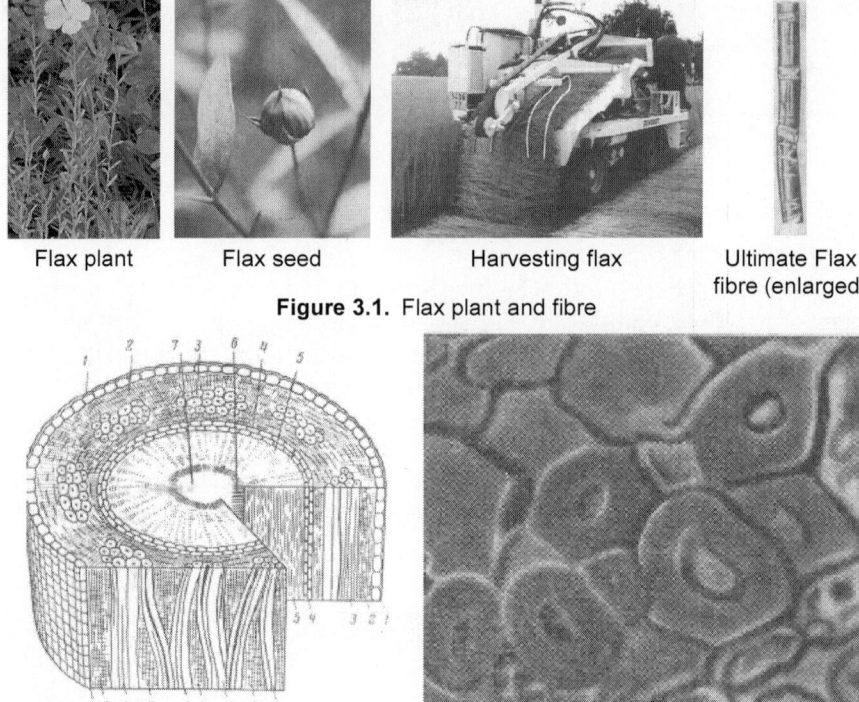

Flax plant   Flax seed   Harvesting flax   Ultimate Flax
                                                                      fibre (enlarged)

Figure 3.1. Flax plant and fibre

Schematic diagram of the cross section   Cross section of flax fibre

Figure 3.2. Flax stem: 1. Epidermis, 2. Parechyma, 3. Bast bundles,
4. Cambium, 5. Woody tissue, 6. Pith, 7. Lumen

Inside the stalk of a flax plant, the fibres carry nutrients and water to the leaf. They are found as individually separated filaments of different lengths, which vary according to the height at which the leaf is situated in the plant. Each filament is composed of a number of single tapered-ended fibre cells evenly joined lengthwise, so that each cell underlays 50% of the preceding one and overlaps 50% of the following one, tightly tied together all along, to form single filaments of the same regular fine thickness, but of different lengths. The fibres are held together by tissues to form the stem of the plant. Like any other bast fibres the main component of the fibres is cellulose, while secondary components are waxes, fats, hemicellulose, lignin and pectin (Figs. 3.2 and 3.3).

During harvesting the flax plant is pulled out from the soil and not cut. Next process is called turning and deseeding which can be done many days after harvesting, for example, after 10 days in case of dew retting, and then bundled or baled.

Like other bast fibres, flax must be separated from the stalks by retting whereby the pectin which binds the fibres together is removed by bacterial action. Water retting, which is essentially bacterial, is practiced in areas such as the Philippines, Taiwan and China. The bacteria take part in this process are *Clostridium* sp., *Clostridium butiricum*, *Granulobacter pectinovorum*, *Clostridium felsineum*, *Clostridium guerfelli* and *Bacillus amylobacter*. Modern times to hasten the bacterial action and decomposition and to stimulate the multiplication of the bacteria, the water heated to 30°C or by adding active bacteria cultures to the water, or even by adding pure enzymes. The process depends very much on the temperature and chemical nature of the water, but takes only 6–8 days under controlled temperature conditions.

Most of the crop grown in Russia and the United States is dew retted, which is predominantly fungal as against bacteria in case of water retting. In this method, the harvested flax straw is left in the field and allowed to remain until the combined action of the moisture from dew and microorganisms makes separation of the fibres possible. The main fungi taking part in the dew retting are *Alternaria alternata*, *Alternaria linicola*, *Cladosporium herbarum*, *Fusarium nivale*, *Cephalosporium* sp. and other, mainly pectinolytic and cellulolytic species. Dew retting has many advantages, even though it is a slow process like, homogeneity of the resulting fibre, preserve the strength of the yarn, fibre separation is easier, even preparation of the fibre helps in further processing like dyeing, etc. Main disadvantages are slow process and difficulties in spreading on the field evenly and properly aligned to get the homogenous retting process which will affect the colour and quality.

After removal of the stalks from the retting medium, thorough drying is necessary to prevent further fermentation. As a result of such retting process, the pectins glueing the bast bundles to the other parts of the stems are disintegrated, and the bundles are easily separated by subsequent treatment in breaking and scutching machines. On the contrary, excessively strong biochemical action may reduce the strength of the lamellae binding the ultimate fibres together and the technical fibre disintegrates into elementary fibres forming the so-called cotton-like flax stock (Fig. 3.4).

The retted and dried fibres are removed from the woody remainder of the stem by the process of scutching, in which the stems are first broken by passage through a series of fluted metal rollers, and the fine pieces of the woody portion of the straw, called shives, are beaten out. About a tenth of the original flax stem is useful fibre.

Cotton-like flax fibre is sometimes used in blends with cotton for manufacturing fabrics. However, it is obtained not by biochemical destruction of valuable technical fibre, which would not be feasible, but from the wastes of the primary treatment of flax (in which a certain part of bast fibres is eliminated together with shive). The bast fibre may be transformed into cotton-like fibre by boiling in alkali solutions. There is also a mechanical process of extracting technical fibre from bast stems. This process is used for obtaining green bast. Chemical analysis of cotton is carried out on natural untreated fibre, while the analysis of flax is usually made on a half-ready product obtained from technical fibre by the wet spinning method (see Table 3.1). The content of lignin in flax fibre can be as high as 5%. It can be seen from the table that the content of cellulose in flax yarn is lower than in cotton, and the amount of impurities is greater.

Table 3.1. Chemical analysis of flax fibre

Yarn thickness tex and (count)	Cellulose	Wax-like substances	Lignin	Proteins	Pectins	Ash	Total	Unidentified impurities
68 (14.5)	73.93	2.39	2.88	2.29	2.04	1.06	84.59	15.41
56 (18)	75.8	2.9	2.7	2.52	2.34	0.86	87.12	12.88
36 (28)	78.82	1.68	2.05	2.1	1.86	0.71	87.22	12.78

The ultimate fibres of flax consist mainly of cellulose. The middle lamellae mainly consist of pectins; some lamellae also contain lignin, which is confirmed in microscopic examination of fibre cross sections by the colouring of lignin by phloroglucinol and hydrochloric acid. Lignin, though in a less amount, is contained in the flax fibre proper, particularly in the lower portions of the stem as has been found by several investigators by means of microchemical reactions and luminescent analysis of fibre and stem cross sections. Flax fibres are multicellular, with each cell having tapering ends and a narrow lumen. The fibres show longitudinal striations and nodes. The ultimate fibres are composed of elementary fibrils (microfibrils), which are spirally arranged.

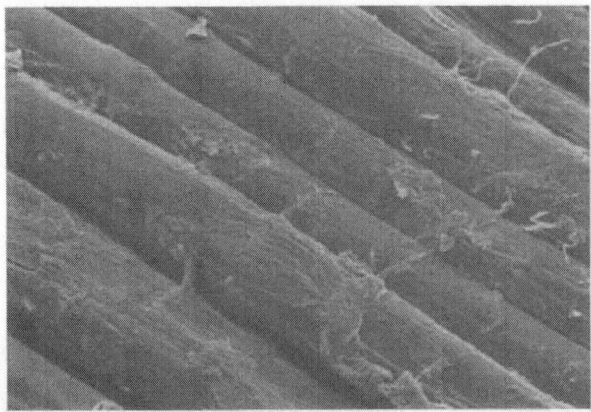

Figure 3.3. Electron microscopic view of the flax fibre (×1000)

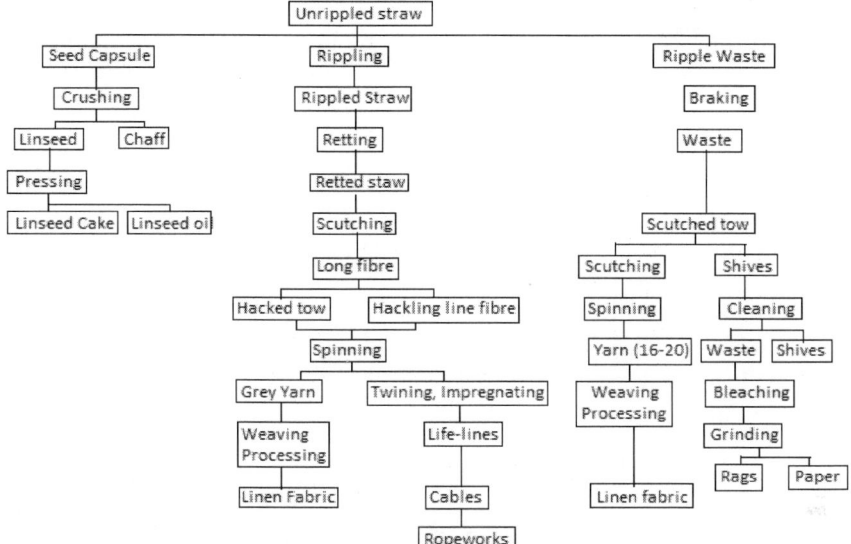

Figure 3.4. Processes involved in flax fibre production

The fibrils are held together by a bonding or gummy substances. The unretted flax contains about 16.7% hemicellulose, 1.8% pectins, 2% lignin and 1.5% fats and waxes. The polymer of flax consists with a degree of polymerization of about 18,000 cellobiose units. Flax is an assembly of ultimates cemented together within the fibres and an assembly of these fibres into bundles. The physical structure of flax ultimate fibres is characterized by displacements or 'breaks' in their surface layers (see Fig. 3.2). These displacements are the result of the multiple deformations to which bast bundles are subjected in processing, beginning with the removal of the flax stem.

These assemblies are prone to discolouration. It is the structure which makes linen feel like linen. Flax has relatively high strength. Disruption of structure by, for example, scouring referred to as cottonization.

A fibre obtained from the stem is of very high quality and is used in making cloth, sails, nets, paper, insulating material, etc. The best quality flax fibre is used for making cloth. It is soft, lustrous and flexible, although not so flexible or elastic as cotton or wool. It is stronger than cotton, rayon or wool, but weaker than ramie. Lower quality fibre is used in manufacturing of towelling, matting, rugs, twines, canvas, bags and for quality papers such as printing currency notes, upholstery tow, insulating material, rugs, twine and paper.

The seed contains 38–40% of a drying oil having wide range of applications. The paint and varnish industries consume about 80% of all the linseed oil produced. The remainder is used in items such as furniture polish, enamels, linoleum, oilcloth, printer's inks, soap making and leather. It is also used as a wood preservative and as a waterproofing for raincoats, slickers and tarpaulins.

3.2   Jute fibre

Jute is one of the cheapest natural fibres and is second only to cotton in amount produced, and variety fibres are composed primarily of the plant materials cellulose and lignin. Jute is a long, soft, shiny vegetable fibre that can be spun into coarse, strong threads.

Jute is an annual herbaceous dicotyledonous plant that grows up to a height of 2.5 m. The stems are about 1–2 cm in diameter with few branches. The colour of the stem, petiole, leaf and pod varies in different forms. Mainly two qualities namely *Corchorus capsularis* and *Corchorus olitorius* are utilized for fibre production on a commercial scale, former one is known as 'white jute' and latter one is known as 'tossa jute'. The two species differ in the quality of fibre they yield. Fibres of *C. olitorius* are frequently softer, stronger and more lustrous than those of *C. capsularis*. Jute fibre is obtained from the bast or phloem layer of the stem. It is thus a ligno-cellulosic fibre that is partially a textile fibre and partially wood. The plant and its fibre length is about 2 m. It is generally used in geo textiles. It has a good resistance to microorganisms and insects. It has low wet strength, low elongation and inexpensive to produce bags (Fig. 3.5).

Jute is an annual plant grown in countries are Bangladesh, India, Myanmar, Nepal, China, Vietnam, Thailand and Brazil. Usually harvested 110–120 after sowing when the plants are at early pod stage. This is important because

Figure 3.5. Jute plant

fibre will be weak if it is harvested before flowering and it becomes coarse and lacks lustre if harvesting is delayed. After harvesting, it is made into bundles of 10–12 kg each and left standing in the field for 3–4 days for helping in defoliation. Next the bundles are allowed to ret in water for 15–18 days after which the fibre can be separated manually. Action of bacteria during retting break down the soft tissues around the fibre bundles and the fibres. Soft water at around 34°C and pH of 6–8 gives good retting. Retting is complete when the bark separates out easily from the core. Stopping retting at the correct point is very important, as it determines the quality of the fibre. Fibres from the retted stalks is done manually (decortication) using knives. Another system of retting followed is ribbon retting, where the outer skins of the stem is removed in the form of a ribbon and retted separately. Retting in slow-flowing clean water produces the best fibre.

Instead of retting stems of the plant, another process called ribbon retting is also followed which has certain advantages. In ribbon retting, the green bark is separated in from the stem after harvesting and retted separately. Decortication of four to six stems is done together manually. This process gives can produce longer fibres and water required is much less compared to conventional process. But other disadvantages are high labour cost, loss of fibre yield (ribboning always leaves some amount of fibre stuck on the stalk 19–30%), loss in strength and entangling of the retted fibre, longer retting period, etc.

After retting, the material is washed well to remove dirt, gum, extraneous plant materials and dried in the sun.

Fibre is situated in the outer skins of the stem of the plant in the shape of spindle of 2.5 m length and 0.02 mm diametre. The fibres are joined into bundles with middle lamellae. Within the ultimate cells of a jute fibre, the ultrafine fibrils, that are purely cellulosic, are the highly ordered regions, while the inter-fibrillar regions are less ordered regions which can make room for the presence of short chain hemicellulose molecules to a larger extent and the bulky lignin molecules to a smaller extent as the bonding material of the middle lamella, providing strong lateral adhesion between the ultimates.

After retting the jute fibre basically is a compound of lignocellulose consisting of calcium, magnesium, aluminium, iron, etc., that are present either in the free state or bonded with functional groups of cellulosic chain. Chemically they are polysaccharides, also called carbohydrates (or holocellulose), and are divided into two groups: alphacellulose (58–63%) and hemicellulose (21–24%). Alphacellulose forms the bulk of the ultimate cell walls with the molecular chains lying broadly parallel to the direction of the fibre (Figs. 3.6 and 3.7). The hemicellulose and lignin, however, are located mainly in the areas between neighbouring cells, where they form the cementing material of the middle lamella, providing strong lateral adhesion between the ultimates.

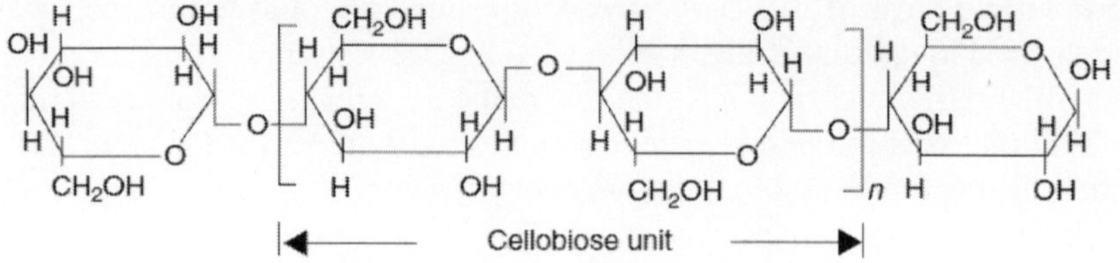

Figure 3.6. Structure of alphacellulose

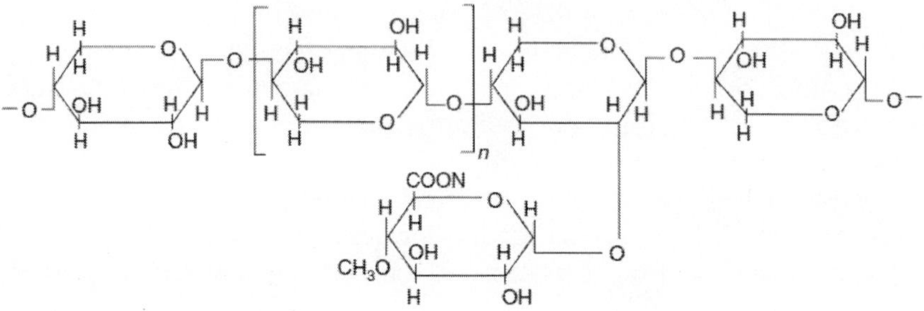

Figure 3.7. Structure of hemicellulose

The hemicellulose consists of polysaccharides of comparatively low molecular weight built up from hexoses, pentoses and uronic acid residues. Thus their three main constituents are hemicellulose-lignin 11.4–12%, alphacellulose 58–63% and hemicellulose 21–24%. In addition, analysis of the hemicellulose isolated from alphacellulose and lignin gives xylan 8–12.5%, glucoronic acid 3–4%, together with traces of araban and rhamnosan. The insoluble residue of alphacellulose has the composition glucosan 55–59%, xylan 1.8–3.0%, glucoronic acid 0.8–1.2%, together with traces of galactan, araban, mannan and rahmnosan. All percentages refer to the weight of dry fibre lignin (11.4–12%) which is not a fibrous matter and is removed by

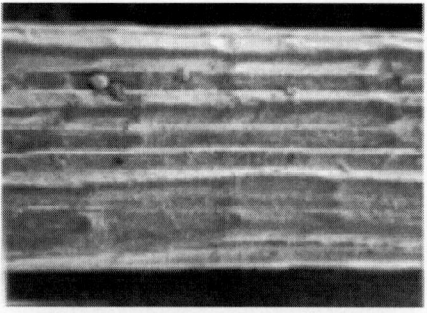

Figure 3.8. SEM view of jute fibres

chlorination by which it is converted to a soluble chlorolignin complex which is removed by a dilute alkali treatment (Fig. 3.8).

Other minor contituents (total of about 2%) are fats and waxes 0.4–0.8%, inorganic matter of 0.6–1.2%, nitrogenous matter 0.8–1.5% and traces of pigments (Table 3.2).

Table 3.2. Average physical properties of a single jute fibre

Property	Average result
Tenacity of single fibre (g/tex)	70
Linear density (tex)	1.8
Elongation at break (%)	1–2
Length breadth ratio	115–140
Specific gravity	1.5–4.0
Moisture regain at standard conditions (%)	12.5

3.2.1   Manufacture

Jute spinning system consists of two stages of carding (Breaker card and Finisher Card) forming the sliver and three stages of drawing and finally spinning stage which can be flyer spinning (or other type spinning for finer quality yarns or blends). After forming the slivers, it is passed through drawing frames in which the faller bars move on spiral screws or the push-bar method for the first stage. Slivers are doubled and fed to each drawing. The output sliver from the final drawing stage then passes to the spinning frame, where its linear density is reduced suitably for the yarn being spun, after which the required twist is inserted using an overhung flyer, with the yarn winding-onto a bobbin rotating on a dead spindle, against a friction drag. Ring or pot-spinning are also used. After the last drawing, the sliver is given a small crimp by passing through a crimp box to hold the fibre together and fed to the spinning frame (Fig. 3.9).

Jute/cotton (50/50) blended spun count of 8/10s yarn is used for making curtain fabrics, blankets, rugs, foot mats, etc. A probable spinning flow chart is given in Fig. 3.10.

Traditionally jute has been used to manufacture packaging materials like hessian, sacking, ropes, twines, carpet, backing cloth, etc., but it has been decreased over the years as it has been replaced by synthetics. Alternative use has been developed and now it is being used for products like home textiles,

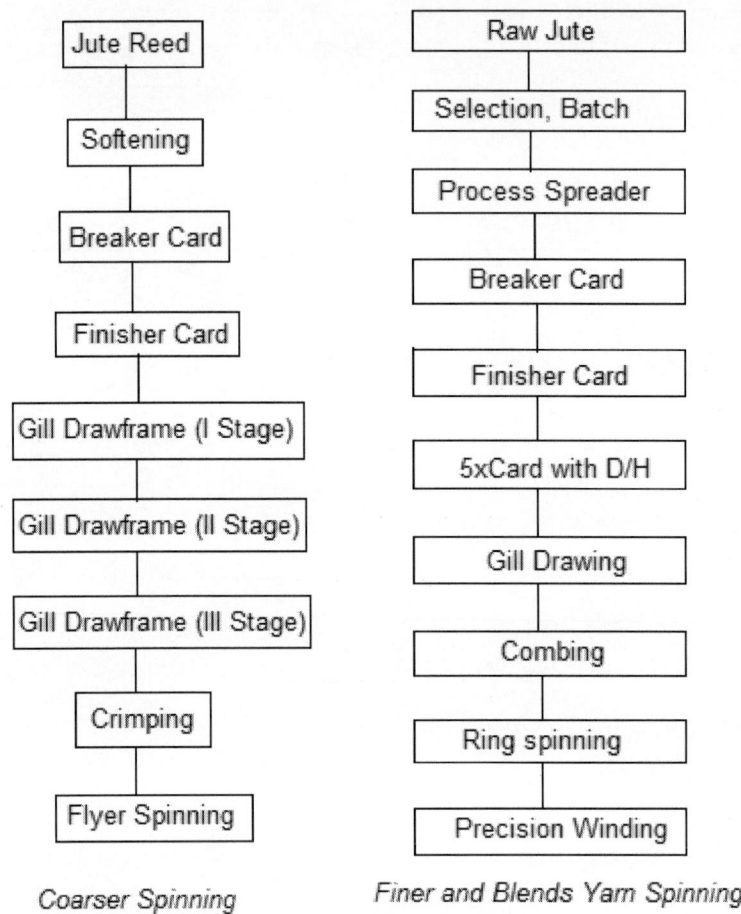

Figure 3.9. Jute spinning system

jute composites, jute geotextiles, paper pulp, technical textiles, chemical products, handicrafts, fashion accessories, etc.

3.3   Ramie fibre

Ramie is one of the oldest fibre crops, having been used for textile. Ramie (*Boehmeria nivea*) is a flowering plant in the nettle family native to eastern Asia. The true ramie or China grass also called Chinese plant or white ramie is the Chinese cultivated plant. A second type is known as green ramie or rhea and is believed to originate from Malay peninsula. It is suitable in tropical climate. Ramie is one of the oldest fibre crops, having been used for at least

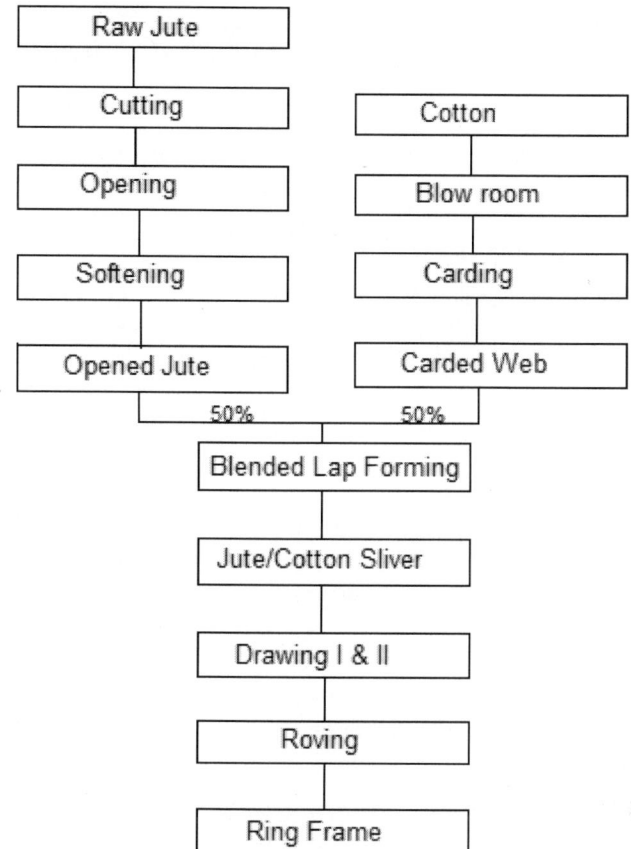

Jute/Cotton (50/50) Blended Yarn Spinning Flow chart (SITRA)

Figure 3.10. A probable spinning flow chart

6000 years, and is principally used for fabric production. It is a bast fibre, and the part used is the bark of the vegetative stalks. Ramie is normally harvested two to three times a year but under good growing conditions can be harvested up to six times per year. Unlike other bast crops, ramie requires chemical processing to de-gum the fibre. This fibre comes from plants *B. nivea* or *B. tenacissema* which belong to a family of stingless nettles. The ramie fibres are removed from the stalks by the process of decortication. It is fine absorbent, quick drying fibre, is slightly stiff and possesses high natural lustre. Its plant height is 2.5 m and its strength is eight times more than cotton. Ramie is one of the strongest natural fibres (Fig. 3.11).

 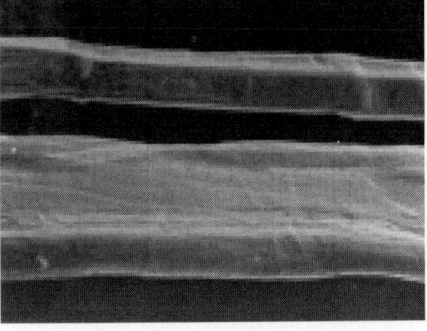

Ramie plant   Electron microscopic view of
                                        the ramie fibre

Figure 3.11. Ramie

Ramie fibre is white and lustrous and often referred to as China grass. Ramie has been grown in Brazil to substitute flax. The fibre is stiff and fairly coarse like canvas. The fibres are released from ribbons or strands in which they are held together by natural gums. Commercial degumming is carried out by treating the fibres with caustic soda solution for 4 h. The use of mixed bacterial cultures and enzymes is also reported to remove the gum from ramie fibre. It exhibits even greater strength when wet. Ramie fibre is known especially for its ability to hold shape, reduce wrinkling and introduce a silky lustre to the fabric appearance. It is not as durable as other fibres, and so is usually used as a blend with other fibres such as cotton or wool. It is similar to flax in absorbency, density and microscopic appearance. However, it will not dye as well as cotton. Because of its high molecular crystallinity, ramie is stiff and brittle and will break if folded repeatedly in the same place; it lacks resiliency and is low in elasticity and elongation potential.

Ramie is a perennial plant having a crop life of 6–20 years and are propagated by the following four methods, in order of importance: (1) rhizome cuttings, (2) division of parent rootstock, (3) laying and (4) stem cuttings. It is harvested two to three times a year, harvesting almost like jute. Harvesting can be done four to six times a year. It is harvested as the stems turn brown. Harvesting is done when the stems brown, just before or soon after the onset of flowering, since at this stage maximum fibre content is achieved.

As in the case of flax or hemp, ramie fibres are found in the bark of the stalk. The skin is removed just after harvesting while the plants are still fresh, by hand or machine. The process of transforming ramie fibre into fabric is similar to producing linen from flax. After soaking the stems in water for a few hours, the inner fibre is stripped away from the skin using a blunt knife or something similar. Having dried the fibre in the shade, it is then split into narrow strips and is washed, dried.

3.3.1   Stripping

Bark and parenchyma are removed from the strips by pulling them between a scraper and a bed plate held in the same hand using a handy tool made of bamboo. The strip is then reversed and the butt end is scraped.

3.3.2   Ribboning

Ribbons from the previous process contain more of the outer parts of the stem than strips. This process is meant to remove the outer bark/epidermis and the bast from the woody core of the stem. The material is passed through fluted rolls whereby the woody core is crushed and removed them from the bast, in fact, the entire cortex is removed. There are many designs of machines for doing this process.

3.3.3   Decortication

Alternatively, decortication process can be done on the green stems or dry stems. This is a direct process to get the stems are fed into the drum where the high velocity blades disintegrate them, causing the fibre to separate, fan out and bend down into the restricted space between the blades and fixed plate. The blades then scrape away the epidermal, parenchymatous and woody tissues from the fibre; much of this is removed between the blades and fixed plate, but the fibre tends to travel around the drum where the licking action of the blades cleans it further. Next, the fibre is washed and dried.

3.3.4   Degumming

The ribbons produced by the hand process or decortication machines contain a large percentage of gums and nonfibrous cells, or parenchyma (30–35%). The water insoluble gums are soluble in alkaline solutions. Hence, a degumming process with alkaline solution has to be performed to get the spinnable fibres from these ribbons.

   The process consists of following steps: (1) boiling of the fibre one or more times in an aqueous alkaline solution with or without pressure and agitation, and with or without penetrants or reducing agents, (2) washing with water and neutralizing, (3) bleaching with dilute hypochlorites or hydrogen peroxide, (4) washing with water and neutralizing and (5) oiling with a sulphonated hydrocarbon.

A typical degumming process is as follows: the material is treated with the recipe given in Table 3.3 two times 1 h under 6 kg/cm^2 pressure at 160°C at 6:1 MLR (wash with water in between).

Table 3.3. Chemicals used in the degumming process of ramie fibre

Quantity	Unit	Additions
6	%	Sodium hydroxide
3	%	Sodium Sulphite
3	%	Sodium tripolyphosphate
3	%	Organo phosphate wetting agent

Wash again with water, if necessary, bleach for 1 h with hydrogen peroxide 1% at 83°C at pH 9, rinse in a dilute solution of acetic acid, apply an oil emulsion such as sulphonated hydrocarbon, between 3% and 4% on dry fibre weight. Remove excess emulsion (by calendering).

3.3.4.1   Bacterial degumming (retting)

Other than traditional method of degumming, microbial degumming has been found promising. In this method, a mixture of bacterial species (they are used as mixed as they cannot survive separately) allowed to act upon the ribbons. The mixed degumming method is simple and economical in that less alkali is required, the treatment is less drastic, and such fibre properties as softness, feel and lustre are also much improved. The combined microbial and chemical method (see above) is also simpler and more economical.

3.3.5   Bleaching

The degummed fibre is bleached in the fibre form or as fabric. Depending on the requirements of the customer, it is bleaches in the fibre form whereby there is a minor loss in strength and is carried out carefully if it is necessary only.

3.3.6   Softening

The harshness and stiffness of the degummed fibre calls for softening treatment to support in the separation of the fibres (which is still held together and spinning operation). This usually done by application of a suitable agent, for example, glycerine, oil, fat, soap, paraffin, wax or tallow, and left for some time to condition. The fibres can be further softened and separated by passing them through a series of paired fluted rollers and then through a pair of smooth rollers; if necessary, they can be passed through these several times.

3.3.7   Spinning

Spinning can be done further with flax or hemp system or specially designed machines for the ramie fibres. The fibre is very fine and silk like, naturally white in colour and has a high lustre. Ramie is a term that is appearing with increased frequency in the labelling of sweaters and some linen-look textiles. Spinning the fibre is made difficult by its brittle quality and low elasticity; and weaving is complicated by the hairy surface of the yarn, resulting from lack of cohesion between the fibres. The greater utilization of ramie depends upon the development of improved processing methods.

Ramie has the highest degree of both polymerization and birefringences. The high strength of ramie is attributed to its highly ordered structure. The crystalline areas increase in lateral size and decrease in crystalline orientation, as the gum is removed from the intercrystalline areas. Ramie fibres tend to have a hairy feel which reduces their cohesion. They are smooth and cylindrical, with thick wall. The surface of the cell is marked by little ridges. The cell of the fibre is long and the cross section is irregular in shape. The lumen narrows and disappears towards the ends of the cell. Ramie fibre absorbs water rapidly and fabrics made from it will launder easily and dry quickly.

3.3.8   Applications

Ramie is used to make suiting, shirting, sheeting, dress materials, table cloths, napkins, towels, handkerchiefs, fine furniture upholstery, draperies, mosquito netting, gas mantles, industrial sewing thread packing materials, fishing nets, fire hose, belting, canvas, marine shaft packing, knitting yarns, hat braids, filter cloths, etc.

Other than textile uses ramie is used for making paper, medicinal uses (antiphlogistic, astringent, demulcent, diuretic, febrifuge, haemostatic, resolvent, vulnerary and women's complaints). Used to prevent miscarriages and promote the drainage of pus. The leaves are astringent and resolvent. They are used in the treatment of fluxes and wounds. The root is antiabortifacient, cooling, demulcent, diuretic, resolvent and uterosedative.

The fabric made of ramie fibre has the following advantages:

1. They are resistant to bacteria, mildew and insect attack.
2. Extremely absorbent, dyes easily.

3. Increases in strength when wet, withstands high water temperatures during laundering. Smooth lustrous appearance improves with washing.

4. Keeps its shape and does not shrink, but can be bleached.

Disadvantages: Low in elasticity, lacks resiliency, low abrasion resistance, wrinkles easily, stiff and brittle.

Other than this, ramie fibre is having lot of competition in the market due to many reasons: (1) High cost due to its long process and many times manual processes which reduces its competitiveness against other textile fibres. (2) The lack of ready supplies of satisfactory quality fibre has discouraged the industrial sector from promoting the crop. (3) There is a traditionally high labour requirement for production, harvesting and decortication. (4) There is a need to degum the fibre prior to processing. (5) The high demand for nutrients and the consequent decline in soil fertility would require special attention to crop rotation. (6) Many alternative crops can be expected to be more profitable hence the cultivation of the ramie is discouraged greatly.

3.4   Hemp fibre

Hemp comes from the plant, *Cannabis sativa*, an annual of the family Moraceae, which grows to a height of 10 feet or more. The most important hemp producing countries are CIS, Yugoslavia, Romania, Poland, Belgium and Hungary. Even though the name 'hemp' refers to the above plant, it is also sometimes refer to other fibre bearing plants like sisal hemp (*Agave rigida*) or Manila hemp (*Musa textalis*), but these plants and fibres are neither related nor associated with *C. sativa*.

Cannbis sativa is a tall, slender annual, with a vigorous growth habit, of the Cannabaceae family. The initial growth is little different than the other plants. As the seedling emerges, it has leaves a single elliptic blade with serrate margins, leaves of the second pair of true leaves have three serrate leaflets emerging from the petiole, a leaf from the third pair of true leaves has five serrate leaflets, and so on. This pattern continues up to between 9 and 12 leaflets (Fig. 3.12).

Hemps are cultivated for fibre and seeds. The varieties which has been developed for extracting fibre have a fibre content of 15–33%, variations being due to seasons, land and other conditions. Fibre hemp varieties require 4–5 frost-free months from planting to produce a harvestable crop, while hemp seed types require 5–6 months to produce mature seed. The fibre quality and yield also depend on the density of the plant sowed. Even though earlier days, 50–500 seeds/m^2 used to be sown, experience and researches

The leaf   The plant   Full grown plant

Figure 3.12. Morphology of hemp plant

have shown that 150–400 seeds/m² gives better yield and quality. The fibre hemp has to be harvested, when the staminate plants have finished flowering but before the seed has ripened, also known as technical maturity (Fig. 3.13).

It is from hemp stem, from which the fibre is extracted. As usual with jute, ramie etc., it has a hard core and a fibrous bark or bast. The bast consists of three layers out of which the inner layer contains the fibre. The fibres are in the form of bundles and run the full length of the plant stem. The length of the ultimate fibres is reported to range from 2 to 60 mm. The breaking strength of hemp fibre is a little higher than that of flax fibre; its elongation is low (2–3%). Its flexibility depends on the fineness of the bundle. The longer bundles require less twist during spinning. Although the elongation of the bundles is low, their flexibility is high and this can cause problems during spinning. Blending flax with hemp improves both the elongation and the flexibility of the yarns, which is low in 100% hemp yarns. However, these blends also decrease the strength of the yarn.

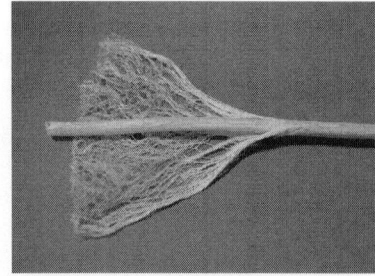

Hemp stem   Photomicrograph of hemp stalk

Figure 3.13. Hemp stem and hemp stalk

The manufacturing process of hemp fibre is almost same as the flax fibre (see processing flow chart). The machineries used are also similar. The fibre yield of unretted hemp straw is about 25%. The yield of long fibre of unretted straw varies from 10% to 13% and of short fibre from 12% to 15%. However, the greater length and thickness of the hemp stalk compared to that of flax need to be taken into account and hemp processing machinery is generally larger and more robust.

The processes to be followed as per the quality requirements of the final fibre or the final product. For example: first class – water retted, used for finer yarns; second class – water retted, for medium count and thicker yarns; third class – unretted, fibres mechanically separated and used for blending or spinning coarse yarns.

3.4.1   Retting

The fibre is first separated from the stalk by mechanical process and further retted. Retting is a biological process that removes the pectic substances that bind the fibres to the other constituents of the hemp stalk. It can be done by ground retting or water retting.

3.4.2   Ground (field) retting

Ground retting takes the advantage of the dew and showers of rain for the growth of the microorganisms on the stalk. After the harvest, the stalks are laid on the ground in thin layers. Ground retting is effective if rainfall reaches 600 mm/month at harvest time. This can take 2–8 weeks.

3.4.3   Water retting

The harvested stalk bundles are steeped in water in concrete tanks of about 50–100 m in length, 5–10 m in width and 1–1.2 m in depth and held beneath the surface of the water by wooden or iron frames which cover the tanks at temperatures of around 15–30°C. The higher the temperature, the shorter the retting period (5–6 days). If temperature is lower, it can take up to 14 days or so (Fig. 3.14).

Retting can be done using enzymes, but it is a costly process due to the cost of enzymes used. But enzyme retting has advantages – it is a very soft treatment like water retting, can be done any time of the year, do not need large field space as in dew/field retting) or chemical retting (can be done within hours using chemicals like NaOH, Na_2CO_3, Na_2SO_3 in the presence or

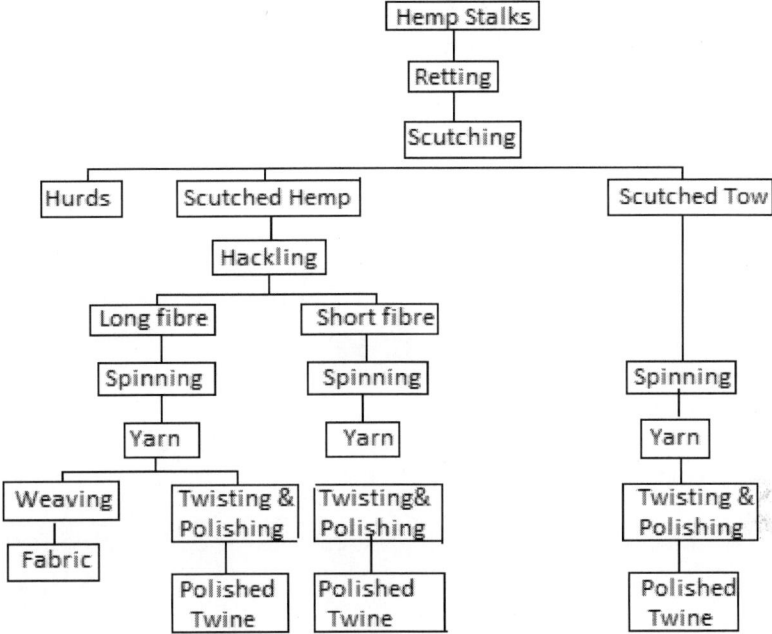

Figure 3.14. Production of hemp fibre

without a chelating agent like EDTA). Fibre is freed from woody matter by a retting process, followed by breaking and scutching.

3.4.4   Breaking

In the next process is to open the sheaves. The stalks are fed into several pairs of horizontal parallel breaking rolls. These rolls split the stalks lengthways and the wooden art is broken into smaller pieces, separated from the fibre and removed suitable method. The effectiveness of the breaking process depends on the quality of the stalks.

3.4.5   Scutching

Scutching helps to remove the wooded parts (herds), dust, earth and other wastes still sticking onto the fibre and make it softer and smooth by removing the lignins. This process can be done better manually but not economical due to very low productivity. The product of scotching is called 'tow'.

Tow is further dried to a required moisture (7–10% moisture) by passing through conveyor drier. Next it passes through various breaking and cleaning rollers and over further reciprocating screens, all aimed at cleaning the fibres and removing impurities but in addition, the high pressure breaker rollers further split and affine the tow fibres.

3.4.6   Spinning

Scutched hemp and tow fibres are further spun into yarn by spinning process using technologies suited for these fibres, which are generally used for twine, ropes, etc. The hemp tow which is meant for fabric production is passed through a process called hackling, whereby the fibres are parallelized, separate and 'affine' (split) the fibres, remove short fibres (tow) and knots of tow and clean the fibre of the hurds and other hurds and other impurities that have not been removed during previous processing.

The output of the hackling machine usually in the form of a sliver, which is further, processed as usual with spinning process but designed for hemp yarn.

The hemp naturally may be creamy white, brown, grey, black or green. This is then softened by pounding it mechanically or by hand. It is yellowish brown fibre. Hemp fibres can be 3–15 feet long, running the length of the plant. Characteristics of hemp fibre are its superior strength and durability, resistance to ultraviolet (UV) light and mould, comfort and good absorbency.

Hemp has cellulose content of about 67% and contains about 16% hemicellulose. Strands of hemp fibre may be 6 feet or more in length. The individual cell is on an average between 0.5 and 1 inch long. They are cylindrical in shape with joints, cracks, swelling and other irregularities on surface. Like flax, the cells of hemp fibres are thick walled and they are polygonal in cross section. The central canal or lumen is broader than that of flax; however, the ends of the cells are blunt.

The fibre will not bleach easily and generally employed in the manufacture of ropes, twines, sacking, carpets, nets and tarpaulins. In addition, hemp also serves today as a raw material for paper industry. Cottonized hemp does not spin easily alone, but it gives useful yarn, when mixed with cotton (up to 50% hemp).

The inner two fibres of hemp are more woody and are more often used in nonwoven items and other industrial applications. Hemp fibres can be 3–15 feet long, running the length of the plant. The hemp naturally may be creamy white, brown, grey, black or green.

The manufacture of cordage of varying tensile strength

- Clothing,
- nutritional products,
- hemp fibres are increasingly used to strengthen cement.

3.5   Coir fibre

Coir fibres are found between the husk and the outer shell of a coconut. The individual fibre cells are narrow and hollow, with thick walls made of cellulose (Fig. 3.15).

Coconut broken into half   A coconut tree   Coir fibre
Figure 3.15. Coir fibres

The coir is of two types:

1. Brown coir
2. White coir

Fibre mechanically extracted from dry mature coconut husk after soaking. It is long, hard and strong fibre but with lower softness, lower water absorption capacity and shorter life than long retted fibre.

3.5.1   Brown coir

Brown coir is harvested from fully ripened coconuts. The fibrous layer of the fruit is then separated from the hard shell (manually) by driving the fruit down onto a spike to split it (*dehusking*). The fibrous husks are soaked in pits or in nets in a slow moving body of water to swell and soften the fibres. The long bristle fibres are separated from the shorter mattress fibres underneath

Figure 3.16. Mannual removal of brown coir from ripened coconut

the skin of the nut, a process known as *wet milling*. The mattress fibres are sifted to remove dirt and other rubbish, dried in the sun and packed into bales (Fig. 3.16).

3.5.2   White coir

White coir fibres are harvested from the coconuts before they are ripe. The immature husks are suspended in a river or water-filled pit for up to 10 months. During this time, microorganisms break down the plant tissues surrounding the fibres to loosen them – a process known as retting. Segments of the husk are then beaten by hand to separate out the long fibres which are subsequently dried and cleaned. Cleaned fibre is ready for spinning into yarn using a simple one-handed system or a spinning wheel.

Mature brown coir fibres contain more lignin and less cellulose than fibres such as flax and cotton and so are stronger but less flexible. They are made up of small threads, each about 1 mm long and 10–20 μm in diameter. The fibres are white or light brown in colour and are smoother and finer, but also weaker.

3.6   Kapok fibre

Kapok (*Ceiba pentandra*) is a tropical tree of the order Malvales and the family Malvaceae (previously separated in the family Bombacaceae), native to Mexico, Central America and the Caribbean, northern South America and (as the variety *C. pentandra* var. *guineensis*) to tropical west Africa. The word is also used for the fibre obtained from its seed pods. The tree is also known as the Java cotton, Java kapok or ceiba. It is a sacred symbol in Maya mythology.

Kapok fibre is a silky cotton-like substance that surrounds the seeds in the pods of the ceiba tree. It can support as much as 30 times its own weight in water and loses only 10% of buoyancy over a 30-day period. It is eight times lighter than cotton. It is extremely good as thermal insulator. It is also lightweight, nonallergic, nontoxic, resistant to rot and odourless. The fibre is very buoyant, resilient, highly flammable and resistant to water. The process of harvesting and separating the fibre is labour intensive and manual. Since it is inelastic and too fragile, it cannot be spun. It has outstanding characteristics of lightness, impermeability, thermal insulation and eco-friendly. It is used as an alternative to down as filling in mattresses, pillows, upholstery, teddy bears, zafus and for insulation. It was previously much used in life jackets and similar devices. The fibre has been largely replaced by man-made materials. The seeds produce an oil used locally in soap and that can be used as fertilizer.

3.7   Pina fibre

Piña is a fibre made from the leaves of a pineapple and is commonly found in the Philippines. It is sometimes combined with silk or polyester to create a textile fabric. The end fabric is lightweight, easy to care for and has an elegant appearance similar to linen piña comes from the leaves of the pineapple plant. Each strand of the hand scrapped piña fibre is knotted one by one to form a continuous filament for hand weaving into the 'piña cloth'. The piña fibre is softer, has a high lustre and is usually white or ivory in colour.

3.7.1   Production methods

Scraping a pineapple leaf to reveal the fibres: Since piña is from a leaf, the leaf has to first be cut from the plant. Then the fibre is pulled or split away from the leaf. Most leaf fibres are long and somewhat stiff. A major use for piña fabric is in the creation of the Barong Tagalog and other formal wear that is common in the Philippines. It is also used for other table linens, bags, mats and other clothing items, or anytime that a lightweight, but stiff and sheer fabric is needed.

3.8   Sunn hemp

The stems of the herbaceous plant *Crotalaria juncea* (legume family, Fabaceae), called sunn or sunn hemp, provide a bast fibre. The plant is native to India, the chief fibre producer, and it is also grown in Bangladesh, Brazil and Pakistan. It has a long tap root and grows to a height of up to 5 m. Harvesting is done manually by pulling or cutting. The plant is defoliated in the field, water retted, and processed similarly to jute. The white fibre is graded by colour, firmness, length, strength, uniformity and extraneous matter content. Sunn is used for canvas, paper, fishing nets, twine and other cordage.

3.9   Urena and abutilon

These are less important vegetable fibres of a jute-like nature. *Urena lobata* (Cadillo) of the mallow family (Malvaceae) is a perennial that grows in Zaire and Brazil to a height of 4–5 m with stems 10–18 mm in diameter. Because of a lignified base, the stems are cut 20 cm above the ground. The plants are defoliated in the field and retted similarly to jute and kenaf. The retted material is stripped and washed and, in some cases, rubbed by hand. The soft, near-white fibre is graded for lustre, colour, uniformity, strength and cleanliness. It is used for sacking, cordage and coarse textiles. Abutilon theophrasti is a herbaceous annual plant producing a jute-like fibre. The plant is native to China and is commercially grown in China and the former Soviet Union. Because of its association with jute in mixtures and export, it is also called China jute. The plant grows to a height of 3–6 in with a stem diameter of 6–16 mm. After harvesting by hand and defoliation, bundles of the stems are water retted and the fibre is extracted by methods similar to those for jute. The fibre is used for twine and ropes.

3.10   Sisal fibre

The sisal plant has a 7–10 years lifespan and typically produces 200–250 commercially usable leaves. Each leaf contains an average of around 1000 fibres. It is extracted by a process known as decortication (Fig. 3.17).

Sisal is used in three grades: (1) The lower grade fibre is processed by the paper industry because of its high content of cellulose and hemicelluloses. (2) The medium grade fibre is used in the cordage industry for making: ropes, baler and binders twine. (3) The higher-grade fibre after treatment is converted into yarns.

Sisal is mainly used for twines and ropes. Rarely used for fabric, other than packing materials and carpets. Hence, the processing of these fibres are not dealt in this book in detail (Tables 3.4–3.6).

Figure 3.17. Sisal plant

Table 3.4. Chemical composition (wt %) of vegetable fibres

Fibre	Cellulose	Hemicellulose	Pectins	Lignin	Extractives
Bast fibres					
Flax	71.2	18.6	2	2.2	6
Hemp	74.9	17.9	0.9	3.7	3.1
Jute	71.5	13.4	0.2	13.1	1.8
Kenaf	63	18		17	2
Ramie	76.2	14.6	2.1	0.7	6.4
Leaf fibres					
Abaca	70.1	21.8	0.6	5.7	1.8
Phormium	71.3				
Sisal	73.1	13.3	0.9	11	1.6
Seed-hair fibres					
Coir	43	0.1		45	
Cotton	92.9	2.6	2.6		1.9
Kapok	64	23	23	13	

Table 3.5. Comparison of the dimensions of long vegetable fibres

Fibre	Ultimate fibre Length (mm)	Ultimate fibre Dia (µm)	Cell cross section Shape	Cell cross section Dia (µm)	Fibre strand Length (cm)	Fibre strand Width (cm)
Bast fibres						
Chinese jute	2–6.5	7–33				
Flax	4–69	8–31	Polygonal	8.8–16.1	25–120	0.04–0.62
Hemp	5–55	16	Polygonal	13.1–23.6	100–400	0.5–5
Jute	0.7–6	15–25	Polygonal to oval	12.3–18.6	150–360	
Kenaf	2–11	13–33	Cylindrical		200–400	
Ramie	60–250	16–120	Hexagonal to oval	6.2–32.4	10–180	
Sunn	2–11	13–64	Irregular	13.6–24.6	108–216	
Nettle	4–70	20–70		50–50		
Leaf fibres						
Abaca	2–12	6–40	Oval to round	14–20	150–360	0.01–0.28
Cantala				13.8–16.4		
Caroa	2–10	3–13		3.2–8.2		
Henequen	1.5–4	8.3–33.2		11.6–22.2		
Istle	9.6–16			11.2–13.4	30–75	
Mauritius	1.3–6	15–32	Cylindrical		124–210	
Phormium	2–11	5.25	Round	10.3–12.1	150–240	
Sansevieria	1–7	13–40				
Sisal	0.8–7.5	8–48	Cylindrical	11–16	60–120	0.1–0.5
Seed-hair fibres						
Coir	0.2–1	6.24				1
Cotton	10–50	12–25	Circular elliptical		1.5–5.6	0.012–0.025
Kapok	15–30	10–30	Round		1.5–3	0.03–0.036
Others						
Broom root					25–40	

Table 3.6. Mechanical properties of vegetable fibres

Fibre	Fineness km/kg	Tensile strength	Elongation %	Elasticity N/tex	Modulus of rupture mN/tex
Flax		24–70	2–3	18–20	8–9
Hemp	139	38–62	1–6	18–22	6–9
Jute	489	25–53	1.5	17–18	2.7–3
Kenaf	180	24	2.7		
Ramie		32–67	4	14–16	11
Urena	342	16	1.9		
Leaf (hard) fibres					
Abaca	32	32–69	2–4.5		6
Cantala	58	30			
Henequen	32	20–42	3.5–5		
Istle	34	22–27	4.8		
Phormium	38	26			
Sansevieria	118	43	4		
Sisal	40	36–45	2–3	25–26	7–8
Seed-hair fibres					
Coir		18	16	4.3	16
Kapok		16–30	1.2	13	10

4
Animal Fibres

Animal fibres are natural fibres which consist largely of proteins such as silk, hair/fur, wool and feathers. The most commonly used type of animal fibre is hair. They can be classified further as hair fibres (staple), secretion fibres (filament), wool, silk, speciality hair fibres and insect fibre (spider silk).

4.1   Silk fibre

4.1.1   General

Natural textile fibres – e.g., cotton, wool and silk – dominated the world textile market until mid-century. The introduction of man-made cellulosic and the synthetic fibres and their sudden availability in the market changed the textile fibre consumption pattern overnight.

Silk is a natural fibre that can be woven into textiles. It is obtained from the cocoon of the silkworm larva, in the process known as *sericulture.* The silkworm as a means of self- protection spins a cocoon around its body by extruding the contents of the two silk glands each filled with a concentrated solution of the silk proteins, fibroin and sericin, the latter forming a sheath around the former. The two glands unite in the spinneret, a minute aperture in the muzzle of the worm. The two filaments solidify on coming in contact with air and form a composite thread. It is a fine continuous strand on wound from the cocoon of a moth caterpillar known as the silkworm. It is the longest and thinnest natural filament fibre with the longest filament around 3000 yards. It is relatively lustrous, smooth, lightweight, strong and elastic. It is essentially composed of protein fibre and is naturally a white-coloured fibre.

4.1.2   Classification of silk

Silk is classified based on origin as Chinese, Indian, Japanese and European silks. Chinese silk is indigenous to China and they produce one of the finest silk and are white, golden yellow, green, red or beige in colour. They highly reelable and are univoltine, bivoltine and multivoltine. Indian races are reared in India and South East Asia. They are mainly multivoltines and the quality of the silk filament is good. The variety grown in Thailand, even though comes under Indian variety, are yellower, coarser and having more silk gum

Figure 4.1. Silkworm cocoon

than the Indian variety. Japanese races are mainly comes from Japan and are mainly white in colour but can be yellowish or greenish also. They are mostly bivoltine or univoltine and comparatively inferior to the above two varieties. European variety is grown in Europe and Central Asia and are suitable for cold dry climates. Cocoon reelability is good since cocoon size is big and without much constrictions (Fig. 4.1).

Silk is also classified as univoltine, bivoltine and multivoltine. This classification is as per generation per year.

It is obvious from the name that univoltine has only one generation in a year, bivoltine two generations and multivoltine more than two generations. It is needless to say that univoltine cocoons are big (so are the larvae) and filament quality is good.

In bivoltine, the cocoons are of less weight and cocoon shell weight, shell ratio and cocoon filament weight are less compared to univoltines, but larvae are more robust and more uniform compared to univoltines. They produce large quantities of thread per cocoon, up to 1600 m or more of good quality, even, lustrous and strong. They are highly vulnerable to disease and hence require very strict rearing conditions in terms of hygiene and temperature and humidity control. They are much better suited to temperate than to subtropical or tropical climates.

Mutivoltines life cycle is short. Larvae are robust and can withstand high temperature and are very hardy and resistant to disease. They can tolerate imperfect rearing conditions and are well suited to subtropical and tropical climates. The cocoons size is small. The cocoon weight, shell ratio and cocoon filament weight are lower compared to bivoltines but filament is fine. They produce relatively low quantities, about 400–800 m per cocoon of thread of fairly poor quality in terms of physical characteristics.

4.1.3   Types of silk

The silkworms are generally cultivated. There are mainly four varieties of silk, e.g. mulberry, tasar, eri and muga, and each variety is produced by feeding on the leaves of certain plants. All the species of silk have four stages in their life cycle, namely, the egg, larva, pupa and moth (Fig. 4.2).

4.1.3.1   Domestic silk (mulberry silk)

The mulberry silkworm belongs to the species *Bombyx mori* and about 95% of the world's production is of this species. This is a white-to-yellow-coloured silk. It is fine and is used mainly for apparel. Major silk-producing countries are China and India – China produces 80–82% of the world production and India about 15–18%. The remaining is produced by countries like Thailand, Brazil, Uzbekistan, Japan, Korea, etc.

4.1.3.2   Wild silk

The other three varieties are produced by the worm *Antheraea mylitta* and are termed as wild silk. They are stiffer and coarser than mulberry silk.

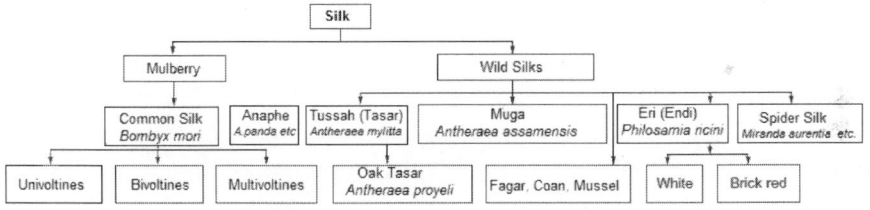

Figure 4.2. Classification of silk

Muga silk: This is a golden-yellow-coloured silk. It is obtained from the semidomesticated silkworm, which feeds on the aromatic leaves of som and soalu plants.

Eri silk: This is got from the domesticated silkworm. It feeds mainly on castor leaves.

Tasar silk: Tasar silk, also known as Kosa silk, is valued for its purity and texture. Kosa silk is drawn from cocoons especially grown on Arjun, Saja or Sal trees. There is another variety called oak tasar. It is a finer variety of tasar. In India, oak tasar is produced by the silkworm *Antheraea proyeli* J. which feeds on oak. They are found in abundance in the sub-Himalayan belt

A   B

Figure 4.3. Electron microscopic images of silk (A) and tusar silk (B)

of India, covering the states of Manipur, Himachal Pradesh, Uttar Pradesh, Assam, Meghalaya and Jammu and Kashmir. A variety found in China is produced by the silkworm *Antheraea pernyi* (Fig. 4.3).

4.1.4   Silk production

Silk is a natural fibre that can be woven into textiles. It is obtained from the cocoon of the silkworm larva, and the process is known as sericulture. The cultivated silkworm (larva of the silkworm moth), belonging to the large order of the Lepidoptera (characterized by the presence of four wings covered with scales, of which 20,000 species are known), goes through four stages of development: egg, larva, pupa in the cocoon and adult moth. Depending on the climate, several generations of silkworm can be raised per year (e.g., from one to eight in China at present). The proper development of the larva at the stage of maturity is extremely important for the successful cultivation of the silkworm (Fig. 4.4).

The female moth lays about 400 eggs to the size of the head of a pin. One enclosure contains about 100 moths, which lay about 40,000–50,000 eggs. After 10–14 days at a temperature above 13°C, the shells are broken through and the caterpillars creep forth. On hatching, the caterpillars are about 3 mm long and weigh 0.47 mg. Eating voraciously (mulberry leaves), they grow very quickly. After 30–33 days, they weigh 3.65–4 g and are 9 cm long, having increased their length about 30 times, their size 400–500 times and their weight 7000–10,000 times. During this development, the larvae change their skin four times, or every 5–6 days. These figures make it clear that silkworm larvae require an enormous amount of fresh mulberry leaves for their food. For example, 50,000 silkworms require 1.5 metric tons of fresh mulberry leaves, necessitating a plantation of about 1080 m² mulberry trees, to produce about 100 kg of cocoons. According to an old Chinese tradition, at the early stage, the silkworm must be fed every half an hour.

Immediately prior to the process of pupation, the silkworm as a means of self-protection first attaches silk fibre to various supports, to form a scaffolding and spins a cocoon around its body by extruding the contents of the two silk glands each filled with a concentrated solution of the silk proteins, fibroin and sericin, the latter forming a sheath around the former. (In cultivation, when the larvae reach maturity, the larvae are selected and gathered. This point is reached when the head of the caterpillar, for example, becomes lighter, more transparent and begins to take on colour. Here the timing and correct selection require considerable experience. The silkworms are placed in bamboo spinning nests which are heated from below and covered with straw. The silkworm seeks the most favourable position for spinning and begins making its pupa bed. Then, moving its head in a figure-8 motion, it begins to spin the silk filament, forming the cocoon itself.)

From the spinning of the cocoon (3 days) to the larva's transformation to the chrysalis stage (3–4 days) and the pause for development (14 days) to the emergence of the moth takes about 20 days. The cocoons from which the silk is taken must be gathered at exactly the right moment. Equally important is the selection of suitable silkworms for further breeding. This requires an

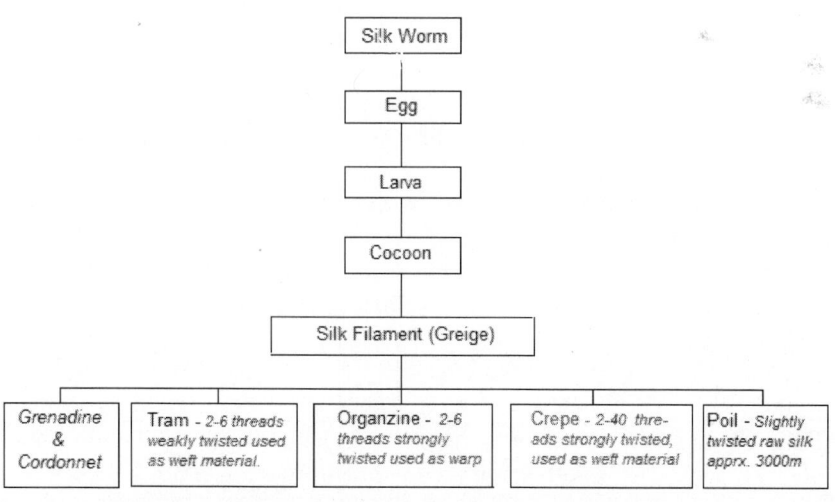

Silk worm lifecycle and the production of the most important silk yarns

Figure 4.4. Silkworm life cycle

especially great amount of experience, since suitable males as well as female moths must be selected. With some types of silkworms, the chrysalis are often hardly distinguishable; the female may either be more rounded at one end or its cocoon somewhat indented. Only skilled breeders can recognize these fine differences, thus also contributing to the production of high-quality silk by careful cross breeding of good moths.

The two glands unite in the spinneret, a minute aperture in the muzzle of the worm. The two filaments solidify on coming in contact with air and form a composite thread. Most cocoons, except those required for propagation purposes, are subsequently heated by a process known as stifling to kill the chrysalis within and prevent the emergence of the moth which otherwise would make the cocoon unreelable. Following the stifling process, the cocoons are inspected and graded, and the defective ones are separated for subsequent treatment as silk waste.

It is a fine continuous strand wound from the cocoon of a moth caterpillar known as the silk filament. It is the longest and thinnest natural filament fibre with the longest filament around 3000 yards. It is relatively lustrous, smooth, lightweight, strong and elastic. It is essentially composed of protein fibre and is naturally a white-coloured fibre.

4.1.5   Silk fibre production

The cocoon, which is the starting material for the processing of silk, varies in colour, size and shape depending on the species of silkworm. Silk cocoons are mostly white and yellow, but may be orange, green or even dark brown. For high-quality silk, the cocoons should be sorted into batches of the same colour. Cocoons vary considerably in their shape. Some cocoons are spherical, egg shaped, peanut-like indentation or elongated and with pointed ends. The exterior is uneven and gritty while the interior – after about per cent of the entire filament has been reeled – is smooth. The rough exterior form of the surface is important for the processing and quality of the silk. Thus, for example, the number of spherules per square centimetre is defined. The quality of the silk is assessed on the basis of the surface graininess.

4.1.6   Stifling

Stifling is the killing of the pupa inside the cocoon avoiding its maturing as moth. There are many methods proctized in stifling:

(a) *Sun drying*: The pupa is killed by prolonged exposure (2–3 days depending upon the intensity of sunlight) of freshly harvested cocoons to hot sun. The disadvantage of this process is that continuous exposure to sun hardens the cocoon shell, so affecting the reelability.

(b) *Steam stifling*: The pupa is killed by exposing fresh cocoons to the action of steam for around 25 min. The process can be done by either basket steaming or chamber steaming.

(c) *Hot-air drying*: This is the most effective method and produces good-quality cocoons such as bivoltine varieties. It facilitates the complete drying of cocoons and ease of storage. Other methods of killing the pupa include the use of infra-red rays, cold air and poisonous gases.

4.1.6.1   Reeling

Reeling consists of unwinding the fibre from several cocoons together and reeling the baves so as to form a composite thread of the required denier (Fig. 4.5).

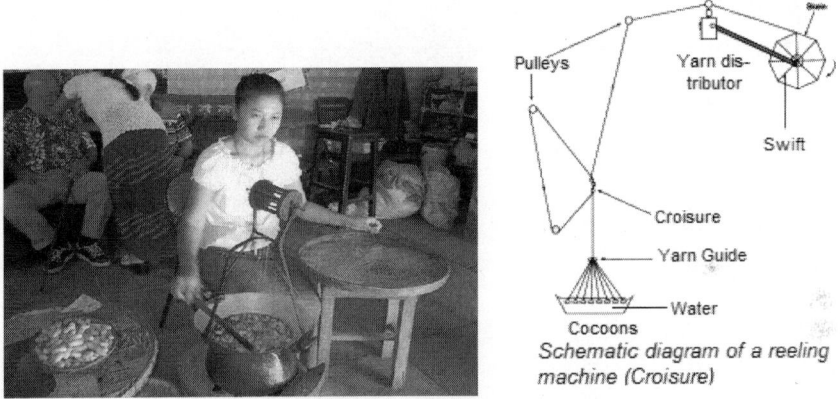

Figure 4.5. Reeling process

4.1.6.2   Handreeling process

A *Bombyx* silk bave is only about 15–25 µ thick and it is too thin and weak to be used singly. The reeling process consists essentially of softening the silk gum, by maceration in hot water, removing the loose outer layers until the free end of the bave has been located and then combining it with the baves from other cocoons. Although much of the world's silk is still reeled from hand-operated basins, 99% of the silk produced in Japan (the world's largest producer and consumer) is reeled on automatic reeling machines. With these machines, the process is considerably less labour intensive.

Silk throwing involves the preparation of the raw silk into a form suitable for knitting and weaving. The first process is that of soaking the hanks in a warm emulsion of various oils and other softening agents in a slightly alkaline solution. During this process, the oils are taken up by the silk gum. The objective is to make the yarns more supple and pliable. Following soaking and drying the hanks are re-wound on to bobbins or cones. It is at this stage that twist is inserted to form different types of yarn. The raw or reeled silk produced in this manner is called 'grege'. The total length of the filament spun by the silkworm for attaching and building its cocoon is ca. 2500–3000 m for the cultivated silkworm. Only about 1200 m of this can be reeled, the rest being used for valuable by-products such as schappe and bourrette silk.

By reeling, different types of fibres are produced. Grege threads can be directly used as weft or warp material. Other special yarns are produced by twisting number of grege filaments as per requirements of the classic silk fabric to be woven. Nonbreaking filament length (NBFL) is the length of silk filament that is present continuously in the cocoon and is important to a reeler with different types of silk (Table 4.1).

Table 4.1. Nature of silk fibre in various types of cocoons

Type of cocoon	Fibre density (g/cm³)	Fibre fineness (den)	NBFL (m)	Total filament length (m)
Mulberry (bivoltine)	1.34	2–3	700–800	1200–1600
Mulberry (multivoltine)	1.34	2–3	400–600	900–1200
Tasar	1.31	8–12	100–250	750–900
Oak tasar	1.31	3–5	300–450	800–1000
Muga	1.3	4–7	150–250	600–800
Eri	1.3	3–4	0.5–2.0	400–500

4.1.6.3   Throwing

Throwing is a process by which the yarn is twisted and supply the weaver or knitter with yarns for a specific purpose. Raw silk yarn is generally too fine to be woven with no twist, except for some special fabrics such as habutae. This means that several yarns have to be assembled and twisted together to form a substantial yarn for weaving or knitting. The continuous filaments reeled from the cocoons have to be joined together as they enter the twisting frame. It is more convenient for the workers to do this operation at waist level and this is why silk is uptwisted rather than down twisted, which is usual for other textile fibres (Fig. 4.6).

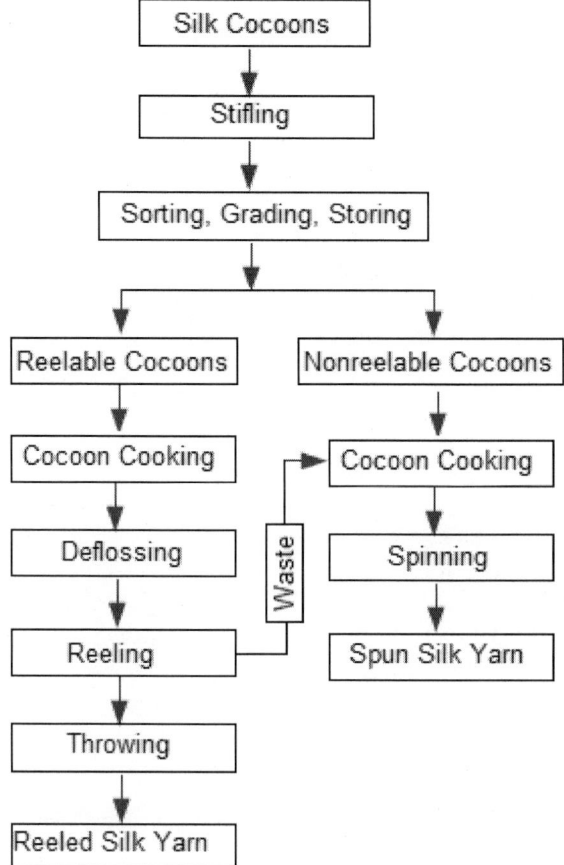

Figure 4.6. Sequence of processes for silk yarn production

The main yarns manufactured by throwing and their uses are as follows:

Voile: They are individual grege threads with 600–3000 twists per metre. Used in the raw state as warp and weft material.

Organzine: Two or more twisted grege single strands are doubled and again twisted into the reverse direction to produce organzine. The preliminary twist consists of about ca. 600–700 and the subsequent one 500–600 twists per metre. Organzine yarn has a stiff handle and produces a tightly closed effect. Used as warp for plain and colour-woven fabrics and knits.

Grenadine: A silk thread formed by doubling two or more ends of voile and twisting them in the opposite direction to that of the individual coil ends. Grenadine is three to four times more tightly twisted.

Tram: Tram is manufactured by doubling two or more silk threads without preliminary twisting and then twisting them slightly, usually between 80 and 150 turns per metre (TPM). They are lofty, voluminous yarn, soft and supple. They are used for fabric with good body with sheen, lustre and smoothness after boiling off. But it lacks the strength of organzine. Dyed as yarn, it is a preferred weft material for colour-woven fabrics. More strongly twisted, it can also be used for knits.

Crepe: Crepe is made by doubling one or more grege threads without a preliminary twist and then twisted with 2500–3500 twists to the metre, producing a hard, strong, tightly closed effect. Probably if the yarn is given between 40 and 80 turns per inch then two or three turns per inch as in tram. The woven fabric made using crepe has a characteristic 'kick up' or crinkle due to the elasticity of these threads which is generally termed as crepe effect. Used in making crepe de chine, crepe charmeuse, crepe meteor, crepe faille, crepe organzine in satin charmeuse and chiffons, as well as other fabrics.

Cordonnets: Coarser double twists consisting of several qreqe threads with a preliminary twist of 400–800 twists per metre and subsequent doubling of two or three threads followed by 300–600 twists per metre. Cordonnet yarns are mostly used for sewing threads but sometimes for knitting yarn as well.

4.1.7   Schappe and bourrette silk processing

Some 60% of the silk extruded by the silkworm is useless for the production of continuous filament yarns, and the spun silk process is based on the utilization of this material and hence it is important. This material, as well as scraps from reeling, twisting and weaving, can be further processed. For a long time, these valuable materials were simply pulped or used for other purposes in the country of origin, due to the intensive work involved and the high investment costs for machinery. The cocoon residues often serve as animal feed, e.g., for chickens or fish.

Specialist terms for processing by-products of grege filaments are as follows:

1. Silk wadding (Spelaia, Blaze) – loose silk mesh around the cocoon

2. Flock silk (Frison, Stursa, Kibizzo) – the outer, irregularly spun cocoon layers (ca. 25% of weight of material), regarded as best raw material for schappe silk

3. Cocons perces (cocons piques) – pierced, damaged cocoons which cannot be normally reeled

4. Bassins (Pelettes, Ricatti, Galettame, Bisou) – the inner, parchment-like layer of the cocoon skin

5. Bourre (Strazza) – scraps produced during reeling the skeins, twisting and cleaning the threads

4.1.8   Dupion

While the silkworms are spinning their cocoons, it sometimes happens that two worms are so close to each other that they spin a double cocoon instead of two separate ones. These double cocoons cannot be reeled in the conventional way but have to be processed on a special machine. This is known as dupion (douppion) reeling, dupion literally meaning 'double'. The two filaments thus produced are intermingled, and when added to normal filaments in fabric manufacturing produce irregular slubs which create an effect very much in demand for making special fabrics, particularly for bridal wear. These fabrics are often mistakenly known as 'wild silk', because they appear natural and irregular.

Many silk fabrics are still produced on handlooms which can produce superior goods for which the purchaser is obviously prepared to pay the extra cost involved. The technique of spun silk production is quite distinct from that of thrown silk yarns, and many of the mechanisms involved in the process are closely related to machines used in the preparation and production of yarns from staple fibres such as cotton, wool, flax and jute. After degumming, cleaning and opening, the fibres are cut into short lengths and then combed into slivers. These are made into yarns by mixing, drafting and spinning.

4.1.9   Properties of silk Fibre

Silk is a protein fibre, and the raw silk is composed of two important proteins: fibroin (the silk filament) and sericin (the gum). Each raw silk strand in the cocoon is known as 'bave'; it is composed of two fibroin filaments (10–14 μm) called 'brin' that are held together by sericin gum (see Fig. 4.7). Silk fibres are biodegradable and highly crystalline with a well-aligned structure. They have a higher tensile strength than glass fibre or synthetic organic fibres, good elasticity and excellent resilience. Silk fibre is normally stable up to 140°C, and the thermal decomposition temperature is greater than 1500°C. The densities of silk fibres are in the range of 1320–1400 kg/m^3 with sericin

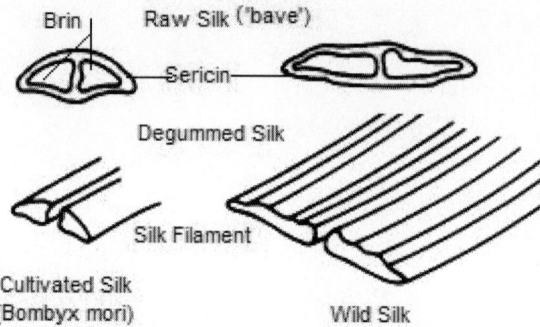

Figure 4.7. Properties of silk fibre

and 1300–1380 kg/m³ without sericin. The mulberry silks show a more or less triangular cross section and a smooth surface. Among the nonmulberry varieties, tasar and muga exhibit an elongated rectangular or a wedge-shape cross section and a large cross-sectional area. Eri silk has a more or less triangular shape. The cocoon fibres of domestic silkworms like mulberry and eri usually have an irregular cross section ranging from triangular to circular. Even within the same fibroin filament, there may be variations in the cross section, depending upon the level of the cocoon layer.

Table 4.2. Density and moisture regain of different types of silk

Silk type	Density, ρ (g/cm³)			Moisture regain (%)		
	Outer layer	Middle layer	Inner layer	Outer layer	Middle layer	Inner layer
Mulberry (bivoltine)	1.35	1.361	1.365	8.52	8.14	4.46
Mulberry (multivoltine)	1.34	1.35	1.35	8.62	8.28	4.06
Tasar	1.3	1.33	1.34	10.76	10.27	4.55
Muga	1.332	1.34	1.348	9.82	9.47	3.56
Eri	1.28	1.29	1.295	10.21	9.79	4.11

The silk is basically a hygroscopic fibre, and in the ungemmed form, the moisture regain is about 11% and the degummed fibre it is only about 9%, from which it is obvious that the sericin is more hygroscopic than fibroin. The moisture regain of wild silks are 1–2% higher than mulberry silks. Layer by layer of the cocoon, the fibre shows less the moisture regain from outside to inside which can be seen (Table 4.2). Silk can absorb up to 30% of its weight in moisture without creating a damp feeling. When moisture is absorbed it generates 'wetting heat' (69 J/g) which helps to explain why silk is comfortable to wear next to the skin. Silk fibres display of heat of wetting from 0°C, which indicate that coupled with high regain of about 10%, silk fibre offers sufficient time for the wearer to acclimatize to the change in weather. It can also swell about 30% of their volume under wet conditions: because of this, silk textile materials have a lower dimensional stability compared to other natural fibres. Due to the swelling action, silk fibres display partial loss of strength under wet condition. Prolonged exposure of silk to steam or boiling water results in hydrolyses of peptide bonds, thereby damaging the silk fibre.

Silk is the only natural fibre which exists as a continuous filament. Each *B. mori* cocoon can yield up to 1600 m of filament. These can be easily joined together using the adhesive qualities of sericin to form a theoretically endless filament. The silk fibre's triangular cross section gives it excellent light reflection capability. The silk fibre is smooth, unlike those of wool, cotton and others. This is one of the reasons why silk fabrics are so lustrous and soft. Silk has a tenacity of approximately 4.8 g per denier, slightly less than that of nylon. Silk has poor resistance to UV light and for this reason is only recommended for those curtains that are lined or not exposed to direct sunlight.

These physical characteristics are determined by the structure of the macromolecule composing the fibroin as shown in (Fig. 4.8). Part of the macromolecule is made up of amino acids with a low molecular weight, offering a series of crystalline regions which confer a high degree of tenacity on the fibre. The rest of the macromolecule is characterized by the presence of amorphous areas enclosing amino acids of a relatively higher molecular weight. The presence of both crystalline and amorphous zones makes for a combination of strength, flexibility and elasticity.

4.1.10   Chemical structure

Silk fibre, which is fibroin, can be described under four headings.

Primary structure: The silk fibroin is composed of many microfibrils (see Fig. 4.8), which are composed of a large number of amino acids in ordered and disordered regions. These amino acids can be represented as $-HNCH_2RCO-$, where R is the side group specific to a different amino acid. The amino acids in the fibroin are joined in a sequential polypeptide chain by the amide linkages (CONH), which are known as polypeptide bonds. The length of the fibroin molecular chain is about 140 nm, and its molecular weight ranges from 300 to 400 kDa. The silk gum called sericin is also protein with same amino acids but varying in percentage.

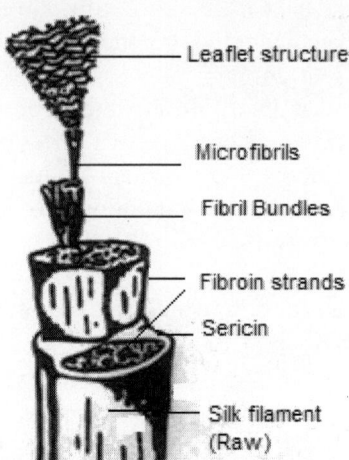

Leaflet structure

Microfibrils

Fibril Bundles

Fibroin strands

Sericin

Silk filament
(Raw)

Figure 4.8. Microstructure of a raw silk filament

Table 4.3. Composition of various amino acids in fibroin and sericin

Amino acid	Side group	Sericin	Fibroin
Glycine	H–	147	445
Alanine	CH_3–	43	293
Leucine	$(CH_3)_2CHCH_2$–	14	5
Isoleucine	CH_3CH_2 CH (CH_3)–	7	7
Valine	$(CH_3)_2$ CH–	36	22
Phenylalanine	$C_6H_5CH_2$–	3	6
Serine	CH_2 (OH)–	373	121
Threonine	CH_3 CH (OH)–	87	9
Tyrosine	HO $C_6H_5CH_2$–	26	52
Aspartic acid	HO OC CH_2–	148	13
Glutamic acid H	HO OC CH_2 CH_2–	34	10
Arginine	H_2N C(NH) NH $(CH_2)_3$	36	5
Cystine (1/2)	–$(S–CH_2)_2$–	5	2
Methionine	CH_3 S CH_2 CH_2-		1
Lysine	H_2N $(CH_2)_4$–	24	3

Amino acid	Side group	Sericin	Fibroin
Proline	CH_2 - CH_2 CH_2 CH - COOH $\diagdown NH \diagup$	7	3
Hystidine	N - CH ‖ $\diagup C$-CH_2- N - CH	12	2
CH_2CH–COOH			
Tryptophane	$C_{11} H_{12} O_2 N_2$		2
NH_3 or amide		86	

Secondary structure: The silk fibroin is composed of simple amino acids, mostly with hydrocarbon side groups; as a result of these side groups strong hydrogen bonding and salt linkages exist between the polypeptide chains of the amino acids, resulting in a β-pleated sheet form of silk fibroin.

Tertiary structure: The tertiary structure of the silk fibroin details the three-dimensional configuration of the polypeptide chains and the β-pleated sheet forms. The crystal structure of silk fibroin, in which four amino acid molecules pass through a rectangular unit cell with $a = 9.37$ Å, $b = 9.49$ Å and $c = 6.98$ Å.

Quaternary structure: It is reports that overall, the silk fibroin structure consist of aggregates of polypeptide chains in β-pleated sheet form, arranged parallel to the silk fibre axis. These *b*-pleated sheet forms are held together by lateral forces with freedom and space in disordered regions. In ordered crystalline regions, close packing of the polypeptide chains and β-pleated sheet forms are assisted by strong hydrogen bonds and further strengthened by the van der Waals forces.

The main types of bonds for the tertiary structure are as follows:

1. Electrostatic salt bonds between positively and negatively charged groups, for instance between $-NH_3+$ and $-COO-$ groups
2. Hydrogen bonds between side chains as, for instance, between hydroxyl groups and free carboxyl groups
3. Bonds due to van der Waals forces between polar groups
4. Bonds between nonpolar side chains of valine, leucine, isoleucine, pheny lalanine residues

Silk fibre is composed of different amino acids displaying amphoteric nature; thus, silk fibres can be dyed with all classes of dyeing agents. Acid dyes, metal complex dyes and reactive dyes are the kinds of dyes most often used for silk fibres.

Sericin is amorphous and dissolves in hot soap solution. Fibroin is the form of a filament thread and dissolves in 5% sodium hydroxide solution at boil. Both fibroin and sericin are protein substances built up of 16–18 amino acids, out of which only glycine, alanine, serine and tyrosine make up the largest part of the silk fibre (see Table 4.3), and the remaining amino acids containing bulky side groups are not significant. The chemical structure of fibroin and sericin for four amino acids – glycine, alanine, serine, tyrosine. Fibroin contains hardly any sulphurous amino acids (cystine) and also only a minimal amount of amino acids as side chains. The low number of large side groups enables silk to be densely packed. The crystalline structure of the polypeptide chains in silk fibroin is shown Fig. 4.9. Silk polymer occurs only in β-configuration. The important chemical groupings of the silk polymer are the peptide groups which give rise to hydrogen bonds, the carboxyl end amine groups which give rise to salt linkages and the van der Waals forces. It has high degree of molecular orientation which accounts for the excellent strength of the silk fibres. However, the polymer system of silk is now considered as being composed of layers of folded, linear polymer as shown in Fig. 4.9. Such a polymer system explains why silk is essential to be about 65–70% crystalline. The DP of silk fibroin is uncertain, with DP of 300–3000 having been measured in different solvents. Silk fibres are smooth surfaced and translucent with some irregularity in diameter along the fibre. The fibres are basically triangular in cross section with rounded corners. The isoelectric point of silk is about 5. Silk fibres are very stable in the face of reducing agents but are easily degraded by oxidizing agents. This means they can be mildly bleached with hydrogen peroxide.

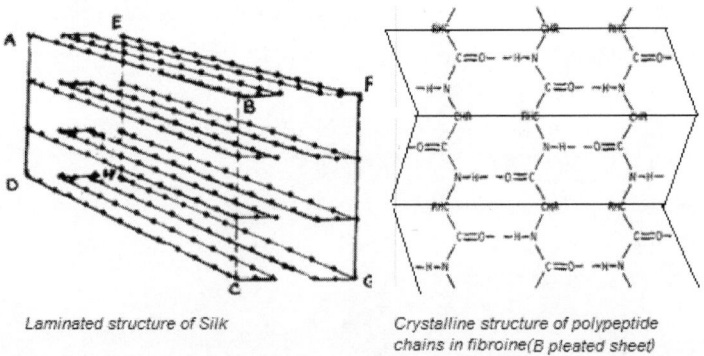

Laminated structure of Silk

Crystalline structure of polypeptide chains in fibroine(B pleated sheet)

Figure 4.9. Polymer system of silk fibre

The actual fibre protein is called fibroin, and the protein sericin is the gummy substance that holds the filament together. The average composition of raw silk is 70–75% fibroin ($C_{15}H_{23}N_5O_8$), 20–25% sericin, 2–3% waxy substances extractable by ether and alcohol and 1–1.7% mineral matter. But the composition changes considerably according the type of silk as shown in Table 4.4.

Table 4.4. Composition of silk fibre in different types

Component	Mulberry	Tasar	Muga	Eri
Fibroin	66–72	78–85	80–86	82–88
Sericin	25–32	14–17	12–16	11–13
Wax	0.3–0.4	1–2	0.5–1.0	1.5–2.2
Minerals, salts and others	0.7–0.8	3–4	2–3	2–3

4.1.11   Chemical properties of silk yarn

Silk fibroin are found in the polypeptide chain with the formula (–CHR–CO–NH–)n, where n = 1100 and R is one of the various amino acid residues. The polypeptide chain has both a backbone and side chains consisting of radicals R of amino acids. The degree of branching of the polypeptide chain depends on the amino acids contained in the protein. Thus the side chains form 19% of the weight in silk fibroin. The side chains can be polar or nonpolar. Amino acids present in the fibre are bifunctional, i.e., they contain both acidic carboxyl (COO–) and basic amino (NH_3+) group in their molecular structure (Fig. 4.10). Hence, the silk fibre can be considered as a zwitterion. Silk has good resistance to acids, but hot concentrated acids break the peptide bonds of the amino acid molecular chains, thereby damaging the fibre. Exposure to weak acids results in the 'scroop effect', which is a famous silk finishing treatment that produces a crackling noise when such silk fibres are rubbed together. It is caused by the creation of a fine skin-like sheath around the silk fibre by reorientation of fibroin on the outermost layer when it is treated with weak acids like acetic acid.

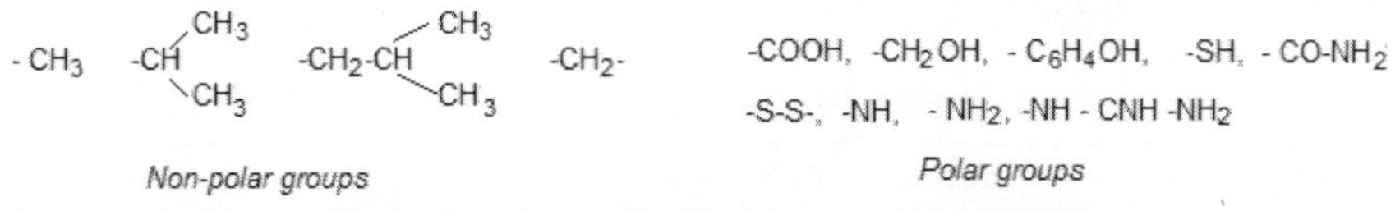

Figure 4.10. Chemical structure of silk yarn.

As mentioned earlier, silk has unusual mechanical properties compared to similar fibres which are because of the of the size and orientation of the crystalline domains, the connectivity of these domains to the less crystalline domains, and the interfaces or transitions between less organized and crystalline domains. They are strong, extensible and mechanically compressible. They also display interesting thermal and electromagnetic responses, particularly in the UV range, and form crystalline phases related to processing. Other properties of silk such as good thermal stability, optical responses, dynamic mechanical behaviour and time-dependent responses have all been used in a number of applications in various fields. In this chapter, details of various properties of different varieties of silk are presented. The tenacity of the silk can be attributed to its linear density. Elongation, on the other hand, increases with an increase in linear density. Various studies in this line has shown that the tenacity ranges between 2.5 and 4.82 g/d (grams per denier), for Japanese and Chinese mulberry varieties, 2.4–4.32 g/d for Indian mulberry varieties, 3.74–4.6 g/d for Indian tasar varieties (Table 4.5).

Table 4.5. Average tenacity, elongation and initial modulus values of silk

Silk type	Tenacity (g/d)	Elongation at break	Initial modulus (g/d)
Mulberry (Bivoltine)	3.75	13.55	95.35
Mulberry (Cross bred)	3.85	16.1	106.8
Tasar	4.5	26.50	84.2
Muga	4.35	22.35	81
Eri	3.70	20.80	89.05

In a study on chemical structure and physical properties of *Antheraea assama* (muga) silk, it has been reported that the tenacity of muga varies between 3.2 and 4.95 g/d. Eri variety silk has shown the lowest tenacity value, ranging between 2.3 and 4.0 g/d. On the other hand, elongation at break showed a higher value for all the nonmulberry silks compared to mulberry varieties and ranges between 19 and 35%. The polypeptide chains of the fibroin may hydrolize in boiling water or in steam. However, this hydrolysis is very slow and only becomes critical in the presence of mineral acids or alkalis. During finishing the fibre has a high absorption capacity for water and the substances dissolved in it. Hard and dirty water must be avoided, as these impair the lustre and handle (Table 4.6).

Table 4.6. Summary of physical and chemical characteristics of silk

Agents	Physical and chemical properties
Moisture regain	Under standard testing conditions, silk absorbs 10.5% moisture. The commercial weight is set at 11% water.
Lustre and handle	On degumming, the silk fibre acquires a soft elegant lustre. This can be because of the triangular structure of the fibre and light reflections due to this. The handle of the degummed silk is scroopy

Agents	Physical and chemical properties
Colour	
Specific gravity	Raw: 1.3–1.37. Degummed: 1.25
Fineness	Fineness of silk ranges 0.9–1. 75 (approx.)
Tenacity	2.5 and 4.82 g/d
Abrasion resistance	The surface of silk is very easily chafed by abrasion, especially when the material is wet and swollen.
Temperature sensitivity	Silk can withstand higher temperature than wool. Prolonged treatments above 170°C can cause damage to the silk fibre.
Biological resistance	When stored in a moist condition, silk may be degraded, i.e., damaged, by bacteria.
Influence of light	Silk fibres may also be damaged by prolonged exposure to direct sunlight.
Action of acids	Silk fibre has good resistance to organic acids but not as resistant as wool. Silk tolerates dilute organic acids well; they enhance its lustre and improve the (scroopy) handle.
Action of alkali	Stronger acid can dissolve silk. Alkalis can attack the chain ends of the peptide. The severity of the degradation also depends on the pH and alkali. However, silk fibroin is considerably more stable in this respect than wool.
Resistance to oxidizing agent and reducing agent	Silk exhibits good stability to the usual concentrations of oxidation and reduction agents in bleaching. The normal reduction agents used in discharge printing also have only a minimal effect on the fibre strength.
Action of Metal salts	Silk has the capacity to form a chemical bond with certain metal salts such as tin chloride, which is used for weighting the silk, mostly with yarn-dyed necktie fabrics.

4.2   Wool

Even though cotton was the first fibre used by human being to make cloth, wool also has an equally old history. The oldest record we can probably get is the mentioning of Phoenicians buying woolen materials from Israelites and shipping to England to barter for tin and raw wool in the seventh century BC. Domestication of sheep for wool may have started around 100 BC. After industrial revolution, sheep rearing and manufacturing of wool and woolen materials have caught up in many countries like Britain, Australia, New Zealand and South Africa.

Wool is the fibre derived from the hair of animals of the Caprinae family, principally sheep, but the hair of certain species of other mammals such as goats, llamas and rabbits may also be called wool. It has the highest moisture regain, i.e., 14%. It exhibits felting property and is easy to spin. Wool has two qualities that distinguish it from hair or fur: it has scales, which overlap like shingles on a roof, and it is crimped; in some fleeces, the wool fibres have more than 20 bends per inch. Due to crimp present in it, it has heat in stored within the length of the fibre is around 3–5 in.

Sheep were one of the first animals domesticated by humans. These sheep looked very different from modern animals. They had a long hair-like outer guard layer with a more downy insulation layer underneath resembling what we think of as wool. They shed this coat in the spring. Primitive people collected the fibre finding it to be of great use, learned to spin. Thus the wool industry was born. At first, sheep were brought to the North American continent in 1493 on the second voyage of Columbus. New England was the first area to establish a spinning and weaving industry. Initially begun in homes, later in small factories, it continued to grow. Eventually the first water-powered textile factories were established in 1788.

The industry was greatly boosted during the Civil War by a great demand for wool to make soldier's uniforms. Breeders use imported merino sheep to improve native stock to help meet the demand. Wool is 5% of the world's textile industry. All domestic wool produced finds use in a variety of products. However, because of defects, contamination and other quality problems, it is of lower value than some imported wool. Main wool producing countries are Australia, Argentina, New Zealand, Russia, Republic of South Africa, Great Britain, China and the United States.

The quality of wool is determined by the breeding, climate, food, general care and health of the sheep.

- Cold weather produced a hard and heavy fibre.
- Poor or insufficient food retards growth.
- It has soft resiliency power, and it is also used to make rugs and blankets.
- The chief wool-producing countries are Australia, Union of Soviet Socialist Republics, New Zealand, Argentina, South Africa and the United States.
- The chief constituent of wool fibre is a protein substance called 'keratin' and it is the only fibre which contains 'sulphur'.

4.2.1   Quality of wool

Generally wool can be classified into different category as per their quality or as per breed of the sheep. As in the case of natural plant fibres like cotton, the quality of wool depends greatly upon the conditions under which it is grown. Wool derives from a living creature, and it is affected not only by the hereditary characteristics of the sheep but by the environment in which the sheep has lived.

There are two types of wool namely clipped or fleece wool taken from live sheep and pulled wool removed from sheep already dead.

Wool may be classified into two types:

1. Classification by sheep

2. Classification by fleece

4.2.1.1   Classification by sheep

There are over 200 grades of wool-producing sheep.

- Merino wool: Produced from merino sheep and the fibres are very fine, strong and elastic. Class I wool (or) merino wool – Merino sheep of Australia, South Africa and South America produce the best quality wool which is strong, fine elastic and has good working properties. Among Australian wool Port Philip wool is reputed to be the finest, Sydney and Adelaide wool are not so fine and a shade yellower, whereas Tasmanian wool has a beautiful whiteness after washing, even though the fineness is not of that level of the above two. South African wool is very crimped or wavy and has a good white colour after washing. Best South American wool comes from Montevideo, Buenos Aires, Punta Arenas, in that order, even though all of them are not of the quality level as those comes from Australia or South Africa.

- It has the greatest amount of crimp of all wool fibres and has a maximum number of scales to give maximum warmth and spinning qualities.

- Class II wools (cross-bred wool) – The effort to rear merino sheep in countries like New Zealand, England, Scotland, etc., was not that successful. Hence cross-bred sheep was tried which can be used for wool and meat was found much successful economically. Wool from these cross-bred sheep is categorized as class II wool. It is obtained from the sheep from England, Scotland, Ireland, Wales, New Zealand, some parts of Australia and South Africa. It is 2–8 in. in length, has a large number of scales per inch and has good crimp. The fibres are strong, fine, elastic and have good working properties. It is not as good as merino wool, but nevertheless it is very good quality wool.

- Class III wools: This class of sheep gives wool of less elastic and resilient. The fibres are 4–8 in. long, are coarser and have fewer scales and less crimp than merino wool and class II wool. They are smoother and have more lustre. They are nevertheless of good enough quality to be used for clothing.

- Class IV wools: This class is actually a group of mongrel sheep sometimes referred to as half breeds and are coarse and hair like. The fibres are 1–16 in. long are coarse and hair like, have relatively few scales and little crimp and therefore smoother and more lustrous. It has least elasticity and strength and used mainly for carpets, rugs and inexpensive low-grade clothing.

- Asian wool: This quality wool comes from sheep bred in China and other parts of Asia, in Turkey and Siberia. They are often long, coarse compared with fibre produced in in Australia, South Africa, etc.

4.2.1.2   *Classification by fleece (based on the shearing of the fleece)*

- Lamb's wool: About 6–8 months.
- Hogget wool: About 12–14 months.
- Weather wool: Any fleece clipped after the first shearing of sheep.
- Pulled wool: The shearing done by a chemical depilatory like lime before it is slaughtered for meet.
- Dead wool: The shearing is done over the dead sheep.
- Cotty wool: The wool obtained from any sheep of severe weather condition.

Merino wool is the best grade of wool. In addition to clothing, wool has been used carpeting, felt, wool insulation and upholstery. There are about 1000 million sheep of nearly 500 different breeds scattered about the world, with large concentrations in New Zealand, Australia, South Africa, South America, China and Russia. The British share of the world's sheep is 3%, with a 2% share of the world's wool, but British sheep have contributed to the world's flocks out of all proportion to the story the statistics alone tell. The basic steps in wool processing have remained unchanged for centuries. Wool obtained from different parts of the sheep varies in fibre fineness, length and crimp. There are variations between locks, or cohering bunches of wool, and between fleece from different areas of the sheep. The quality of wool obtained from the belly is short and burry; the shoulders yield the finest wool. Sorting is the separation of the different qualities of wool from the fleece. The most important property is usually fibre fineness, closely related to softness of handle. The sorter unrolls the fleece on the sorters board, usually waist high and cuts away any wool carrying tar or paint marks. Next removes the coarsest wool, placing it in a separate basket and finally reaches the fine wool on the shoulders. A sorter can generally sort up to 4500 kg of Australian fleece in a week.

4.2.2   Wool production

Wool fibres grow from small sacs or follicles in the skin of the sheep. The wool fibres grow in groups of 5–80 hairs, and there are 1550–3410 per sq. cm (10,000–22,000 per sq. in). A typical sheep of the class I group sheep may carry as many as 120 million wool fibre in their fleece, and class II group will have some 16–40 million fibres. These fibres grow on an average at the rate of 2.5 cm (1 in.) in 2 months.

4.2.3   Shearing

Sheep are normally shorn of their fleece every year (in some countries like South Africa up to twice a year). On large stations the fleece is removed in one piece by power operated clippers. By skilled hands, the sheep fleece is clipped in 2 ½ min. This type of wool is known as 'fleece or clop wool'. Immediately after it has been removed, the fleece is 'skirted'. This involves pulling away the soiled wool around the edges. Then the whole fleeces are graded by experts who judge the fineness, length, colour and other characteristics. Finally, the various grades are packed into large sacks and then sewn up into large sacks and the sewn up into the bales. Each bale contains about 136 (300 lb) or more of wool (Fig. 4.11).

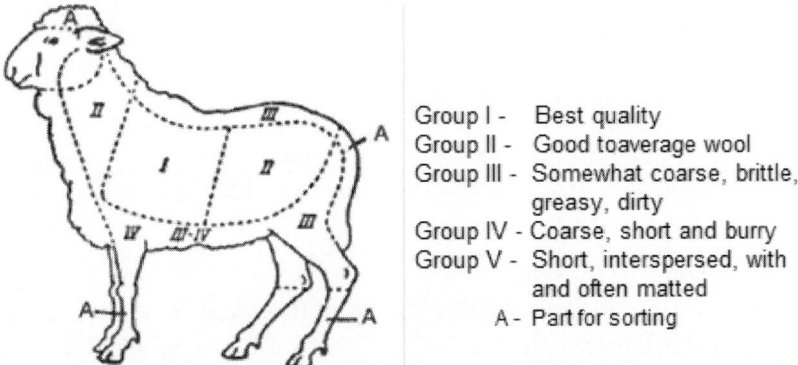

Group I -   Best quality
Group II -   Good to average wool
Group III -   Somewhat coarse, brittle, greasy, dirty
Group IV - Coarse, short and burry
Group V - Short, interspersed, with and often matted
A - Part for sorting

Figure 4.11. Various grades of sheep fleece based on the quality

4.2.4   Slipe wool (Mazamet)

Wool is also removed from the pelts of slaughtered sheep. The pelts are treated with lime and sodium sulphide. This loosens the wool, which can be pulled away without damaging the wool. This wool is called 'slipe wool'. The wool from hide is also removed by bacterial action and such wool is called 'Mazamet'. Wool removed from hide is inferior to clip wool, but used to mix with fleece wool.

4.2.5   Manufacture

Manufacturing process involves following:

1. Preparation
2. Sorting and grading
3. Garneting
4. Scouring
5. Carbonizing
6. Drying
7. Oiling
8. Dyeing
9. Blending
10. Carding
11. Gilling and combing
12. Drawing
13. Roving
14. Spinning

1.   Preparation
   (a) The hair of the sheep is trimmed first.
   (b) The raw wool or newly sheared wool called grease wool because it contains the natural oil of the sheep. Fleeces vary from 6 to 18 lb (3–8 kg) in weight and provides 3 lbs (1.5 kg) of scoured wool.

2.   Sorting and grading
   (a) Wool sorting is done by skilled workers who are experts in distinguishing by touch and sight. Twenty different grades are obtained from one fleece. Each grade is determined by type, strength,

length, fineness, elasticity and strength determined by type, length, fineness, elasticity and strength.

(b) Wool is sorted and graded according to the quality, then trimmed, rolled up, tied and packed in sacks weighing about 225–350 lb (100–160 kg). Superior wool comes from the sides and shoulders of sheep where it is longer, finer and softer and it is treated as one fleece. Wool from the head, chest, belly and shanks is treated as a second fleece. Then wool reaches the mill in bags. The raw wool or newly sheared wool is called grease wool as it contains the natural oil of the sheep (Fig. 4.12).

(c) Counts: The price of raw wool depends on the buyers assessment of its fineness and length. Quality is defined by numbers which at one time described the limiting fineness of count of yarn into which it could be spun. An 80s wool, for example, was considered capable of being spun into a yarn of 80s count. This meant that 1 lb (454 g) of wool would yield hanks each containing a fixed length of yarn. In the worsted industry, the standard length of a hank is 540 yards (504 m). Under modern conditions, these spinning limits are no longer valid but the traditional numbers continue to be used to describe wool quality fineness. Merino wool can be used to spin 60s to 100s, wool from cross breeds

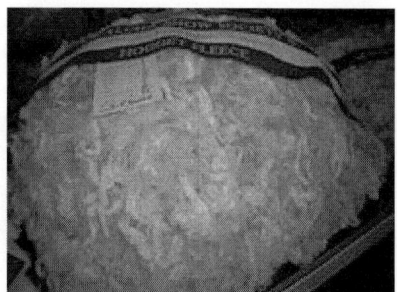

Figure 4.12. Raw sheep wool

from 36s to 60s and coarse wool such as that is used for carpets can spin 44s or coarser.

(d) Fibre length: The average length of wool fibres is described by special terms such as 'combing' or clothing. When the fibres are long enough to undergo combing and be made into worsteds – 65 mm (21/2 in.) or more – they are combing wools; 176 mm (7 in.), they are long wools. Fine fibres, 38–65 mm (1.5 in.), can be combed in the French comb and are 'French combings'. Short wools or less than about 32 mm (1.25 in.) are described as 'carding or clothing wools'.

(e) Classifying: When the bales of wool are opened in the mill, the fleeces are skirted if this has not been done earlier. The fleece may be classified as a whole or, if variable in quality, separated into sections such as shoulders, sides, back, thighs, britch and belly. In general, the shoulders provide the best wool and the flanks a slightly lower quality wool. The belly, tail and legs yield the poorest quality of wool. In medium and long wool breeds of sheep, the head, legs and britch usually produce the highest proportion of hair and kemps. Some breeds, such as the scotch blackface, produce hair and kemp fibres in all parts of their fleece.

3. *Garnetting*: Separating the used and unused materials to a fibrous mass by picking and shredding. Recycled wool fibres are obtained by separately reducing the unused and used materials to a fibrous mass by a picking and shredding process called garneting. The fibres are then put through a dilute solution of sulphuric acid or hydrochloric acid, which destroys any vegetable fibre that may be contained in the raw stock. This process is known as 'carbonizing' and the resultant wool fibres are called 'extracts'. The new staple ranges from ¼ to 1½ in. in length.

4. *Scouring*: 'A thorough washing of raw wool in an alkaline solution is called scouring'. The scouring machine contains warm water, soap and a mild solution of soda ash or other alkali and is equipped with automatic rakes which stir the wool. Rollers between the vats squeeze out the water. Valuable by-products are obtained from the spent liquors in the scouring of wool. The most important by-product

(A)   (B)

Figure 4.13. (A) Wool washing and (B) wool carbonizing

is lanolin, which is largely used in manufacture of cosmetics, adhesive plasters, disinfectants, ointments, etc. This removes the grease and oil in wool and makes it absorbent to dye.

5. *Carbonizing*: If the fibres are having still vegetable matters, the fibres are treated with a dilute solution of sulphuric acid to remove vegetable matters. This process is called carbonizing (Fig. 4.13).

6. *Drying*: Wool is not allowed to dry completely, 12–16% of moisture is left over.

7. *Oiling*: Wool becomes unmanageable after scouring, the fibre is usually treated with various oils, including animal, vegetable and mineral or a blend of these to keep it from becoming brittle and to lubricate(oil) it for the spinning operation.

8. *Blending*: Wool of different grades may be blended or mixed at this stage. Inferior grade of wool mixed with better grades of wool or a small amount of cotton is blended with a raw wool and an increase of twist can contribute in greater the strength in the fabric. Man-made fibres such as nylon, polyester or acrylic may be blended with wool and the wool helps in contribution of warmth, absorbency, drape and handling (Fig. 4.14).

Figure 4.14. Blending of wool

9. *Carding*: The carding process introduces the classification of woolen yarns and worsted yarns. At this point, of manufacturing process, it should be decided whether wool fibre is to be made into a woolen or a worsted product, because manufacturing of woolen and worsted is different. Fibres are passed through rollers covered with thousands of wire teeth to orient the fibres parallel. This separates the woolen yarns and worsted yarns (Fig. 4.15).

Figure 4.15. Carding

10. *Gilling and combing*: The carded wool which is to be made into worsted yarn is put through gilling and combing operations. The gilling process removes the shorter fibres (called as combing noils and of 1–4 in. in length) and places the longer fibres (tops) as parallel as possible and further cleans the fibres by removing any remaining loose impurities. Combing noils (shorter fibres) are used for ordinary and less expensive fabrics and tops (longer fibres) for manufacturing worsted fabrics as gabardine, whipcord, serge and convert and produce fabrics with good colour, feel and strength.

11. *Drawing*: It is done only to worsted yarns. It is an advanced operation which doubles and redoubles slivers of wool fibres. The process draws and twists the fibres and makes the slivers more compact and thin.

12. *Roving*: It is a process to hold the thin slubbers intact. The fibre passes between the rollers, over the coarse wire teeth of the first card clothing, and over progressively finer toothed card clothing. And the fibre that leaves the machine are in the form of untwisted ropes known as 'rovings' (Fig. 4.16).

13. *Spinning*: Here the wool roving is drawn out and twisted into yarn. Woolen yarns are spun on mule spinning machine. Worsted yarns are spun on mule, ring, cap or flyer kind of spinning. Two different systems to spin worsted yarns – is English System and French System.

Figure 4.16. Wool after roving

Wool fibre is made up of two parts: the follicle, which is located below the surface of the sheep skin and produces the fibre material, and the shaft, which grows from the follicle and is composed mainly of the fibrous protein α-keratin (see Fig. 4.17). The fibre can be categorized in terms of zone 1 – bulb zone (proliferation and differentiation); zone 2 – elongation (fibril formation); zone 3 – prekeratinization (lateral aggregation); zone 4 – hardening (keratinization); zone 5 – posthardening (hard keratin). Follicle is a living cell producing a continuous substance which emerges as the fibre shaft (wool). The rate of fibre shaft production is around 1 cm/month for each follicle. The number of follicles per sheepskin (the follicle density) varies according to

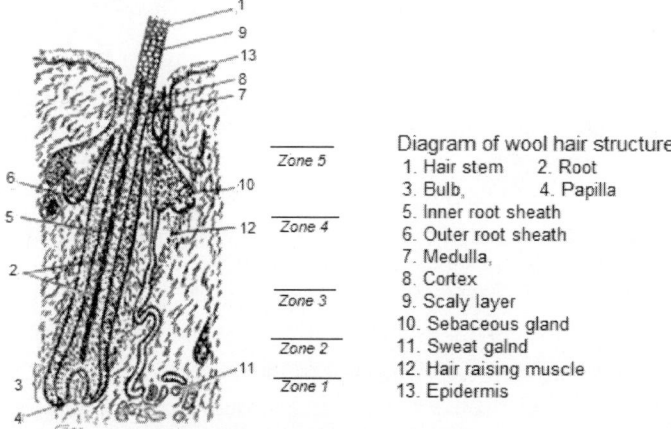

Diagram of wool hair structure
1. Hair stem   2. Root
3. Bulb,   4. Papilla
5. Inner root sheath
6. Outer root sheath
7. Medulla,
8. Cortex
9. Scaly layer
10. Sebaceous gland
11. Sweat galnd
12. Hair raising muscle
13. Epidermis

Figure 4.17. Structure of a wool hair

breed and dictates the amount of wool produced per animal. This explains why the annual yield of one sheep can vary from 1 kg (most sheep breeds) to 4 kg (Australian merino) clean wool fibre. The diameter of the fibre is also genetically determined, although this can also be slightly affected by weather and food. Consequently, wool fibre diameters range from 11 μ (fine Australian merino) to 100 μ (sheep of northern hemisphere).

4.2.6   Different types of wool

Merino wool: One of the finest and soft wool. Produced by merino sheep which are grown in Australia, South Africa and South America (Argentina is one of the highest wool producing country in South America). In Australia itself, the wool produced in different area are of different quality – Port Philip produces the finest Australian quality, Sydney and Adelaide wools not so fine and a shade yellower. Tasmania produces first rate quality and on washing gives a beautiful white. South African wool is very crimped or wavy and has a good white colour like Tasmanian wool. The South American wool is not of such good quality as Australian and south African qualities. Merino sheep is also reared in Europe (Germany, France, Spain, etc.) and gives good quality wools.

Cross-bred wool: Some areas where merino wool were not able to rear cross-bred wool were tried successfully – New Zealand, some parts of Australia and South Africa, Britain, Ireland, Argentina, etc. These wools are mainly used for knitwear, tweeds, worsteads, carpets, cheviot suitings, etc. Urugway produce wool which are named after the country's principal port and are marketed as 'Montevideo wools', but are also a cross-bred wool.

Englands sheep population is relatively high, and it produces types of wool which are not found anywhere else:

1. Cheviot wools: harsh, robust, interspersed with kemps

2. Shetland wools: coarse, long yet soft wools, rare. Shetland wool is from sheep bred and reared in the Shetland Islands. The wool from the Shetland sheep is particularly fine and imparts a soft handle to the textiles produced from it. The shades are principally mottled grey over blue and reddish brown. They are used in men's jackets, light to medium weight between-season coats as well as winter coats in the form of double cloth; lighter weight qualities are also used for women's jackets, costumes and coats.

3. Lustre wools: wools with high light reflectance, rare. It is up to 30.5 cm long and used for making lustrous dress fabrics, buntings and linings.

4. Down wools: Curly, crisp handle, medium length 7–100 mm. Used to make hosiery, cheviot suitings and flannels.

5. Mountain wools: Great quality, coarse, long. Used for making tweeds and carpets.

Asian wool: China, Turkey and Siberia also produce wool from different breeds of sheep. The wool produced are often long and coarse compared to the above varieties.

4.2.7   Fineness of wool

Wool always vary in thickness. Generally finer wools are shorter and coarse wools are longer. Fineness of wool refers to the diameter of a fibre which is expressed in microns (1 μ = 1/1000 mm), and on this basis, wools are classified as summarized in the Table 4.7.

Table 4.7. Classification of wool based on the fineness of the yarn

Variety	Fineness in microns
Fine merino wools	17–20
Merino wools	20–24
Fine cross bred	24–28
Medium cross bred	28–37
Coarse cross bred	over 37

To express the fineness of wool sometimes are classified as per 'quality fineness number' Under English system, the wool is assigned a number corresponding to the finest English count of yarn which can be spun from it. German system uses capital letters from the alphabet which is more arbitrary (e.g., AA, A, etc.; see Table 4.8).

Table 4.8. Expression of fineness of wool by English and German systems

Wools	English number	German system
Merino wools	80s	AAA
	70s	
	64s–70s	AA
	64s	A
	60s	A/B
Cross-bred wools	58s–60s	B

Wools	English number	German system
	56s–60s	B/C1
	50s–56s	C1
	50s	C2
	48s–50s	D1
Coarse cross bred/Asian	48s	D1/D2
	46s	D2/E1
	40s–46s	E1/E2
	36s	F1

4.2.8   Dimensions of the wool fibre

The dimensions of wool fibre vary according to their fineness coarseness, length, etc. A fine wool may be about 38–128 mm (1½–5 in.), medium wools 65–150 mm (2 ½–6 in.) and long wools 125–375 mm (5–15 in.). The average diameter of a merino high quality wool is about 17 μ, a medium quality wool 24–34 μ and along wool about 40 μ. They have an oval cross section.

4.2.9   Morphology of wool

Morphology of wool is comparable to human hair and other hair fibres. The structure principle of a multiple composite system can be found with all keratin fibres. The outer cover is formed by the cuticle, which surrounds the inner strand, the cortex, in the centre of which an axis of medullar cells can be found.

The wool fibre under microscope will show four distinctive regions – outer sheath or epicuticle, scale cell layer, cortex and medulla (see Fig. 4.18).

1. *Outer sheath or epicuticle*: The epicuticle covers the fibre like a covering made of wax. This is the only part of the fibre, which is not protein and is the outer most water repellant thin membrane. Even though it is water repellent, it has small pores, which allows small amount water entry. This is why a wool clothing can repel a light shower whereas the water vapour from sweat the body can be absorbed by the interior of the clothing without felling damp. During wet processing or by mechanical means, if this sheath is damaged the wool gets wetted faster. It is about 100 Å thick.

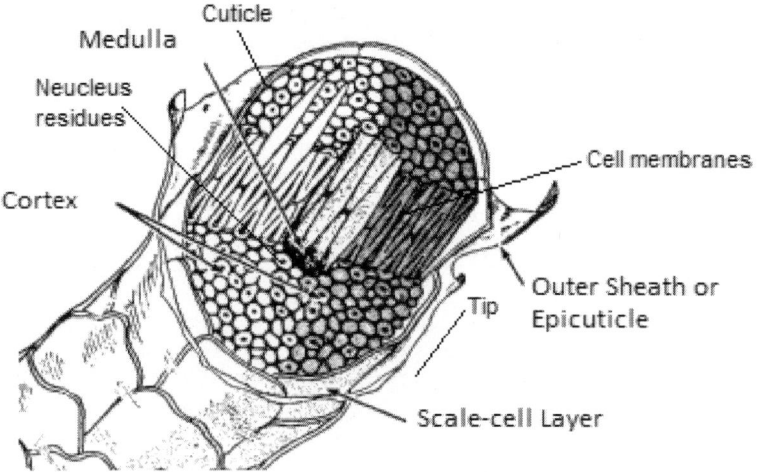

Figure 4.18. Four regions of a wool fibre

2. *Scale cell layer (epithelial scales)*: Epithelial scales come below the epicuticle and it is a flat, scale-like cells which overlap like the shingles on a roof of tiles. The free ends of the scales point towards the tip of the fibre. The epicuticle and scaly layer together form the cuticle of the fibre. In finer fibres, the scales encircle the fibre, giving it the appearance of a stack of flowerpots sitting one inside the other. In coarser wools, the scales may not encircle the fibre, but overlap in two directions. These epithelial scales are, on an average, 0.05–0.5 μ thick.

The scale structure of the outer covering layer of wool has an effect on the felting and dyeing behaviour. If the wool fibres are nonfelted, the original hydrophobic scales are either rendered hydrophilic by a polymer or by oxidation (or a combination of the two) in such a way that a water film significantly alters the friction between the fibres.

The scale layer consists of

(a) *Outer epicuticle*: Hydrophobic, low cystine content, i.e., low degree of cross linking.

(b) *Exocuticle*: Hydrophobic, high cystine content or degree of cross linking, division into a-layer (extremely high in cystine) and b-layer (high in cystine). This is a layer resistant to enzymatic decomposition and heavily disulphide bonded. Since it contains relatively few polar groups, it works as an effective barrier for the diffusion of hydrophilic substances.

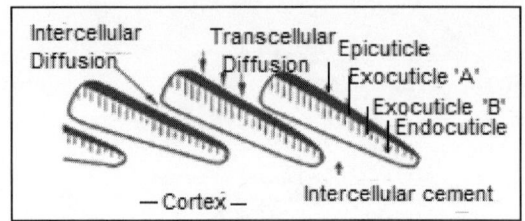

Figure 4.19. Diffusion routes into a wool fibre (Zollinger)

(c) *Endocuticle*: Low in cystine, i.e., low degree of cross linking. It is not as resistant as exocuticle and is solubilized by proteolytic

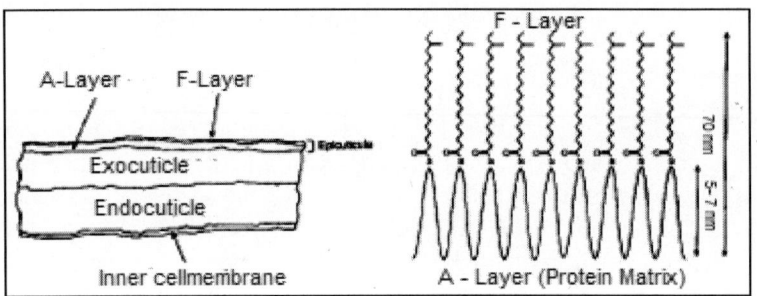

Figure 4.20. Migration of intercellular cement and cellular membrane complex

enzymes. In it is found cell detritus from the cytoplasm of the formerly living cells.

(d) *Intercellular cement*: globular proteins, low in cystine. The material between the macrofibrillae, the intermacrofibrillar cement, consists of residue from cytoplasm and nuclei of earlier active cells.

The high sulphur content of keratins, i.e., a high cysteine content and high degree of cross linking, in addition to the increased content of hydrophobic amino acids, is regarded as a cause for the hydrophobic nature of the scales. For this reason, the exocuticle (with approx. 35% cystine content) is regarded as a diffusion barrier for dyestuff molecules during the dyeing process. The high sulphur content causes reduced swelling capacity of this layer of fibres. Consequently, changes to the wool fibre by splitting the cystine links causes acceleration of the dye absorption. If the wool is untreated, the dyestuff does not diffuse vertically (Fig. 4.19) through the exocuticle (so-called 'transcellular diffusion') into the wool fibre, but migrates via the intercellular cement/cellular membrane complex (Fig. 4.20), i.e., between the scales, into the fibre (so-called 'intercellular diffusion'). This intercellular diffusion path can be observed at the beginning of the dyeing process.

In fine wools, only 10 μ or less of each scale may be exposed but actually it is about 30 μ long and 36 μ wide.

3. *The cortex*: The cortex is 90% of the bulk of the wool fibre and consists of dead cells and intercellular connective material. This region consists of millions of long spindle-shaped cells, thick in the centre and tapering to points at each end. These cortical cells are 100–200 μ in length and 2–5 μ wide. The cortical cells are themselves built up from fibrous components called fibrils, which are in turn constructed from protofibrils. These may be seen through the electron microscope as globular particles, which possibly consist of the keratin molecules themselves. The cortex of the wool fibre is composed of two/three distinct sections, ortho cortex and para cortex and a the meso cortex cells in which the proteins differ slightly in chemical and physical properties. Ortho- and para-cortex cells are found in merino wool fibres arranged as two strands twisted together, which causes the crimp in the fibre, exposing the ortho-cortex cell strand to the exterior and the para-cortex cell strand to the interior of the curl of the fibre. The fibre can be regarded as being formed from two half fibres of semicircular cross section which are joined lengthwise. The average length of the cortical cells is about 80–110 μ, the average width 2–5 μ and the thickness 1.2–2.6 μ.

Pigment granules containing the hair colouring melanin are found both in the cuticle and between the macrofibrillae. The macrofibrillae themselves are made up of microfibrillae, highly organized fibrillar proteins and matrix, a supposedly less organized structure, which sheathes each macrofilament. The great resistance and flexibility of the fibre is due in the main to this microfibrillae–matrix complex. Its strength is chiefly guaranteed by the number of disulphide bridges, which (it is believed) develop within and between the proteins of the macrofibrillae. These proteins are described as keratins.

4. *Medulla*: Many wool fibres, especially the coarser ones, have a hollow space running through the central axis of the fibre, called medulla. It can be hollow or may contain a loose network of open cell walls.

4.2.10   Characteristics of wool

4.2.10.1   Crimp

Wool fibres have a special wavy structure, which enables the fibres to hold together when twisted into yarns. This crimp of the fibre is pronounced in the fine wool fibres. The crimp gives a special elasticity because of the springy structure, which stretches under tension and return when tension is released. The crimp does not consist of waviness in a single plane, but takes the form of a three-dimensional waviness. It is related to the spiral form of two core sections of different constitution (ortho and para cortex), which twist spirally around one another in phase with twists of the crimp. The crimpiness varies with the origin and breed of the animal. The relevant criteria of crimpiness are:

1. Type of crimpiness
2. Number of crimps or waves along a given length
3. Permanence

Important characteristics such as heat retention, handle, loftiness and lustre are largely governed by the crimpiness. Although the waves or crimps of wool takes a three-dimensional form, we can distinguish the following types of crimpiness on the basis of the shape of the curve:

1. Completely flat: Coarse cross bred (carpet wools), (depth:length of staple = 1:1.1).
2. Flattened curves: Cheviot wool where the curve represents less than a semicircle, ratio 1:1.3.
3. Normal curves: Merino wools, fine and medium cross bred, curve in the shape of a semi circle, ratio 1:1.6.
4. Pronounced curves: Fine merino wools from special stock, curve exceeds a semicircle. Ratio 1:1.9.
5. Exaggerated curves: Defective wools, so-called 'hunger-fine wools'.

Crimpiness is expressed as the number of waves per inch or sometimes per centimetre. The crimp is of great importance and can only be removed by the simultaneous application of moisture, heat and tension.

4.2.10.2   Length

One should distinguish between the natural length of the fibre in the crimped form and the length of the straightened but not stretched fibre. The species of the sheep and the period of growth of the fleece are the determining factors of the length of the fibre. The increase in length after a year growth is however insignificant.

Table 4.9. Type of wool and length of the fibre

Type of wool	Length of the fibre (mm)
Merino and fine cross bred	25–80
Merino wool with less crimp	80–130
Straight smooth cross bred	120–300
Long, very lustrous carpet wools	Up to 550

4.2.10.3   Elasticity

Elasticity is the property of wool to recover from a fixed deformation. Elasticity is a property which is a must for a textile fibre to cope with the stress and strain to which it is subjected to during manufacturing, processing and wear. Hence we have to think of

Extensibility

Compressibility

Flexural rigidity

As explained earlier the wool has an inherent limited elasticity like rubber. This may be explained in relation to the chemical structure of the fibre. The molecules of wool, like rubber, are highly folded. In the fine structure of wool, the long protein molecules are lying alongside one another, held together at intervals by chemical cross links. When the fibre is stretched, the chain unfolds and when the force is released, they return to the folded state. Figure 4.21 shows how the folded molecules of α-keratin changing into the stretched molecules of β-keratin under tension.

The elasticity of a fibre play an important part in the shape retention and dimensional stability of the textile manufactured from it. Wool is having highest elasticity to any natural fibres. Even at load equal to half of the breaking load the elasticity of wool is almost 80%.

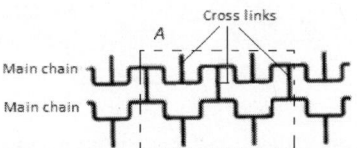
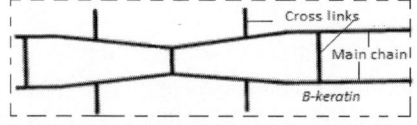

Folded molecule of alpha-keratin

Dotted portion A of the folded molecule stretched

Figure 4.21. Folded molecules of α-keratin and stretched molecules of β-keratin

4.2.10.4   Colour

Almost 90% of the wool traded today is having a cream shade which is often referred as 'natural white' in wool trade. Paler the shade, higher the value in the wool market as even pales shades can be dyed on these fibres without prior bleaching.

4.2.10.5   Lustre

Wool has a natural lustre pleasing to the eye. Natural wool have a more subdued reflection while coarse wool are highly lustrous. Lustre of a fibre is determined by its crimpiness, scale structure and the moisture content of the fibre.

4.2.10.6   Handle

Wool has a unique special handle and it is needless to say that the fine wool have a softer feel while coarser wools are harsher.

4.2.10.7   Tenacity

Wool is comparatively a weak fibre. The low tensile strength of wool is due to the relatively few hydrogen bonds that are formed. When wool absorbs moisture, the water molecules gradually force enough polymers apart cause a significant number of hydrogen molecules to break. Breakage and hydrolysis of these inter polymer forces of attraction are apparent as swelling of fibre and result in further loss of tenacity, when wool is wet.

4.2.10.8   Resilience

Wool has very good elastic recovery and excellent resiliency (ability to withstand elastic force). This is because of helical structure of arrangement of monomer or due to the helical polymeric configuration of wool, which is caused by the unique chemical and physical properties of wool. The fibre tends to bend and turn into a resilient three-dimensional structure.

4.2.10.9   Hygroscopicity

This is defined as the capacity of a fibre to absorb moisture from the atmosphere without feeling wet. Wool is an absorbent fibre due to the amorphous nature of_polymer system. In relatively dry weather, wool develops static electricity. This is because there are not enough water molecules in the polymeric system to dissipate any static electricity, which is being developed. Under ordinary atmospheric conditions, wool will absorb and hold 16–18% of its weight of moisture. Under higher humid conditions, wool will hold about 35% of its weight of water. Without feeling damp. Because of this phenomenon of wool, it is important to include definitions and the moisture conditions in the purchase or selling contract of wool. This makes wool good for all climates since it aids in the body's cooling mechanisms to keep moisture away from the skin.

This property is extremely valuable in apparel textile, where it is important not to interfere with one of our essential body functions – respiration through the skin. The absorbent nature of wool tends rather to assist this function and to equalize extremes of temperature. In marketing, however, represents more of a drawback since the commercial weight of the wool depends on the moisture content (regain) is 17% for loose wool, 18.25% for tops, 17% for woolen yarns and 18.25% for worsted yarns. Adjusting the moisture content to these values is known as 'conditioning'.

4.2.10.10   Heat of wetting

Wool is renowned for its ability to give off small but steady amount of heat while absorbing moisture. This is known as 'heat of wetting'. It is considered to be due to energy given off by the collision between water molecules and the groups in the wool polymer system. This friction is severe and thereby liberates energy. Thus, wearer feels slightly warm when the wool fabric absorbs moisture. It holds in air to insulate the wearer.

4.2.10.11   Settability

Wool can be 'moulded' or 'set' under the influence of heat, moisture and pressure. Due to this property the processor can permanently impose on the wool fabric the characteristics and effects required for certain end use. Using a warm iron over a damp cloth, the care accorded to wool fabrics in most households, you can shape the garment as required, is the simplest example of the exploitation of this property. The action of heat and moisture breaks up certain molecular linkages in the wool which under pressure, reform into the new shape desired. On cooling down, they automatically retain the new formation giving the wool a permanent new shape.

4.2.10.12   Felting

Wool is the only fibre with the capacity of felt. Felting is brought is brought about by the interaction of various properties and can be deliberately promoted by applying special mechanical treatments in association with chemicals (alkalis, acids). In felting, the individual fibres start to move within the fibre mass and creep towards their root end because the schalke structure prevents movement towards the tip. As a result, the fibres consolidate into a dense mass which can be used for industrial felts and hat bodies without first having to be converted into thread or fabric. Unlike setting, felting cannot be reversed.

4.2.10.13   Uniformity

Uniformity refers to the uniform fineness of the wool fibre. The diameter of the fibre should be the same along the while length of the fibre. The variation in the uniformity is a sign of lower quality.

4.2.10.14   Yield

Yield is defined as the percentage of clean wool obtained from the original grease wool. Fine wool have a high grease content giving a low yield, approximately 80–90%, while coarse wools yield about 80–60%.

4.2.10.15   Specific weight

Of the natural fibres, wool has the lowest specific weight, 1.31 g/cm^3, and is thus also lighter than some of the man-made fibres, e.g., polyester with a specific weight of 1.4 g/cm^3.

4.2.10.16   Resistance to fire

Wool contains moisture in every fibre allowing it to resist flame without any additional chemical treatment. The wool will just char and self-extinguish.

4.2.10.17   Dyeability

Wool absorbs many dyes deeply, uniformly and directly without the use of chemicals. This characteristic allows wool to achieve very beautiful and rich colours when dyed.

4.2.10.18   Durability

The flexibility of wool makes it very durable. A single wool fibre can be bent back on itself more than 20,000 times without breaking. Compare this to the only 3000 times of cotton and 2000 times of silk. Its elasticity makes it very resistant to tearing. Wool also has an outer film making it resistant to abrasion. Wool fibre can be stretched up to 50% of its length when dry and up to 30% of its length when wet without breaking. It will return to its original length when released.

4.2.10.19   Effect of sunlight

Sunlight can decompose kerating. This process starts even before it is removed from the sheep. Long exposure to strong sunlight can discolour the fibre, sulphur in the wool is converted to sulphuric acid, makes the fibre feel harsh and becomes sensitive to alkali and even soap.

4.2.11   Fine structure of wool

Wool structure are formed from a type of low sulphur proteins which are referred to as intermediate filament (IF) proteins. The matrix consists of proteins containing glycine and tyrosine, as well as proteins containing sulphur, known as IF-associated proteins (IFAP). The smallest link in the hierarchy of the wool structure is the α-helix. These combine to form protofibrils. These in turn combine to form microfibrils, which then result in macrofibrils, surrounded by the matrix. The α-keratins, as the main chemical constituent of wool are insoluble, hard substances which are highly elastic. They are present in the macrofibril in the form of two components, the microfibril and the matrix. The matrix consists of tyrosine and sulphur proteins, whereas the low sulphur proteins are assigned to the microfibrils. The macrofibrils on the other hand form the cortex cell, embedded in the intermacrofibrillar cement. Chemically, the intermacrofibrillar cement is a nonkeratin protein consisting of cytoplasm and residual nuclei.

The X-ray studies has have given an idea of the α-helix structure to account for the secondary structure of the keratin fibre. Fig 4.22 gives an approximate arrangement of the α-helices in protofilaments, protofibrils to microfibrils and intermediary filaments follows a pairing rule: two α-helices are joined together by hydrophobic bonds into a double helix structure: a dimmer of 50 nm length; two dimmers twist together into a coiled-coil tetramer; two chains of tetramers self-assemble to become a protofilament; two protofilaments form a protofibril and four protofibrils arrange into a microfibril or IF, the crystalline structure of wool fibre.

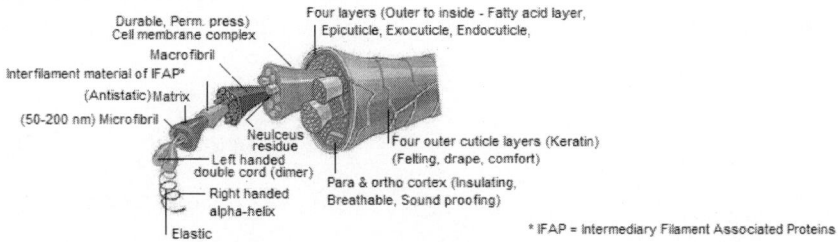

Figure 4.22. Wool structure and their effect on the properties of the fibre

4.2.12   Chemical structure and properties

Wool fibre contain following elements – carbon (around 50 wt%), hydrogen (7 wt%), oxygen (22 wt%), nitrogen (16 wt%) and sulphur (5 wt%). The ash content of wool is around 0.3–0.9%, and the metal detected in traces are Ca, Cd, Cr, Cu, Hg, Zn, Pb, Fe, As and Si, and these may be incorporated into keratin from extraneous sources.

4.2.12.1   Keratin

The main component of wool fibre is keratin – wool protein. Keratin, like any other proteins, is an extremely complex chemical containing carbon, hydrogen, oxygen, nitrogen and sulphuric. The major characteristics of wool, especially chemical, depend on their action on keratin. For example, the action of peroxide, alkali on wool is exactly as its action on keratin.

In general, the properties of a fibre is attributed to the primary structure of the linear macromolecules, their conformation (secondary structure) and the interactions between the certain parts of the macromolecules, the side-chain groups in particular (tertiary structure). The tertiary structure is effectively fixed by the intra- and interchenaric bonds (cross links). As in the case of silk, wool is a polymer made of amino acids and the peptide bonds keep them together. By hydrolysis of these bonds yields the 20 common natural α-amino acids with the general structure (H_2N–CHR–COOH) but one can expect more than 100 amino acids bound together to form the protein chains which make up wool fibre. Due to this wool show many different aspects in their cross-link chemistry. Half of these amino acid residues as monomers act as a source of reactive side chain groups. There is a difference between primary and secondary valence links but, in many cases, it is not possible to differentiate clearly between the two. A further distinction can be made between natural cross links produced in the wool by the biosynthesis and synthetic links, which do not appear until they are formed during the reactions which occur later. The insolubility of the cell membrane is explained by the presence of glycoproteins and isopeptide link (Fig. 4.23).

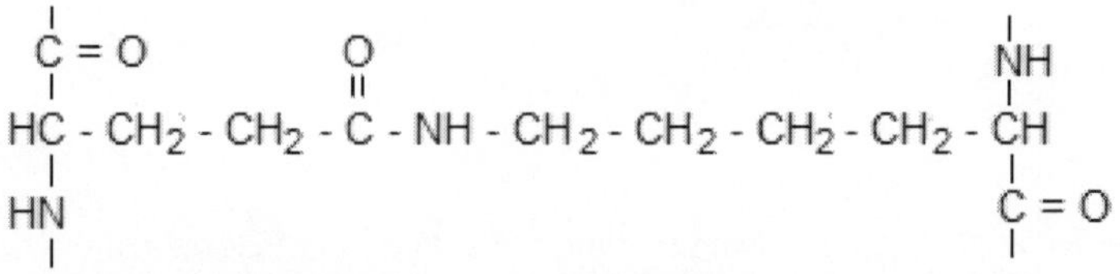

Figure 4.23. Isopeptide link

Peptide condensation product made from amino acids, in which the amino group of one molecule links with the carboxyl group of another, discharging water.

In the same way, any number of different amino acids can react with each other (polypeptides), whereby the acid amide group (COHN) represents the characteristic binding link (peptide bond). Wool is chemical polypeptides made up of about 20 different amino acids. The peptide bond determines the chemical properties and dyeing behaviour of the protein fibres (wool, silk) and also polyamide fibres (Table 4.10).

Table 4.10. Major amino acids found in wool keratin

Group of amino acid	Name	Side chain R
'Acidic' amino acids and their ω-amides	Aspartic acid	$-CH_2-COOH$
	Glutamic acid	$-(CH2)_2-COOH$
	Asparagine	$-CH_2-CONH_2$
	Glutamine	$-(CH_2)_2-CONH_2$
'Basic' amino acids and tryptophan	Arginine	$-(CH_2)_3-NH-C(NH_2)=NH$
	Lysine	$-(CH_2)_4-NH_2$
	Histidine	
	Tryptophan	
Amino acids with hydroxyl groups in the side chain	Serine	$-CH_2-OH$
	Threonine	$-CH(CH_2)-OH$
	Tyrosine	$-CH_2-C6H_4-OH$
Sulphur-containing amino acids	Cysteine	$-CH_2-SH$
	Thiocysteine	$-CH_2-S-SH$
	Cystine	$-CH_2-S-S-CH_2-$
	Methionine	$-(CH_2)_2-S-CH_3$

Group of amino acid	Name	Side chain R
Amino acids without reactive groups in the side chain	Proline	$-CH_2$ $\diagdown$ CH_2 $-CH_2$ $\diagup$
	Alanine	$-CH_3$
	Valine	$-CH(CH_3)_2$
	Leucine	$-CH_2-CH(CH_2)_2$
	Isoleucine	$-CH(CH_2)-CH_2-CH_3$
	Phenylalanine	$-CH_2-C_6H_5$
	Glycine	$-H$

Wool can stretch 100% when it has swollen in hot water. On relaxation, it returns almost to the original length. X-ray studies has shown that in unstretched form the wool keratin is in α-keratin form where polypeptide chains are folded together in a particular way and are therefore shortened, and the β-keratin form when it is stretched (see Silk section). In α-keratin, the polypeptide chain is wound in the shape of a spiral or screw (see Fig. 4.23). The side chains are oriented towards the outside and each amino acid has an intramolecular hydrogen bond between it and the third neighbouring group in the direction of the helix, thus stabilizing the molecule. When stretched intramolecular hydrogen bonds break and reform as intermolecular links between two different polypeptide chains 'pleated sheet' forming the structure of β-keratin and this process is reversible.

Table 4.11. Various fibrils in wool and the dimensions

Fibril type	No. in cross section	Diameter, nm	Length, nm
Macrofibril	Approx. 700 microfibrils	300	10,000
Microfibril (intermediate filaments)	8 protofibrils	10	1000
Protofibrils	4 polypeptide chains	2	1000
Individual (intermediate molecule)	2 polypeptide chains		50
Polypeptide chain (α-helical)	1 polypeptide chains	1	50

The fibre is amphoteric due to the presence of both cationic and anionic groups. The cationic aspect is due to the protonated side groups of arginine, lysine and histidine and the free terminal amino groups. Anionic groups are present in the form of dissociated side groups of aspartic and glutamic acid residues and carboxyl end groups. The terms 'acid' and 'basic' amino acids only refer to the chemical structure of the nondissociated and dissociated carboxyl groups –COOH of the aspartic acid and glutamic acids or nonprotonated groups in the lysine, histidine and arginine.

Since both 'acid' and 'basic' groups exist simultaneously in amino acids, they react with each other to form 'salt linkages (Fig. 4.24).

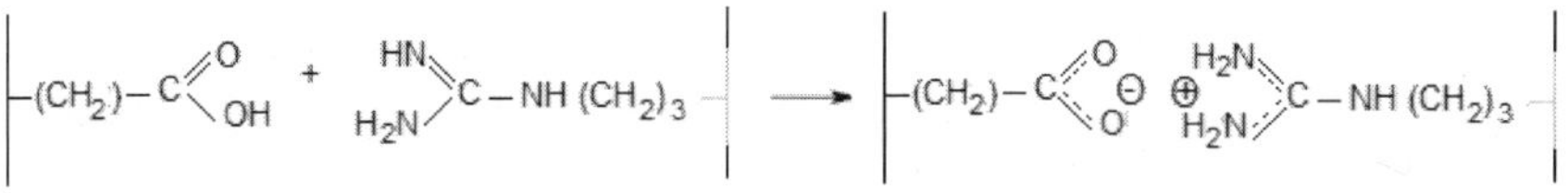

Figure 4.24. Acid and basic groups in amino acids react and form salt linkages

Apart from the carboxyl and amino end groups, the amphoteric character of the keratin molecule depends on the side chain functions. The side chains are characterized by their content of bifunctional acidic and basic amino acids. At the isoionic point, the carboxyl groups in the protein have donated protons to the amino groups so that exactly equal numbers of anions

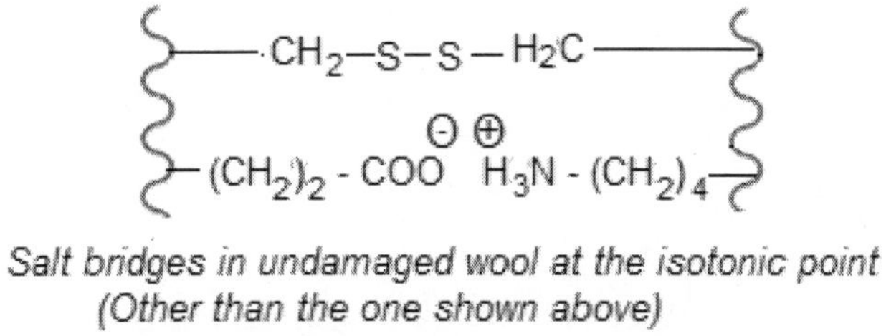

Salt bridges in undamaged wool at the isotonic point (Other than the one shown above)

Figure 4.25. Disulphide bridges in wool fibre

and cations are present. The isoionic point was found to be pH 4.9. The value measured by electrochemical methods is called the isoelectric point and is only slightly different from the isoionic point. Wool fibres are stabilized by salt bridges as well as by disulphide bridges (Fig. 4.25 and Table 4.12).

Table 4.12. Summary of major properties of wool (keratin)

Property	Wool
Colour	Most wool is white or near white in colour. Some breeds give brown or black wool but mainly in coarse quality
Lustre	Wool has an inherent lustre which varies as per type of wool
Chemical composition	80% keratin proteins; 19% nonkeratin proteins; 1% internal lipids

Property	Wool
Molecular masses:	9000 to 60,000
Intermolecular forces	Cystine bonds, hydrogen bonds and salt linkages; hydrophobic interactions
Isoionic point	pH 4.9
Moisture regain	18%
Crystal structure (30%)	α-Helix
Amorphous sections	Nonhelical segments in microfibrils; matrix in macrofibrils; nonkeratin proteins
Microfibrils	7–11 nm
Macrofibrils	50–200 nm
Pore volumes	35%
Biological composite structures	Cortex, cuticle
Glass temperature	Dry: +174°C; wet: –5°C
Decomposition temperature	250°C
Tenacity	Dry: 8.8–15.0 cN/tax (1.0–1.7 g/den); wet: 7.0–14.0cN/tax (0.8–1.6 g/den)
Tensile Strength	1190–2030 kg/cm^2 (17,000–29,000 lb/in^2)
Elongation	Wool has an elongation at break of 25–35% at standard condition and 25–50% when wet
Elasticity	Wool is a resilient fibre. It has elastic recovery up to 99% at 2% extension and 63% at 20% extension
Specific gravity	1.32
Organic solvent	Stable to dry-cleaning solvents
Effect of acids	Decomposed by conc., H_2SO_4, stable to most of the mineral acids even at higher temperature except HNO_4, which may oxidize the wool
Effect of alkalis	Wool is sensitive to alkali like NaOH. Soda ash will tender wool and turn yellow, if used in too concentrated solution, especially when hot. Ammonium carbonate, borax and sod. Phosphate is mild alkalis and has minimum effect on wool. Ammonia under normal conditions does not damage wool

Property	Wool
Insects	Wool is attacked by moths and other insects
Microorganisms	Wool has poor resistance to mildew, bacteria, etc.
Effect of sunlight	Wool keratin can decompose by the action of sunlight. Under the action of sunlight, the sulphur is converted to H_2SO_4 and the wool turns discoloured and harsh. It becomes more susceptible to many chemicals, which they are resistant otherwise, like alkalis, soapy water, dyeing gets affected.
Effect of heat	At 130°C wool becomes yellowish and starts to decompose, at 300°C it chars and forms black knob. As it decomposed, it gives the characteristic smell of burning feathers, wool does not continue to burn if removed from the flame
Microscopic view of wool	
Cross section of wool	

4.2.13   Benefits of wool

Wool fibre/garments have the following benefits:

1. It is breathable: Wool has moisture absorbing capacity of up to 35% of its own weight due to its hydrophilic core and this property makes the fabric breathable – the perspiration is readily absorbed by the garment and released as vapour into the air and provides comfort for the wearer throughout the day. No other fabric offers this amazing wearer sensitive comfort.

2. Insulation: Wool is natural insulation capacity. Even though synthetic also has the same ability, but has health hazards due the very fine fibres are used for this purpose. (As against wool, the synthetic fibres are fine and brittle which can become airborne and cause respiratory and allergy problems during usage.)

3. It can control humidity: Humidity control is due to the moisture regain capacity of wool fibre.

4. It is less allergic: Wool does not harbour dangerous chemicals, dust or mould that can lead to allergic or other reactions.

5. It can absorb toxic chemicals.

6. It can reduce sound: Wool is a perfect sound-insulating material, with the capacity to dampen or absorb both high and low frequency sound.

7. It is flame resistant: Wool is naturally nonflammable, unlike almost all alternatives. It simply requires more oxygen to burn than is available in the air, making it a superior fibre for fire safety.

8. It provide UV protection: Testing of various textiles made of wool shows that wool has a natural UV protection factor (UPF) of 30+ in more than 70% of cases – this is much higher than most synthetics and cotton.

9. It reduces static electricity: Due to its moisture absorption capabilities, wool is far less likely to cling to the body when worn.

10. It has high thermal resistance.

11. Since wool fibres are flexible the garments are more durable: Wool fibres can be bent 20,000 times without breaking, which explains why wool garments are so long lasting.

12. Compared synthetics wool is active in response to fluctuations in body temperature means that it offers warmth when it is cold, but will self-adjust when temperatures increase by releasing heat and moisture to maintain a comfortable clothing climate.

13. Woolen garments have health benefits including improved sleep for the general population, reduced risk of SIDS (sudden infant death syndrome) for babies and lower incidence of microbial infection for hospital patients.

14. Wool is a better perspiration controlled fabric than synthetic: It has the capacity to absorb perspiration caused by exertion. The natural absorption of wool moves moisture away from the skin so that it can evaporate, ensuring coolness and dryness.

15. Modern woolen garments are having inherent easy care property: Wool has had a reputation for being difficult to machine wash in the past, but modern wool fabrics are treated to be truly easy-care. Shrink resistant finishing helps the woolen fabric to be crease resistant and shrink proof.

16. It is stain resistant: Wool fibre has a protective layer that prevents stains from being absorbed. As it is also anti-static, it picks up less dust and lint from the air.

17. Woolen fabrics are quick drying: It was formerly thought that synthetics dry at a faster rate than wool; however, the latest technology enables wool to be as quick drying.

18. Wool has the ability to reduce the smell of sweat: Wool will absorb moisture, reducing sweat on the body; this in turn reduces the amount of resulting body odour caused by sweat and its contact with any bacteria on the skin.

4.2.14   Difference between woolen and worsted yarns

4.2.14.1   Woolen yarns

1. Fibres are short (2 in. length)
2. Woolen yarns are only carded, less twisted and hence weak in strength.
3. Woolen fabrics are woven with plain weave and sometimes twill weave. Weaving is not compact and fabric is not durable or strong.
4. Woolen fabrics are soft, fuzzy, thick and warm but not durable.
5. Woolen can be easily adulterated and napping finish is given to produce soft surface.
6. Less expensive than worsted.
7. These fibres are warmer than worsted. It has no lustre and is less durable. The napped surface tends to catch and hold dirt but stains can be easily removed.

4.2.14.2   Worsted yarns

1. Fibres are long (2–8 in.)
2. Worsted yarns are carded and combed, highly twisted and stronger.
3. Worsted fabrics are woven chiefly with twill weave and weaving construction is close, compact, and as such more strong fabric.
4. Worsted fabrics are flat, rough, harsh when worn next to the skin, but more durable.
5. Worsted cannot be easily adulterated as it has a hard finish on the surface.
6. Costlier or expensive than woollens.

7. It wrinkles less than woolen, holds creases and shape and become shiny with use.

4.2.15   Wool blends

Wool is blended with a wide variety of fibres.

4.2.15.1   Wool and cotton

Wool is blended with cotton in various ratios. The properties of the yarns and fabrics will be affected by the proportions of the fibres blended. Wool contributes warmth, resilience, abrasion resistance and drapability. Cotton adds strength and reduces the cost of yarn and fabric. Both fibres are absorbent and can be blended to make a comfortable, durable fabric with a nice hand.

4.2.15.2   Wool and linen

Wool is sometimes blended with linen. Linen may be used in such a blend, which is stronger than a pure wool fabric, but is more resilient and drapable than a pure linen fabric.

In order for wool to be sold in International Trade, methods of expressing grades for raw wool have been devised. On the world market, grade is expressed by a number system (Table 4.13).

The finer the wool, the higher the number.

Table 4.13. Comparative grading systems

Fine	80s, 70s, 64s
Half blood	62s, 60s, 58s
Three-eighths blood	56s
Quarter blood	50s, 48s
Low-quarter blood	46s
Common	44s
Braid	40s, 36s

Large wool producing and exporting nations are: Australia (50s–80s), New Zealand (40s–60s and carpet grade), South Africa (60s–70s and carpet grade), Argentina (40s–60s, 64s and carpet grade).

Countries contributing to greatest amount of carpet wool to world market are: Argentina, India, Pakistan, New Zealand, Syria, Iraq and Iran.

Wool is used in the manufacture of various products like: boots, carpet, blankets, sweaters, coats, seat covers, bed sheets, cushion covers, curtains (Table 4.14).

In garment form, wool is beneficial because it

- resists wrinkles
- retains shape
- resists soiling
- resists flames
- wool is durable
- wool is comfortable in all seasons
- repels moisture

Table 4.14. Major uses of wool and woolen materials

Property of wool	Area of usage due to this property
Moisture absorption	Police/military uniforms, socks, gloves, nightwear, firefighters uniforms, infant apparels
Absorption of odours, filtration of gas, chemicals, air purifying	Aviation – Interior trimmings, flight attendants garments, aircraft interiors and interior sound proofing, general - air-conditioning, sound and vibration controls, toxic chemical, dust, electrostatic and odour filters, heat exchangers
Elasticity, controlled felting	Sheep skin boots and garments, hats, pullovers, uniform, fashion garments, water proof garments, machine washable suits
Breathability	Woven garments, nonwoven garments, accessories, millinery, flannels, thermal underwear. Sports clothing, baseball filling, billiard clothing, skiwear
Comfort, softness, handle and drape	Vital signs vest, molecular templating, intelligent kneesleeve
Sound and vibration absorption	Piano felts, wool filters for sound and chemicals, gaskets and washers, buffering pads, baby blankets, seat covers
Heat and cold insulation, less allergic - medical uses	Second skin injury prevention, bandages, medical sheep skins, wound dressings, pressure bandages
Fire resistance, antistatic	Roof insulation, upholstery, quits, blankets, wall coverings, carpets

4.3   Other animal hair fibres

Hair fibres obtained from different kinds of animals also contribute to the fabric formation. Animal hairs are adapted by nature for the climate they live in. These fibres can be used alone or mixed with wool (Fig. 4.26).

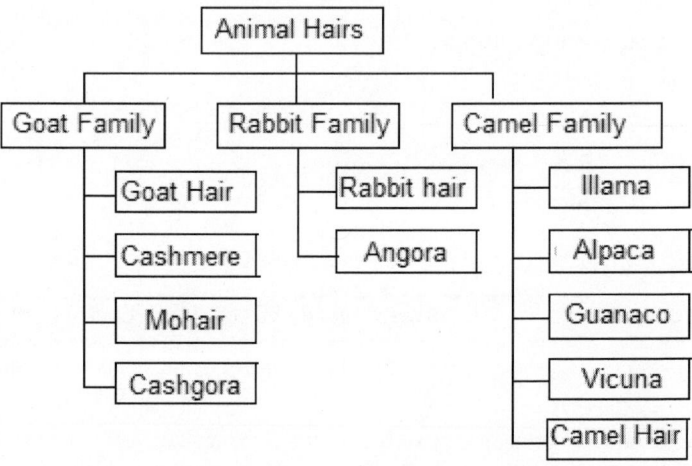

Figure 4.26. Classification of animal hair

There are many animal hairs which are used for textile purposes: camel, mohair (angora goat), cashmere, llama, angora rabbit, guanaco, vicuna, alpaca, etc.

Table 4.15 summarizes the fibre diameter comparison of various animal hairs.

Table 4.15. Comparison of fibre diameters of various animal hairs

Animal	Fibre diameter (micrometres)
Merino	12–20
Vicuna	6–10
Alpaca (Suri)	10–15
Angora rabbit	13
Cashmere	15–19
Camel (Down)	16–25
Llama	20–30
Mohair	20–30

4.3.1   Camel hair

Also known as high quality coat fibre and apt protect both heat and cold. It is a very expensive fibre and hence mixed with other fibres like wool.

Camel hair is collected from dromedaries (one hump) and mainly from two-humped Bactrian camel (Fig. 4.27), found from Turkey east to China and north to Siberia. Significant supplier countries of camel hair are Mongolia, Iran, Afghanistan, Russia, New Zealand, Tibet and Australia. It is the undercoat (down hair) which is principally used. Animals from the Gobi desert in Central Asia yield particularly fine and therefore valuable hair.

There are five primary steps in the production: (1) *Collection* – by shearing or moulting. The hair is not usually gathered by shearing or plucking; it

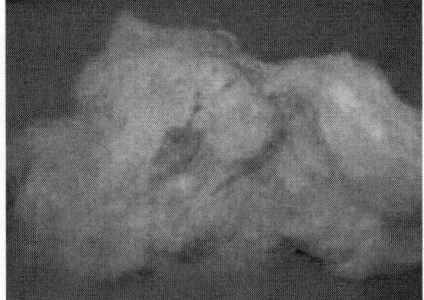

Figure 4.27. Two-humped Bactrian camel

is most often collected as the animal sheds its coat. (2) *Sorting* – coarse hair separated from soft hair. Both the outer coat and the undercoat are shed at the same time, and combing, frequently by machine, separates the desirable down from the coarse outer hairs. (3) Dehairing. (4) Spinning. (5) Weaving or knitting. It has got good strength, lustre, smoothness and warmth. It is lightweight and thermostatic.

Bactrian camels have protective outer coats of coarse fibre that may grow as long as 15 inches (40 cm). The fine, shorter fibre of the insulating undercoat, 1.5–5 inches (4–13 cm) long, is the product generally called camel hair, or camel hair wool. The resultant fine fibre has a tiny diameter of 5–40 µ and is usually a reddish tan colour. The natural colours of the camel hair give the fabric its characteristic colours, which ranges from light tan to dark brown (Table 4.16).

Table 4.16. Major properties of camel hair

Properties	Camel hair
Tensile strength	15.7 cN/tex, (16 g/tex)
Elongation	39–40
Elastic recovery	50% Breaking load 0.8 50% Breaking extension 0.7
Work of rupture	4.6 cN/tex, (16 g/tex)
Initial modulus	294 cN/tex
Specific gravity	1.32
Moisture regain	13%
Effect of heat, sunlight, chemicals, organic solvents, insects, etc	Similar to wool
Enlarged cross section	

Camel-hair fibre has greater sensitivity to chemicals than does wool fibre. Its strength is similar to that of wool having a similar diameter but is less than that of mohair. Fabric made of camel hair has excellent insulating properties and is warm and comfortable. Camel hair is mainly used for high-grade overcoat fabrics and is also made into knitting yarn. The coarse outer fibre is strong and is used in industrial fabrics such as machine beltings and press cloths, Over coats and jackets, blazers and sweaters, skirts, paint brushes, gloves and hosiery items, scarves, mufflers, caps carpets, knitwear, waterproof coats, blankets, rugs, ropes and industrial beltings.

4.3.2   Mohair

Mohair refers to a silk-like fabric or yarn made from the hair of the docile angora goat (also called mohair goat). The length of its fibre is 9–12 inches. Found in mountains of Tibet, Turkey, the United States and South Africa. Today South Africa and the United States are the largest mohair producers, with the majority of American mohair being produced in Texas. The hair is long, brightly lustrous and relatively thick and stiff. The undercoat is particularly finer and therefore particularly coveted for the use in garments manufacture where it finds varied applications, very often in blends with wool.

4.3.2.1   Manufacturing

1. Shearing – Removing fur from animal body. Angora goat can be clipped twice in a year. One goat may give 1.8–2.4 kg (4–5 lb) mohair per clipping. As in the case of wool, the quality of the fibre varies, depending on its source under which it lived. The fleece is graded into tight lock, flat lock and fluffy types. Tight lock is characterized by its ringlets and is usually very fine. Flat lock is wavy and of medium quality and fluffy or open fleece is of the lowest grade. The dead fibres and dull ones in the fleece are called kemps, much the same as in the case of wool.
2. Scouring – treating fibre with detergents
3. Dehaired – separating coarse and soft hair
4. Spinning

The microfibrils (keratin IFs, KIFs) are chiefly parallel to the fibre axis in the cortex of mammal hair. Mohair is characteristically wear resistant, and hence mohair fabric is used when durability is the prime requirement. Mohair can be dyed easily and has a natural lustre. The surface of the fibre is covered in epidermal scales, which are much more tightly connected with the fibre stock than wool scales. There are only approximately half as many scales as on the wool fibre (see Fig. 4.28).

(A)   (B)

Figure 4.28. (A) Angora goat (mohair goat). (B) SEM view of a mohair fibre

The overlapping is slight so that the fibre has a soft handle. Light is easily reflected from the surface so that mohair has a characteristic lustre. Two typical textile properties of mohair are the lustre and soft handle as a result of the scale structure of the fibre surface and good breaking strength and wear resistance.

The scale edge height of mohair, in comparison to 1.07 μm for Australian wool, is 0.4 μm. The individual cells consist of approximately 0.2 mm wide macrofibrils in the cross section. The macrofibrils are in turn bundles of 10 nm fine microfibrils in a hexagonal pack. This structural arrangement is found in the low-sulphur keratin fibres, which supply highly informative radiographs with many radio reflexes. The microfibrils or KIFs represent approximately 60–70% of the fibre mass. These KIFs are mainly responsible for the stability of keratin fibres against elongation and draw stretching. Each keratin consists of a central alphahelical rod domain of 311–314 amino acid residues. The central rod domain is flanked at both ends by regions, which vary both in their size and in their chemical character (Table 4.17).

Table 4.17. Properties of mohair fibre

Properties	Mohair
Tenacity	11.8–12.8 cN/tex (12–13 g/tex)
Elongation	30%
Elastic recovery from	
50% breaking load	0.8
50% breaking extension	0.6
Specific gravity	1.32
Moisture regain	13%
Effect of sunlight, chemicals, solvent, insects, microorganisms	Similar to wool
Cross-section appearance	
Microscopic appearance	

The mohair chemically and physically behaves almost like wool only. Hence, the processing routes are all similar to wool processing.

Mohair is composed of keratin similar to wool. It is resilient, dust repellant, absorbent, lustrous and silky. It is also antiwrinkle, flame resistant, moth and mildew proof and gives warmth. The fibre is used in making carpets, sweaters, coats, home furnishings, doll wigs, etc.

4.3.3   Cashmere

Cashmere wool is obtained from the Cashmere goat and is also known as available in East India and northern Himalayan mountains of Kashmir, Tibet (Tibetan goat) Mongolia and China (Fig. 4.29). These animals are found in high altitudes where the temperature is extremely low. The cashmere hair is obtained at the time of moulting by pulling, combing out or shearing. The wool hair of the length up to 300 mm long, coarse but smooth and soft. The colour is usually grey to white and classified as per colour, lighter the colour it is more valuable. The downy hair has a silky lustre, and is wavy and soft. The colour is rarely white, usually grey, brownish and blackish. The length is 40–90 mm, the fineness 10–20 μm, which corresponds to the finest merino wools but tensile strength is relatively lower. The handle is particularly soft and pleasant. Cashmere is spun in pure form and in blends, primarily of finest

Figure 4.29. Cashmere goats

merino wool, natural silk, cotton and synthetics. Worsted yarns are used for knit goods, necktie fabrics and hand-woven shawls; carded yarns are more frequently produced and used for fine cloths and fabrics for ladies' wear, with a napped or fleece-like surface.

Down wool is about 2.5–9 cm long, while the coarse fibres are 5.0–12.5 cm long. Like wool, the hair is having an outer layer of scales, about 5–7 numbers per 100 μm length on an average. These scales cause an irregular surface. Cashmere fibre has cortex but differs in structure from the wool in the fact that it does not have a distinct medulla. The cortex cells are spindle shaped with some long narrow spaces between the cells forming striations in the fibre. The fibre is usually grey, buff coloured or white. The coloured fibres are full of tiny granules of pigment. The cross section of the fibre is either circular or slightly oval and is about 15 μ in diameter. Some coarser fibres which are thicker (60 μ) are used to mix with merino wool.

Chemical properties are similar to wool, easily affected by acids and alkali. Cashmere can easily be dissolved in alkali and even in washing soda (soda ash). Since these fibres are not produced in large quantities, it very expensive. It is soft and wool like, lightweight, warm and finer as compared to normal wool fibre and less durable than wool. Their undercoat is particularly fine and soft and therefore has very good insulating properties. Cashmere is inherently comfortable and have a beautiful drape.

The fibre is used for making high-quality expensive garments, shawls, hosiery, scarves, sweaters, jackets, gloves, etc., either as pure cashmere or mixed with merino wool and other fibres.

4.3.4   Llama

It is obtained from animal which looks like camel (one-third of its size) native of Andes Mountains in Ecuador, Argentina, Peru and Bolivia (Fig. 4.30).

The fleece of llama is a mixture of coarse and fine fibres which is not very elastic like other hair fibres. Natural colours range from white, brown to black. They are insulative, light weight, wrinkle resistant, colour fast and extremely durable. Fibres are strong of about 30 cm or more long. It is almost like camel hair, with very soft scales which are often difficult to find with naked eye. Pigment granules are found in medulla.

Figure 4.30. Llama

The fibre from llama used in making of hand-made socks, sweaters, jackets, carpets, rugs, etc.

4.3.5   Alpaca

Related to camel breed, is a close relative of llama. It is also grown in the same regions where llama grows. It has a soft fleece having about 60 cm in length. It has a glossy appearance when it is spun and woven into woven goods. It is finer than mohair but not so shiny. Suri is highly selected type of alpaca (Fig. 4.31).

Found in Andes mountains in South America. Main trading centre is Arequipa, Peru. Fibres are finer than llama fibres. The fibres are normally black, brown, fawn or white. The cortical layers are striated and often medulla is absent. The scales are indistinct as in the case of llama (Fig. 4.32).

Used for women's dresses and high grade lining fabrics and is also added to wool yarns for a variety of end uses.

Figure 4.31. Alpaca

They are distinguished by a silky lustre, strong, water repellant, highly insulative, obtained in many colours but reddish brown is most valuable. They are lightly crimped, medium fineness and long in staple.

The fibre is used for making dress fabric, linings, plushes and tropical suitings.

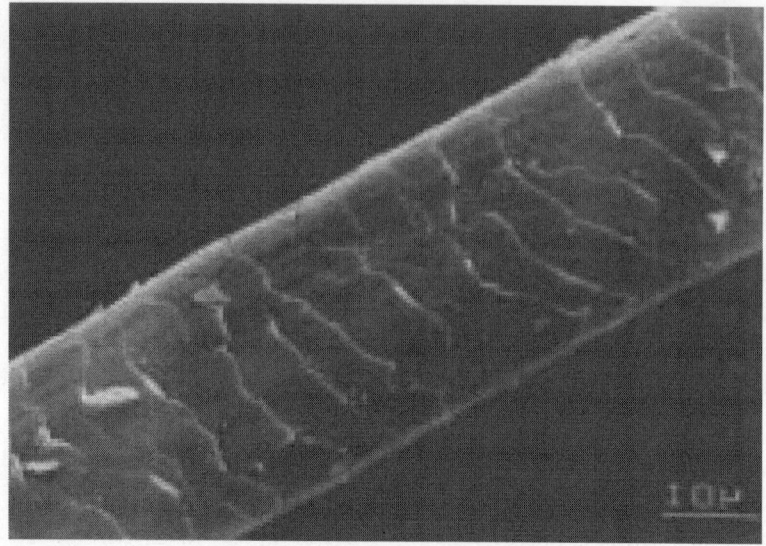

Figure 4.32. Alpaca hair (SEM)

4.3.6   Angora (rabbit)

It is obtained from rabbit and is a fur fibre, even though sometimes erroneously described as angora wool. These rabbits are grown in European countries like France, Italy and Asian countries (Japan, China), Chile and the United States. The fibre is very expensive.

The hair is clipped every 3 months and are about 7.5 cm long. The annual yield amounts to 250 g of hair per rabbit from three to four shearings. Outer guard hairs and fine fur are separated by blowing the fibre in a stream of hair. Both the fibres have different properties and are used for making yarn. The fur is more soft and fine, but will not give enough strength if used alone. Hence the hair also are mixed with fur in different proportions as per end use and the effect required. The hair is mostly pure white, lustrous and silk like (specific gravity 1.1–1.2). The outer guard hairs, which is responsible for the fluff, amounts to approx. 2.2%. Hair thickness approx. 12.7 µm, tensile strength approx. 20 cN/tex, elongation at break approx. 38%, thermal retentivity approx. 50–70%. The fibres are obtained by shearing or combing. Angora hair has a high thermal retentivity due to the internal air-containing medulla and is also very soft, light and free of pigment.

The collection of fur may be done by

1. Plucking loose hair

2. Shearing

3. Collection of moulting fur

Figure 4.33. Angora fibre

The fur has an attractive appearance and soft luxurious handle (Fig. 4.33). Angora fibres consist of three main components (Fig. 4.34):

1. *Cuticle*: 0.3–0.4 μm thick, makes up 10% of the fibre weight. The typical arrangement of scales on the fibre surface can be described as chevron structured; they have a protective function, i.e., to prevent the penetration of foreign matter into the fibre interior.

2. *Cortex*: Composed of microfibrils and matrix similar to wool.

3. *Medulla*: Depending on fibre type, the medulla is represented more or less frequently. It can be continuous or discontinuous and occurs only in fibres with a diameter of >10 μm. In correlation with the diameter, the medullation increases in a form so that the fibre can be marked with one or several rows. The longitudinal section reveals the scale structure of the angora hair.

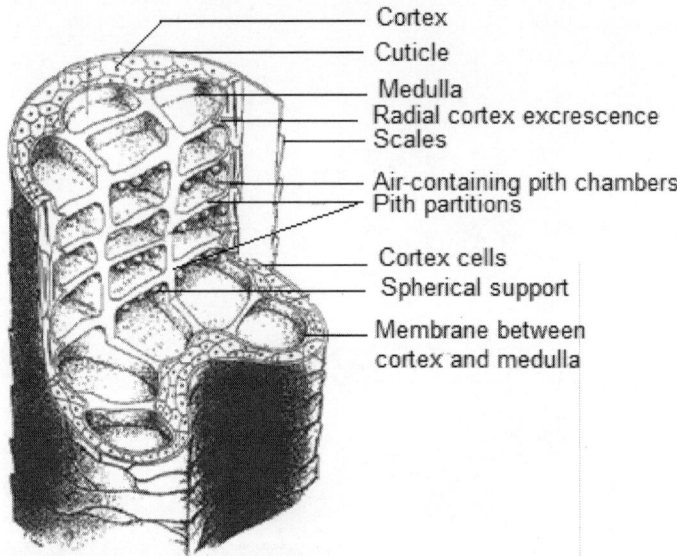

Cortex
Cuticle
Medulla
Radial cortex excrescence
Scales
Air-containing pith chambers
Pith partitions
Cortex cells
Spherical support
Membrane between cortex and medulla

Figure 4.34. Morphology of Angora fibres

4.3.6.1   *Properties*

The hairs are longer compared to the fur, which will be around 20 mm. On examining the cross section the fur fibres are round, oval or even rectangular. The coarse hairs are often dumb-bell shaped or in the form of a sharp-edged oval. These fibres are also having scales on the outer surface, where each scale covers almost half around the fibre. Scales on the guard hairs are little different, they have serrated edges and the edges often run slanting across the fibre (see Fig. 4.35).

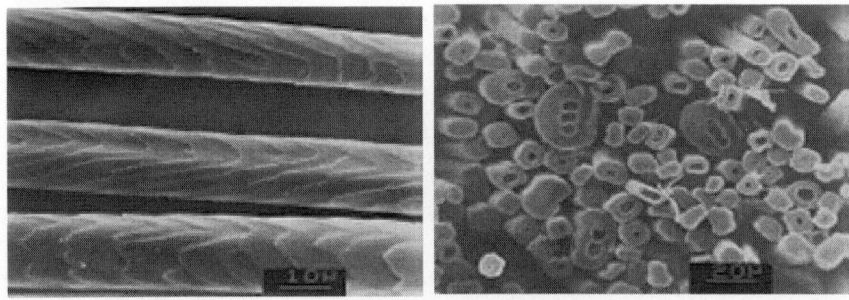

Figure 4.35. Angora rabbit hair enlarged (left: longitudinal view x1000; right: cross section x500)

They have thick medullas which contain many pockets of air in it.

Chemically this is also a keratin fibre and may be a mixture of several proteins. The polypeptide chains of angora keratin are composed of approx. 20 different amino acids (Table 4.18).

Table 4.18. Amino acid composition of angora hair fibre

Amino acid	General formula	Formula of the residue, R	Amino acid content, %
Cystine		$-CH_2-S-S-CH_2-CH(NH_2)-COOH$	14.96
Glutamic acid		$-CH_2-CH_2-COOH$	13.56
Arginine		$-CH_2-CH_2-CH_2-NH-\underset{\underset{NH}{\|\|}}{C}-NH_2$	10.33
Serine		$-CH_2-OH$	7.62
Leucine		$-CH_2-CH-(CH_3)_2$	7.46
Proline	$\underset{\underset{\overset{\|\|}{O}}{HOO}}{HOO} - CH - \underset{NH_2}{\|}$ R	$HOOC-CH\underset{NH}{\overset{CH_2-CH_2}{\diagup\diagdown}}CH_2$	6.97
Aspartic acid		$-CH_2-COOH$	5.89
Threonine		$-CH(OH)-CH_3$	4.85
Glycine		$-H$	4.79
Lysine		$-CH_2-CH_2-CH_2-CH_2-NH_2$	4.59
Tyrosine		$-CH_2-C_6H_4-OH$	3.78
Valine		$-CH-(CH_3)_2$	3.5
Phenylalanine		$-CH_2-C_6H_5$	3.08
Alanine		$-CH_3$	3.01

Amino acid	General formula	Formula of the residue, R	Amino acid content, %
Isoleucine		$-CH(CH_3)-CH_2-CH_3$	2.71
Histidine		$-CH_2-C-NH$ $\begin{array}{c}\parallel \quad \diagdown \\ CH-N \diagup CH\end{array}$	2.65
Citrulline		$-CH_2-CH_2-CH_2-NHCONH2$	0.27
Methionine		$-CH_2-CH_2-S-CH_3$	Negligible

They behave almost same way as wool chemically, but absorbs less moisture. Hot water makes the fibre soft and plasticize them. Usually with the keratin fibres, they dissolve in alkali.

For technological reasons, these fibres are frequently spun with wool, a proportion of synthetic fibre or with cotton. Like wool the angora fur gets felted very easily and hence used for manufacturing felts.

The fibre is used in making sweaters, cardigans, suiting, baby clothes, gloves and berets. Used for knitting after mixing with wool in spinning.

5.1   Classification of synthetic fibres

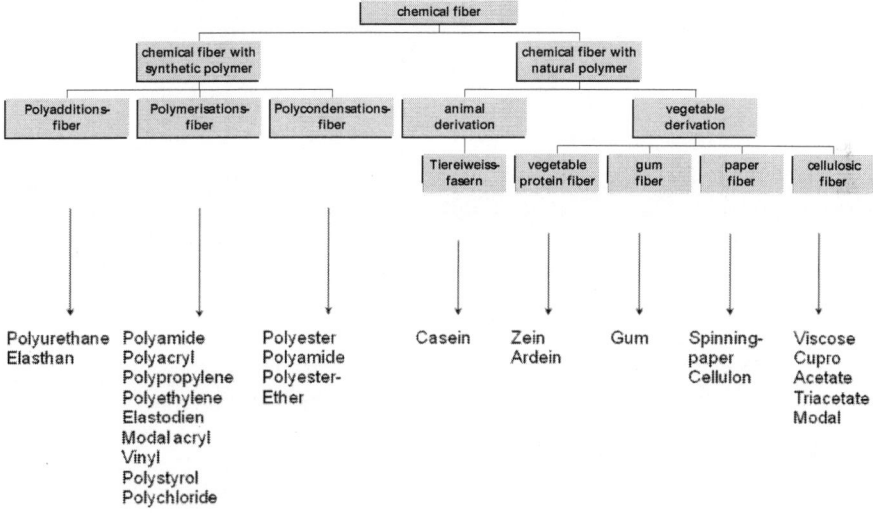

Figure 5.1. Classification of synthetic fibres

5.2   Man-made regenerated fibres

Cellulosic-regenerated man-made fibres include rayon, viscose, modal and recently developed lyocel, cuprammonium, cellulose acetate (secondary and triacetate), polynosic, high wet modulus (HWM). Protein-based fibres are obtained from natural sources such as casein fibre from milk, groundnut fibre from groundnut, zein and azlon fibres from corn. Fibres whose chemical composition, structure and properties are significantly modified during the manufacturing process. Man-made fibres are spun and woven into a huge number of consumer and industrial products, including garments such as shirts, scarves and hosiery; home furnishing such as upholstery, carpets and drapes; and industrial parts such as tire cord, etc.

Natural man-made fibres can be classified into cellulosic fibres and non-cellulosic fibres.

1. Cellulosic fibres: Cellulose is one of many polymers found in nature. Wood, paper and cotton all contain cellulose. Cellulose is an excellent fibre. It is made up of repeat units of the monomer glucose. The three types of regenerated cellulosic fibres are rayon, acetate and triacetate which are derived from the cell walls of short cotton fibres called linters. Paper for instance is almost pure cellulose.

2. Noncellulosic fibres:

   (a) Protein-based fibres: Azlon fibre from soya and corn, casein of milk

   (b) From other sources: Mineral – glass, ceramic and graphite; metallic fibres – by mining and refining of metals like silver, gold, aluminium and steel; rubber fibres – sap tapped from the rubber tree.

Fibre forming polymer is either natural or synthetic.

Man-made fibres can be classified according to the fibre length, their mechanical properties and the material of which they are composed. The following overview shows the most important types, sorted according to length.

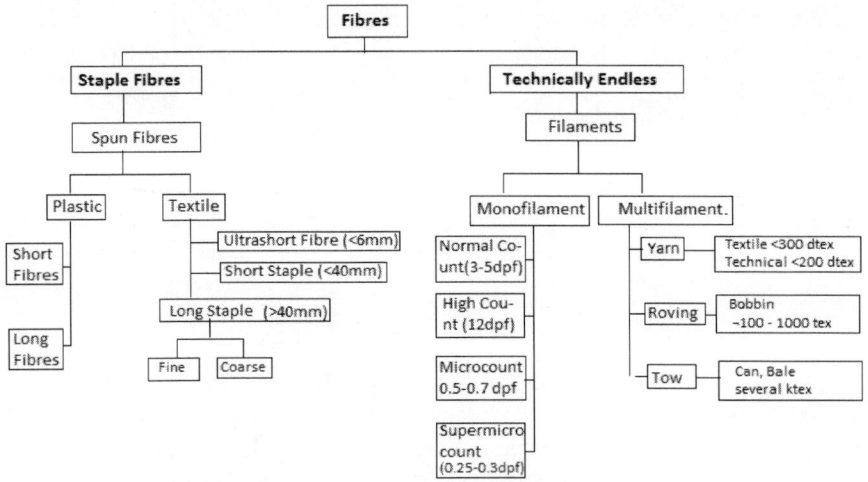

Figure 5.2. Classification of man-made fibres according to length

5.3   Man-made cellulosic fibres

5.3.1   Rayon

Rayon fibre is composed of pure cellulose, the substance of which the cell walls of such woody plants as trees and cotton are largely composed of. They are made from cellulose that has been reformed or regenerated; consequently,

these fibres are identified as regenerated cellulose fibres. Rayon is a manufactured regenerated cellulosic fibre. It is the first man-made fibre. It has a
serrated round shape with smooth surface. It loses 30–50% of its strength
when it is wet. Rayon is produced from naturally occurring polymers and
therefore it is not a synthetic cellulosic fibre. The fibre is sold as artificial silk.
There are two principal varieties of rayon namely viscose and cuprammonium rayon.

In 1884, Count Hilaire de Chardonnet produced the first natural man-
made textile fibre from nitrocellulose and it was named rayon. In 1890, I.H.
Despaisses of France developed the cuprammonium process for making
rayon. Although it was not economically competitive to manufacture, improvements in the spinning techniques developed in Germany resulted in successful commercial production there in 1919.

5.3.1.1   Manufacturing process

In the production of Rayon purified cellulose is chemically converted into a
soluble compound. A solution of this compound is passed through the spinneret to form soft filaments that are converted or almost regenerated into
almost pure cellulose. Because of the reconversion of soluble compound to
cellulose, rayon is referred to as a regenerated cellulose fibre.

There are several types of rayon fibre in commercial use today, named
according to the process by which the cellulose is converted to soluble form
and then regenerated rayon fibres are wet spun, which means that the filaments emerging from the spinneret pass directly into the chemical baths for
solidifying or regeneration. Viscose rayon is made by converting purified
cellulose to xanthate, dissolving the xanthate in dilute caustic soda and then
regenerating the cellulose from the product as it emerges from the spinneret.
Most rayon is made by the viscose process. The various steps involved in the
viscose process are given below.

1. *Cellulose* – Purified cellulose for rayon usually comes from specially
   processed wood pulp (Fig. 5.3).

Figure 5.3. Wood for the production of pure cellulose

2. *Steeping* – Firstly, sheets of cellulose in the form of purified wood pulp are steeped in 18% caustic soda (or sodium hydroxide) and allowed to steep for enough time for the caustic solution to penetrate the cellulose and convert some of it into 'soda cellulose', the sodium salt of cellulose. This is necessary to facilitate controlled oxidation of the cellulose chains.

$$- O - \underset{\underset{OH}{|}}{C_6H_7O} - (OH)_2 + NaOH \longrightarrow \quad - O - \underset{\underset{ONa}{|}}{C_6H_7O} - (OH)_2 + H_2O$$

3. *Pressing* – The soda cellulose is squeezed mechanically to remove excess caustic soda solution (Fig. 5.4).

Figure 5.4. Pressing

Hemicelluloses left over from the cellulose pulping process tend to accumulate in the steep soda, and levels of up to 3.0% are tolerable, depending on the type of fibre being produced. Equilibrium hemicellulose levels may prove to be too high for the production of stronger fibres, and in this case, steps must be taken to purify the soda, e.g., by dialysis, prior to reuse. In any case, a proportion of the soda from the presses is filtered for later use in dissolving and washing; the steep liquor itself is corrected to the required concentration by the addition of fresh hemicellulose-free strong soda.

4. *Shredding* – The soda cellulose is mechanically shredded to increase surface area and make the cellulose easier to process. This shredded cellulose is often referred to as 'white crumb' (Fig. 5.5).

Figure 5.5. Shredding

5. *Ageing (also called mercerizing)* – Alk-cell mercerizing (or ageing) is ideally carried out on a conveyor belt running through a temperature- and humidity-controlled tunnel. Towers, silos, cylinders or even extended treatment in shredders are used as alternatives to belts. During mercerizing, oxidative depolymerization of the cellulose occurs, and this reduces the molecular weight to a level where

Figure 5.6. Ageing

the final viscosity of viscose and cellulose concentration are within acceptable ranges. The optimum conditions for a given pulp depend on the details of the equipment and the cost–quality balance required by the business. The ageing temperature is generally around 50°C, and the humidity is adjusted to prevent any drying of the alk cell. Catalysts such as manganese (a few parts per million in steep liquor) or cobalt have been added to the steep soda to accelerate the ageing process or to allow lower temperatures to be used. Catalyzed ageing on a conveyor takes 3–4 h, and at the end of the belt, the alk cell is blown or conveyed to a fluid-bed cooler. Whatever system used, tight control of the ageing process is essential for good quality of viscose solutions, because the viscosity of the viscose depends on it. Varying viscosity can play havoc with all subsequent operations up to and including spinning. Productivity gains achieved by high temperature or catalyzed ageing tends to be accompanied by increased viscosity variability and reduced reactivity in xanthation. The white crumbs are allowed to stand for a time in contact with the oxygen of the ambient air during which 'ageing' occurs, the very long cellulose molecules are thus reduced in length to allow a satisfactory spinning solution to be prepared later.

6. *Xanthation* – The properly aged product which is called 'alkali cellulose' and in the form of white crumb is placed into a churn, or other mixing vessel, and treated with gaseous carbon disulphide to form a soluble derivative of cellulose, cellulose xanthate. In theory, for regular rayon manufacture only one of the hydroxyl groups on each pair of anhydroglucose units needs to be replaced by a xanthate group, i.e., the target degree of substitution (DS) is 0.5, which to achieve without waste would need 23% CS_2 on cellulose. The desired reaction is

$$- O - \underset{\underset{\displaystyle ONa}{|}}{C_6H_7O} - (OH)_2 - \; + \; CS_2 \; \longrightarrow \; - O - \underset{\underset{\displaystyle O - C - SNa}{\underset{\displaystyle \overset{\|}{S}}{|}}}{C_6H_7O} - (OH)_2 - \tag{1}$$

but side reactions between the CS_2 and the NaOH also occur, one of which is

$$2CS_2 + 6NaOH \; \rightarrow \; Na_2CO_3 + Na_2CS_3 + Na_2S + 3H_2O \tag{2}$$

It is the sodium trithiocarbonate from this side reaction that gives the viscose dope its characteristic orange colour. In addition to the cellulose xanthate forming reaction, equation (1) being reversible, cellulose can reform by

$$\text{Cell-OCS}_2\text{Na} + 2\,\text{NaOH} \rightarrow Na_2CO_2S + NaSH + \text{Cell OH} \tag{3}$$

Clearly, free sodium hydroxide in alkali cellulose leaving the slurry presses causes problems in xanthation, and the presses have to be operated to minimize reactions (2) and (3). At the end of xanthation, any remaining traces of CS_2 are flushed from the wet churn prior to, or in some cases by, admitting a charge of the dissolving or mixer soda in order to commence dissolution. For a dry churn operation, the vessel is opened to allow the golden xanthate crumbs to be discharged into a separate mixer.

7. *Dissolving* – The yellow crumb is dissolved in aqueous caustic solution. Cellulose and its derivatives dissolve more easily in cold alkali than in hot, and the initial contact between cellulose xanthate and the mixer soda should occur at the lowest temperature possible. Two per cent NaOH at no more than 10°C is admitted to the wet churn and the paddle speeded up to wet out all the crumb prior to discharge. Some of the required mixing soda charge can be retained to wash out the churn before restarting the cycle. The large xanthate substituents on the cellulose force the chains apart, reducing the inter chain hydrogen bonds and allowing water molecules to solvate and separate the chains, leading to solution of the otherwise insoluble cellulose.

It is possible to add modifiers or delustrants at the dissolving stage. However, modern viscose dope plants feed several spinning machines, which are often expected to make different grades of fibre. It is therefore now more common to add the materials needed to make special fibres by injection close to the spinning machines.

8. *Ripening* – The viscose is allowed to stand for a period of time to 'ripen'. Two important process occur during ripening: redistribution and loss of xanthate groups. The chemical and physical character of the solution changes slowly with time until an optimum spinning condition is reached. Meanwhile it is subjected to vacuum to remove gas bubbles. Dexanthation, and the redistribution of xanthate groups into the most favourable positions on the cellulose molecules, occurs automatically as the viscose ripens. For regular staple fibres, the process takes about 18 h, but for the stronger fibres, shorter times and more difficult spinning conditions have to be tolerated. During this ageing process, reaction (2) goes into reverse and the xanthate groups attached to the 2 and 3 positions on the anhydroglucose units hydrolyze 15–20 times as fast as those on C-6. Transxanthation occurs; the unoccupied C-6 hydroxyl competing with reaction (2) for the CS_2 that is leaving C-2 and C-3. The reactions favouring more spinnable viscose are in turn favoured by ageing for longer times at the lowest possible temperature. Cold viscoses are unfortunately much more viscous, and therefore harder to filter and deaerate.

9. *Filtering* – The viscose is filtered to remove undissolved materials that might disrupt the spinning process or cause defects in the rayon filament. Filtration of viscose is not a straightforward chemical engineering process. The solution of cellulose xanthate contains some easy-to-deal-with undissolved pulp fibres, and some gel-like material that is retarded rather than removed by the filters. The viscose is unstable and tends to form more gel as it ages. Its flow characteristics make the material close to the walls of any vessel or pipe move more slowly, get older and gel more than the mainstream viscose. Therefore, while filtration can hold back gels arising from incomplete mixing, new gels can form in the pipework after the filters.

The removal of particles that would block the spinneret holes occurs in several stages. Durable nylon needle felts filters that can be cleaned by automatic backwashing are used. Second filtration, usually after deaeration, is also changing to fully automatic systems with sintered steel elements that do not need manual cleaning. Third-stage filtration, close to the spinning machines, is used to provide a final polishing of viscose quality, but is only justifiable for the premium quality fibres. All processes nevertheless use small filters in each spinneret to catch any particulate matter that may have eluded, or been formed after, the main filter systems.

10. *Degassing* – Bubbles of air entrapped in the viscose must be removed prior to extrusion or they would cause voids, in the fine rayon filaments.

The correct viscose age or ripeness for spinning varies according to the type of fibre being made. Ripeness can be assessed by establishing the salt concentration necessary to just coagulate the viscose dope. The preferred test uses sodium chloride (salt figure), although the alternative method (Hottenroth number) is based on ammonium chloride.

5.3.1.2   Spinning (wet spinning)

The filtered deaerates viscose the right salt figure is extruded at a measured rate through the holes of spinnerets, in which each hole produces a fine filament of viscose that are immersed in a bath containing water, sulphuric acid and salts. The emerging filaments are coagulated and chemically changed back to cellulose. The forming filaments are pulled through the bath by the first godets, which in some systems also serve as an anchor against which stretch can be applied by the second godets or traction units. The assembly of pumps, spin baths, jets, godets, stretch baths, traction units and ventilation systems is known as the spinning machine.

$$- O - C_6H_7O - (OH)_2 - \quad \xrightarrow{+H_2SO_4} \quad - O - C_6H_7O - (OH)_3 - \ + NaHSO_4 \ + CS_2$$
$$O - C - SNa$$
$$\overset{\|}{S}$$

Jets for continuous filament textile yarn are typically 1-cm-diameter gold–platinum alloy structures with 20–500 holes of 50–200 μm in diameter. Tire yarn jets are also 1 cm in diameter but typically use 1000–2000 holes to give the required balance of filament and yarn denier. Staple fibre jets can have as many as 70,000 holes and can be made from a single dome of alloy or from clusters of the smaller textile or tire yarn jets. The precious metal alloy is one of the few materials that can resist the harsh chemical environment of a rayon machine and yet be ductile enough to be perforated with precision. Glass jets have been used for filament production, and tantalum metal is a low cost but less durable alternative to gold–platinum (Fig. 5.7).

The basic chemistry of fibre formation is however independent of the hardware. Spin-bath liquors are mixtures of sulphuric acid (5–15%), zinc sulphate (0.05–7%) and sodium sulphate (10–28%) controlled at temperatures ranging from 30 to 60°C. They are circulated past the jets at carefully controlled rates and fully recycled. The liquid filament emerging from the jet

Figure 5.7. A spinneret

coagulates at the interface between acid bath and alkaline viscose to form a cuticle, and later a skin, through which the rest of the coagulation and regeneration is controlled. This regeneration mechanism causes the skin–core nonuniformity, which is characteristic of regular rayon fibre cross sections. Zinc salts are added to the spin bath to slow down the regeneration reaction by forming a less easily decomposed zinc cellulose xanthate intermediate. This allows greater stretch levels to be applied and results in fibres with thicker skins. There is still uncertainty as to whether the zinc cellulose xanthate gel acts by hindering acid ingress or water loss. High levels of zinc in the spin bath allow the production of tough fibres for tire reinforcement and industrial use.

Drawing – The rayon filaments are stretched while the cellulose chains are still relatively mobile. The yarns are kept separate and collected either as acid cakes for off-line cake washing, or wound onto bobbins after on-line washing and drying. This causes the chains to stretch out and orient along the fibre axis. In staple fibre production, the filaments from all the jets on the spinning machine are combined into a few large tows and cut to length prior to washing.

Washing – The freshly regenerated rayon is contaminated with sulphuric acid, zinc sulphate, sodium sulphate, carbon disulphide and the numerous incompletely decomposed by-products of the xanthation reactions which need to be removed. Several different washing techniques may be used. When the filaments are being formed in the bath, simultaneous processes take place during which sulphur is evolved and deposited on the surface of the filaments. In processing, the filaments are subjected to stretching with the purpose of obtaining a supermolecular structure in the fibre and imparting the required physical and mechanical properties to it. The filaments thus obtained are grey-coloured, hard to the touch and without gloss.

A typical washing sequence is follows: a hot acid wash (2% H_2SO_4 at 90°C) decomposes and washes out most of the insoluble zinc salts. This wash completes the regeneration of xanthate and removes as much sulphur as possible in the form of recoverable CS_2 and H_2S (A). An alkaline sodium sulphide desulphurization bath solubilizes sulphurous by-products and converts them into easily removed sulphides (B). A sulphide wash to remove the sulphides created in bath B (C). A bleach bath (optional) uses very dilute hypochlorite or peroxide to improve fibre whiteness (D). A dilute acid or sour bath removes any remaining traces of metal ions and guarantees that any residual bleaching chemicals are destroyed (E). A controlled-pH freshwater final wash removes the last traces of acid and salt prior to drying (F). At last, a finish bath gives the fibre a soft handle for easy drying and subsequent processing (G) (see Fig. 5.8).

Water consumption in washing is minimized by using a countercurrent liquor flow, clean water entering the system at the point where the fibre leaves, and dirty liquor leaving for chemical recovery at the point where the fibre enters.

In case of continuous filament yarn, a continuous washing and drying process based on self-advancing reels is used. Whereas the processing of cake yarn can take many days to complete, the continuous process is over in

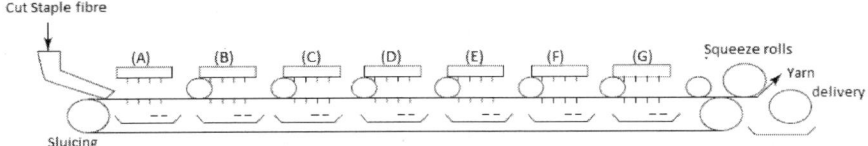

Figure 5.8. Staple fibre washing sequence

minutes. The process is less labour intensive and more productive than the cake system. It is most used for industrial yarns – still the reinforcement of choice for radial tire carcasses. During subsequent finishing, the filaments are subjected to washing and desulphurization, i.e., to a treatment with a solution of sodium sulphide or sulphite for dissolving the sulphur deposited on the filaments. As a result of this treatment, the sulphur content may be reduced to 0.05–0.1%.

Bleaching, soaping, drying – Further bleaching (with hypochlorite, hydrogen peroxide or sodium chlorite), soaping or oiling to soften the filaments and to improve the conditions of subsequent mechanical treatment and final drying (Fig. 5.9).

While the cellulose is regenerated, sodium sulphate and carbon disulphide are formed, the latter being released in the form of a gas. Some of this carbon disulphide is condensed and recovered. The carbon disulphide which is not condensed is washed out and the components with an unpleasant smell are removed. During the stage when the xanthate is formed, a side reaction takes place during which the carbon disulphide partly reacts with the sodium hydroxide in the alkali cellulose. This produces sodium trithiocarbonate via various intermediate steps.

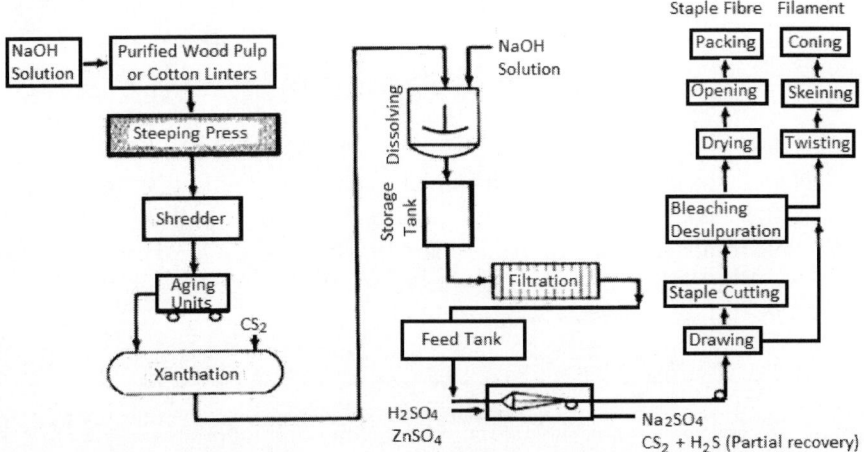

Figure 5.9. Viscose manufacturing process - Filament and staple fibres

$$3CS_2 + 6\ NaOH \rightarrow Na_2CO_3 + 2\ Na_2CS_3 + 3H_2O$$

The characteristic orange colour of viscose is explained by the presence of the sodium trithiocarbonate.

The manufacturing process has many disadvantages including the poor ecology due to the emission CS_2 and H_2S. Many researches has been done in this regard and new processes has been put forward including alternative solvents for cellulose and developed new processes having better safety and more productivity. One of such processes is the manufacture of modal with N-methyl-morpholine-N-oxide (NMMO) process which has been explained in this book elsewhere. Other processes developed in this regard are CS_2-free processes for the production of regenerated cellulose fibre are:

- Process for producing tencel fibres which was developed by Courtaulds,
- Akzo process for spinning newcell fibres and
- Lenzing process for spinning lyocell fibres.

The average results of chemical analysis of viscose fibre are given below:

α-Cellulose	80%
Hemicellulose	15%
Pentosans	3.5%
Resin	0.5%
Soaps	0.5–0.7%
Sulphur	0.1%
Ash	0.4–0.5%
Lignin-like substances	0.3%

α-Cellulose is that part of the technical product, which is insoluble in a 17.5% NaOH solution after treatment for 1 h at a temperature of 20°C. Hemicellulose is the name for low-molecular polysaccharides with a DP below 150. Resin, pentosans and lignin residues pass into the viscose fibre from the original wood pulp. Sulphur remains due to incomplete desulphurization, and soap is left on the filaments after finishing.

Generally, viscose fibre also contains titanium dioxide (up to 2% of the fibre weight) which has been added to impart a mat finish to the fibre. A part of manufactured viscose fibres is dyed in stock and therefore contains dyes in its composition.

The X-ray pattern of viscose fibre is similar to that of mercerized cotton, which is cellulose hydrate. However, in mercerization of cotton, the conversion of cellulose to cellulose hydrate is hardly at all accompanied by hydrolytic splitting of chains, whereas during the process of manufacturing man-made fibres the chains become considerably shorter and the DP of cellulose hydrate in ordinary viscose fibre is within 280–320. Another substantial difference between man-made cellulose fibres and cotton fibres is that the structure of viscose fibres is looser due the less orderly arrangement of the main chains. Usually, in the formation of filaments the main chains are not so compactly arranged as in cotton or flax fibres. Owing to this, viscose fibres are endowed with the following properties: their hygroscopicity is higher than that of cotton and their swelling in water is much greater, so that in the (wet condition they lose a considerable part of their strength (50–60%), as compared with cotton, they are less resistant to the action of acids and much less resistant to alkalies (caustic alkali solutions cause strong swelling of viscose, which can dissolve them).

Other forms of regenerated cellulose fibres that are classified as rayon without separate, distinctive names include HWM rayon, cuprammonium rayon and saponified rayon. HWM rayon is highly modified viscose rayon that has greater dimensional stability in washing. Cuprammonium rayon is usually made in fine filaments that are used in lightweight summer dresses and blouses, sometimes in combination with cotton to make textured fabrics with clubbed, uneven surfaces. When extruded filaments of cellulose acetate are reconverted to cellulose, they are described as saponified rayon, which dyes like rayon instead of acetate.

5.3.1.3   Rayon fibre characteristics

The drawing process applied in spinning may be adjusted to produce rayon fibres of extra strength and reduced elongation. Such fibres are designated as high tenacity rayons, which have about twice the strength and two-thirds of the stretch of regular rayon. An intermediate grade, known as medium tenacity rayon, is also made (Fig. 5.10).

It is well known that rayons are cellulose fibres. Cellulose exists in different crystalline configurations of which the most important are cellulose I and cellulose II. Natural cellulose contains more of cellulose I, whereas cellulose II is major constituent of regenerated fibres and hence the difference in properties. The number of intramolecular hydrogen bonds in cellulose II is found to be half that found in cellulose I which also is the reason for the rayons less stronger when compared to natural cellulosic fibres like cotton, cellulose, hemp, ramie, flax, jute, etc. X-ray analyses have shown that 40% of the fibre substance in the case of viscose fibres and 70% in the case of cotton is found

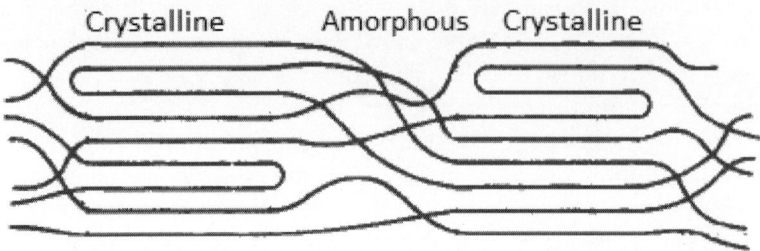

Figure 5.10. Schematic representation of crystalline and amorphous zones in cellulose

in crystalline zones which alternate between amorphous segments. The molecular chains are partly folded and form a so-called fringed fibrilar structure. The drawn fibre becomes arranged but the amorphous portion remains. Even though the amorphous portion contributes negatively towards the strength of the fibre all the reactions with chemicals, dyes, finishing agents, etc., are generally restricted these areas only where the water molecule penetrates and the hydrogen bond splits. Even though the –OH groups in this areas are responsible for all these reactions but not all OH groups are available to react because, for example, some of the dye molecules are too large and are so sterically inhibited that reaction is impossible. During the spinning process when the fibre enters the precipitation bath a tiny skin is first formed by coagulation and this acts as a membrane of the fibre. Thus the cross section of viscose can show a sheath and core structure. These structural characteristics are responsible for most of the characteristics of the viscose fibres. Thus moisture absorption and swelling behaviour of viscose fibres: it rapidly absorbs moisture from the surrounding air and reacts more sensitively to fluctuations in climatic conditions. Under standard atmospheric conditions (20°C and 65% relative humidity), the moisture absorption of viscose 13–14%, while of cotton is only 7–8%. The swelling during moisture absorption and consequent loss in strength is one of the most important properties of viscose. This is because of the effect of the hydrogen bonds between the chains are removed by their bonding with water. The relative wet strengths for viscose and cotton are 42–50% and 100–113%, respectively. During the drawing since the crystalline structure percentage increases, yarn with fine filaments has normally, a higher relative wet strength (Fig. 5.11).

The action of alkalis are much more intense than water. Alkalis swell the viscose to higher degree than water as intermicellar swelling turns into intramicellar swelling. The reactivity of viscose is increases because of the accessibility almost all hydroxyl groups due to swelling. Stronger caustic soda can dissolve the cellulose converting it to soda cellulose. Viscose can be

Figure 5.11. Scanning electron microscopic (SEM) view of the viscose (×2000)

safely treated with 4–6% caustic soda solution with a higher affinity for dyes. Thus viscose can be causticised with caustic soda of strength between 6–8%.

5.3.1.4   Major uses of rayon fibre

Apparel: Accessories, blouses, dresses, jackets, lingerie, linings, millinery, slacks, sport shirts, sportswear, suits, ties, work clothes.

Home furnishings: Bedspreads, blankets, curtains, draperies, sheets, slipcovers, tablecloths, upholstery.

Industrial uses: Industrial products, medical surgical products, nonwoven products, tire cord.

Other uses: Feminine hygiene products.

5.3.1.5   Modified filament rayons

In an attempt to improve the functionality or user friendly characteristics of rayon filaments, many new versions of filaments are developed. In most cases, the manipulation and modification of the spinning process, the physical structure and form of the rayon filament can be changed.

5.3.1.6   Varying cross section

Modification of filament cross section is becoming of increasing importance today, as it can cause profound changes in the characteristics of yarns and fabrics. Circular cross-section filaments, for example, are poorer in covering power than lobed cross sections typical of the normal viscose filament. Many synthetic fibres are now being produced in noncircular forms, such as dog bone and trilobal cross sections as explained in polyester manufacturing process in this book. As an easiest way, the cross-sectional shape of the filament may be varied by extruding through spinneret holes of suitable shape. Viscose rayon has been made experimentally in a variety of cross-sectional shapes, and some have become of commercial importance. Straw filaments, for example, are produced by some manufacturers. Flat filaments are made by extrusion of viscose through slit orifices instead of circular ones; these filaments have improved covering power, but tend to be of harsh handle. Another method is varying the diameter of a filament continuously between thick and thin, providing rayon filaments which make up into special effect fabrics (Fig. 5.12).

Figure 5.12. Multilobal cross section of wet-spun viscose

5.3.1.7   Bubble filled filaments

One of the major disadvantages of the common round cross-sectional filament yarn is its poor covering power. One of the method in improving the covering power is to introduce air or gas bubbles inside the fibre while spinning. This may accomplished by spinning an agitated viscose solution to produce a foam in which air bubbles are entrapped. In 1976, Courtaulds Ltd marketed a hollow viscose fibre 'Viloft' which is made by generating carbon dioxide inside the filament. This has greatly increased bulk and high moisture absorption of the fibre. In blends with polyester fibres, it provides increased covering power. Blends of hollow viscose with cotton are used in shirtings and dress fabrics and for terry towel pile. Hollow viscose fibre is widely used in nonwovens, particularly in fields such as surgical and medical fabrics where high moisture absorption and moisture holding properties are important. Another method of introducing air bubbles inside filaments developed in the United States was to inject air into the filament as it was extruded. This produced a continuous filament containing discrete bubbles 3–6 mm (1/8–1/4 in.) long, which was used as a substitute for kapok in life jackets, pontoons, insulated clothing, etc.

5.3.1.8   Spun dyed filaments and staple yarns

It was practice to add titanium dioxide in rayon for dullening the natural sheen of the rayon if not required. As an extension of this process, finely dispersed pigment can be added to viscose solution before spinning. As you spin the pigments get blocked inside to give a light and wash fast dyed filament yarn. Spun dyed filaments are used only for making fabric for bulk requirements. In fashion garments, it may not be very much useful.

5.3.1.9   Crimped rayon

The serrated structure of wool gives certain special characters to it, in addition to spinning qualities which are advantages as a fibre. There was attempt to create such waviness or crimp by one way or other during spinning which enhanced the qualities of rayons. Since its spinning quality was better it was more required for filaments which are meant for staple yarn. This was accomplished mechanically, for example, by passing the filament between gear-like rollers, or chemically by controlling the coagulation of the filament in such a way as to create a fibre of asymmetrical cross section. The chemical method is to spin into a coagulating bath containing less acid and more salt than usual and then carefully controlling the stretch, which gives a crimp to the resultant fibre. There are disadvantage for this method. The filaments produced this way have an asymmetrical cross section, one side being thick-skinned and almost smooth, the other side being thin-skinned and highly serrated (see Fig. 5.13). When the fibres are wet, they swell much more on the thin-skinned side than on the thick-skinned side, so that there is a tendency to curl. A similar effect may be introduced into rayon by using the 'bicomponent' technique which has been developed successfully in the production of some synthetic fibres. This method consists in the extrusion of twin filaments through orifices set side by side, in such a way that the two filaments join as they coagulate. The composite filament is made from viscose solutions of different characteristics, and the two portions of the filament have different

Figure 5.13. Cross section crimped viscose produced by chemical method which shows one side being thick skinned and the other side being thin skinned

swelling properties. In water, the filament tends to curl as one side swells more than the other. Surface variations instead of the usual striated surface of the regular rayon can be produced means of finishing treatments, and by using vibrating spinnerets, which gives better appearance and handle of the fabric made using this type of yarn.

5.3.1.10   Polynosic fibre

Viscose manufactured in the starting had good lustre and was being compared to natural silk, but many of the qualities like wet strength, was much inferior to silk or any other fibres as such. Since it was cotton like with qualities of cotton and cheaper, it was accepted in the market but always had a nonacceptance due to the poor strength qualities. Researches continued to make a better viscose, even though the production of viscose remained higher than any other man-made fibres due to the cheapness, abundant availability of raw material (wood).

Since it is a cellulosic fibre, comparison has been made to cotton. Look in this angle, viscose has some better qualities like since produced in filament form it will have uniform density and composition, staple fibre can be cut of any length from filament or mixture of length, which can solve to spinning faced by cotton and above all the cost and the evenness. Despite this remarkable progress made by viscose rayon and the continuous improvement in the quality of the fibre, viscose retains unattractive characteristics which have been associated with it since the earliest times. These shortcomings have prevented viscose rayon from competing as effectively as it might with the natural fibre it most nearly resembles – cotton.

Viscose rayon is sensitive to the effects of moisture. When rayon is wet, it absorbs water and swells, the diameter of the filament increasing by more than 25%. At the same time, the tenacity falls by about 50%, and the extensibility increases by some 20%. The initial modulus of the rayon falls, and the filament will stretch in response to only a small tensile stress. Elastic recovery from such stretching is poor. Researches has given more impressive progress in this direction has taken place with the development of the new types of viscose rayon which have become known as HWM modal and polynosic rayons. The main structural differences between viscose rayon and cotton can be summarized as (a) differences in the DP of the cellulose molecules, and (b) differences in the arrangement of these molecules in the filament.

The effort towards achieving the above structural qualities manufactured rayons was being made in which the DP of the cellulose was increased by modification of the viscose production process and the orientation of the cellulose was improved by slowing the regeneration and coagulation of the cellulose filaments. The DP was increased from 250 of regular rayon to 500 by different methods by 1950s. Thus, an interesting variety of viscose staple fibre is the so-called polynosic fibre, which in its properties is closer to cotton fibre. Due to some special processing carried out in manufacturing. Polynosic fibres are distinguished from viscose fibres by a higher degree of cellulose polymerization (450–500), a more orderly structure, and better uniformity. Owing to this, polynosic fibre is characterized by higher strength, reduced elongation in the dry state (these properties are similar to those of cotton), less swelling in water, smaller loss of strength in the wet state (25–35%) and reduced solubility in alkalies. In practice, the term HWM is commonly used to describe a broad range of fibres of this type, the term 'polynosic' being used for those with the HWM. 'Modal' is widely used as a generic term for regenerated cellulose fibres obtained by processes giving a high tenacity and a HWM.

5.3.1.11   Modal

Difference between polynosic and modal as per Textile Institute (United Kingdom):

Polynosic fibre: A regenerated cellulose fibre that is characterized by a initial HWM of elasticity and a relatively low degree of swelling in sodium hydroxide solution.

Modal fibre: A generic name for regenerated cellulose fibres obtained by processes giving a high tenacity and a HWM.

Modal fibres have a higher DP (350–600), higher resistance, wet modulus, dimensional stability and alkaline resistance compared with normal viscose fibres. HWM fibres have lower alkaline resistance and lower brittleness compared with polynosic fibres.

Modal is a pure cellulose fibre made of 100% natural origin wood. The development of modal fibre began in the 1930s. The Japanese introduced it in 1942 in the form we still use it today. Both deciduous and coniferous trees can be used in its production. Only the highest quality of cellulose is used to be finished into modal fibre. Viscose is a basic cellulose fibre, whereas modal is the more refined fibre which is achieved through different finishing techniques. Modal is more durable, compared to viscose and does not wrinkle or shrink as viscose does. It has an optimum wear comfort – it 'breathes'. It has good moisture absorption, a soft feel, high colour brilliance. It is easy to care.

The improved properties compared with the classic cellulose regenerated fibres are due to the clearly higher mole mass or chain length of the individual molecules and the more compact internal structure of the fibres, which come closer to the properties of cotton. During the manufacturing stage, a deceleration of the xanthate decomposition, caused by the addition of a modifier and the precipitation bath composition by variable spinning speeds (wet elongation 15% at 2.25 cN/dtex load), results in an extension of the plastic state and in the possibility of a strengthened longitudinal orientation and the development of better internal structures. These developed structures play a crucial role in the finishing process. The deviating behaviour of classic regenerated and modal fibres is well known. One problem is the tendency for uneven dye uptake, which can often be recognized as colour streakiness. Local or temporary differences in the colloidal precipitation processes, delay and orientation processes, chemical decomposition of the xanthate as well as subsequent thermal effects, may constitute the cause of this.

Modal fibres are typically of round cross section and do not display any skin effect. The microstructure is fibrillar, the filament breaking up into smaller and smaller fibrils when disintegrated, for example, by nitric acid. The fibrils are distributed uniformly throughout the filament cross section, producing a homogeneous structure. The DP is in the region of 500. The degree of crystallinity of HWM modal fibres is in the region of 55%, compared with 40–45% for ordinary rayon and 70–80% for cotton. The crystallites in HWM modal fibres are larger than those in ordinary rayon. The degree of orientation of the long cellulose molecules, in both the amorphous and the crystalline regions of the fibre, is higher in HWM modal fibres than in ordinary rayon.

Modal is manufactured mainly in three varieties to meet the requirements in the market – high strength, standard, high elongation. Their characteristics are given in Table 5.1.

Table 5.1. Characteristics of three types of modal fibre

Characteristics	High strength		Standard		High elongation	
	Dry	Wet	Dry	Wet	Dry	Wet
Tenacity	41–46 cN/tex (4.6–5.2 g/den)	3 0–3 5 cN/tex (3.4–4.0 g/den)	28–35 cN/tex (3.2 – 4.0 g/den)	18–26 cN/tex (2–3 g/den)	34–42 cN/tex (3.8–4.8 g/den)	21–30 cN/tex (2.4–3.4 g/den)
Elongation	6–10%	8–14%	8–12%	9–16%	12–14%	16–20%
Initial modulus, wet per 100% ext'n at 2% ext'n		132–221 cN/tex (15–25 g/den)		88–159 cN/tex (10–18 g/den)		53–79 cN/tex (6–9 g/den)
Initial modulus, wet per 100% ext'n at 5% ext'n		221–353 cN/tex (25–40 g/den)		124–247 cN/tex (14–28 g/den)		88–115 cN/tex (10–13 g/den)
Water imbibition	65–75%		55–70%		65–75%	

Elastic recovery is higher for HWM modals than for cotton or rayon staple, especially in the wet state.

5.3.1.12   *Advantage of modal over cotton*

Modal is soft, smooth and breathes well. Its texture is similar to that of cotton or silk. It is cool to the touch and very absorbent. Like cotton, modal dyes easily and becomes colour-fast after submersion in warm water. One of the advantages of modal over cotton is its resistance to shrinkage, a notorious problem with cotton. Modal is also less likely to fade or to form pills as a result of friction. Its smoothness also makes hard water deposits less likely to adhere to the surface, so the fabric stays soft through repeated washings.

5.3.2   Lyocell/tencel

Lyocell is a fabric most know better by its brand name tencel. It has a soft finish, packs light and is made from cellulose (vegetable matter), or wood pulp, typically a mix of hardwood trees like oak and birch. This makes it a natural fabric, and it is noted for its durability and strength, in addition to its ecofriendly manufacturing techniques. In the evolution of man-made cellulosic fibres, viscose fibre was made first with many advantages like good moisture absorption, anti-static properties, good dyeing properties, easily made flame-retardant and produces comfortable clothing. On the contrary, it causes pollution problems in the spinning process and has many disadvantages like it has a lower wet tenacity, the production cycle is longer, etc. Further, modal, polynosic, etc., were introduced by improving some of the disadvantages of the previous fibre (a comparison of their properties are given in Table 5.2). But with the further development of lyocell fibre, it adopted a solvent-spinning process whereby pollution was highly controlled and the fibre has many advantages: higher tenacity, shorter production cycle and no pollution problem.

Table 5.2. Comparison of properties of different fibres

Fibre type -->	Cotton	Viscose	Modal	Polynosic	Cupro	Lyocell	Lyocell LF
Titre dtex	1.8	1 - 100	1.0-3.3	1.3-4.2	1.4-2.2	0.9-3.3	1.3
Tear strength cond. cN.dtex	24-28	20-24	34-36	35-40	15-2-	40-44	34-36
Elongation cond. %	7.0-9.0	20.0-25.0	13.0-15.0	10.0-15.0	7.0-23.0	14.0-16.0	10.0-12.0
Wet tear strength cN/tex	25 -30	10.0 - 15.0	19.0-21.0	27.0-30.0	9.0-12.0	34-38	28-31
Relative wet strength %	105	55	57	75	60	85	
We elongation %	12.0 - 14.0	25 - 30	13.0-15.0	10.0-15.0	16.0- 43	16 - 18	13.0-15.0
Degree of Polymerisation	2000-3000	250-350	300-600	500-600	450-550	550-600	
Loop strength cN/tex		10.0 - 14.0	12.0 - 16.0	8.0- 12.0		18.0-20.0	15.0 - 17.0
Fibrillation tendency	2	1	1	3	2-3	4-6	2
Water retentivity %	45-55	90-100	75-80	55-70	100	65-70	63
Crystallinity %		25	25	40	25	40	45

5.3.2.1   Manufacturing process of lyocell

It was well known that the amine oxides dissolve cellulose, but it was not possible to use the same in a manufacturing process of viscose they decompose easily, particularly at raised temperatures, in contact with reducing groups – i.e., with end groups of cellulose – or under the influence of metal ions. But with the discovery of NMMO, which is considerably more stable, less toxic than even ethanol and clinically harmless the chances of it being used for cellulose solution for spinning has highly increased drastically.

A process for the manufacture of lyocell fibre with an increased tendency to fibrillation, comprising the following steps:

1. Dissolving cellulose in NMMO solvent to form a solution

2. Extruding the solution through a die to form a plurality of filaments

3. Washing the filaments to remove the solvent, thereby forming lyocell fibre

4. Drying the lyocell fibre, wherein the DP of the cellulose is not more than about 450 and the concentration of cellulose in the solution is at least 16% by weight.

The NMMO is recovered by separate recovery process.

As shown in Fig. 5.14, the main raw material for lyocell manufacture is wood pulp. The wood pulp grade used is similar to the dissolving pulp used for viscose rayon but has a slightly lower DP. The cut pulp is conveyed to vessels where it is mixed with a 76–78% amine oxide solution in water. A small quantity of a degradation inhibitor is also added to the mixer; other additives such as titanium dioxide (for producing matt fibre) can also be added. The mixing is achieved at 70–90°C in a ploughshare mixer that contains a number of high-speed refiners to break the pulp down and aid solvent wetting. The resultant slurry consists of swollen pulp fibres and has the consistency of dough. This premix is dropped into an agitated storage hopper from which it is accurately metered to the next stage of the process.

5.3.2.2   Solution making

Premix is heated under vacuum to remove sufficient water to give a clear, dark amber-coloured viscous solution of the cellulose which may contain 10

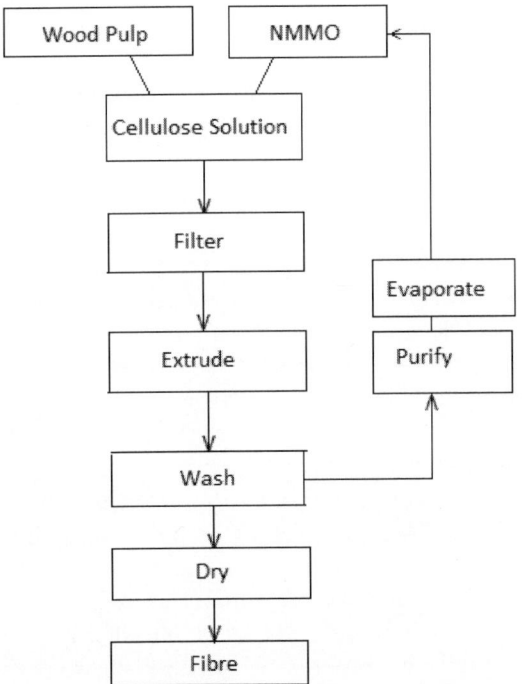

Figure 5.14. Flow diagram of lyocell manufacture

to 18% cellulose. The evaporator vessel is operated under vacuum to reduce the temperature (circa 90–120°C) at which the water evaporates. This is important because the amine oxide solvent (NMMO) in solution can undergo an exothermic degradation process if it is overheated.

The solution is pumped by a number of specialized pumps in series through the transport system. The transport system consists of a solution cooler and a hydraulic ram buffer tank, which feed into the solution primary filters. Due to the viscosity of the solution, the pressures involved in pumping solution can be as high as 180 bar. A complication to the process design is caused by the tendency of the amine oxide in solution to exothermically degrade due to maloperation of equipment or chemical contamination of the solution. In such a case, the pressure inside the vessel can increase exorbitantly to a point, which can rupture the equipment causing safety hazards. To allow for this possibility, bursting discs are provided at strategic positions throughout the plant to relieve pressure in the event of an exotherm.

5.3.2.3   Filtration

It is necessary to filter the solution out of various impurities like undissolved pulp fibres, inorganic compounds such as sand and ash before filtration. Filtration is effected in two stages, one as a central filter and second just before each spinning machine position. The filters use candle filter elements, which is reused after cleaning with hot NMMO and ultrasonic washing after a certain usage.

5.3.2.4   Spinning

The solution is accurately pumped to several spinning positions after filtration by metering pumps. Each spinneret consists of thousands of tiny holes through which the solution is extruded into fibres into a small air gap across which air is blown by the cross-draught system to condition the fibres. Next the fibres or tow are pulled down through the spin bath where the cellulose is regenerated in dilute solvent. The fibres are drawn, or stretched, in the air gap by the pull of traction units, or godets.

For spinning, the solution is split into sub-streams, which serve a number of spinning positions. The solution is then supplied to each jet, via a filter, by a metering pump. It is then extruded and spun through an air gap into a spin bath containing dilute amine oxide solution. Each jet consists of thousands of tiny holes through which the solution is extruded into fibres. Just below each jet face is a small air gap across which air is blown by the cross-draught system to condition the fibres. After passing through the air gap, the fibres, or tow, are pulled down through the spin bath where the cellulose is regenerated in dilute solvent. The fibres are drawn, or stretched, in the air gap by the pull of traction units, or godets.

5.3.2.5   Washing and after treatments

The fibres are brought together into a large tow and is washed as a single large continuous tow through a series of wash troughs, each of which consists of a wide, shallow bath containing a number of wedges which deflect the tow band alternately up and down as it is pulled along the trough. This serves to allow dilute solvent into the tow band and then squeeze it out. Wash liquor leaving the wash line goes into the spin bath system. In some manufacturing process, the fibres are cut before washing.

After this, the fibres can be bleached if required, finished with softeners and antistatic agents to facilitate the further processes and dried on a drum drier. Crimping, cutting and baling can be followed.

5.3.2.6   Solvent recovery

An important part of the economy of the manufacturing programme is the efficient recovery of the relatively expensive solvent. There are chances of the NMMO to oxidize to N-methylmorpholine by catalytic action of transition metals like copper and iron. To control this, it is essential that a stabilizer such as propylgallate is added in the solution which acts both as an antioxidant and a chelating agent. Solvent recovery consists of two main processes, ion exchange of the dilute solvent then evaporation of the excess water to a concentration required in premixing. The NMMO reconcentrated in a steam heated multiple effect falling thin film evaporator.

5.3.2.7   Characteristics

It is soft, strong and absorbent and has excellent comfort. It has good drape and more comfortable than cotton. It is softer and smoother after repeated washing. It can be fibrillated during wet processing to produce special textures, excellent wet strength and wrinkle resistant. Very versatile fabric dyeable to vibrant colours, with a variety of effects and textures. Can be hand washable simulates silk, suede or leather touch. It has good drapability and is biodegradable.

5.3.2.8   Fibrillation

Fibrillation is an inherent characteristic of many fibres including lyocell, but it varies in different fibres. Fibrillation is the longitudinal splitting of a single fibre into microfibres of typically less than 1–4 µ in diameter. The splitting occurs as a result of wet abrasion against fabric or metal. The fibrils are so fine that they can become almost transparent, giving a white or 'frosty' appearance to the finished fabric. In cases of extreme fibrillation, the microfibrils become entangled, giving a pilled appearance. Any process that abrades the fibre in a wet condition will generate fibrillation to some extent. Therefore processing of a fabric on rope dyeing equipment can lead to fibrillation, for example, on a winch or jet where the fabric rubs against itself and metal.

In fabric manufacture, lyocell fibres are available at a fineness of 1.3 or 1.7 decitex and finer. The lyocell is mostly finished in traditional process route, in yarn form or fabric form. But the two decisive machine-related factors that determine the possibilities and limits of a lyocell finish are mechanical energy input and a smooth running of the goods. Free movement of the goods is essential for a uniform fibrillation and largely depends on the machinery used. Due to the high swelling of the fibres, the spaces between the fibres, especially when woven, are reduced and the textile becomes less flexible. Smooth running of the textile becomes more difficult. With rising square metre weight, the risk of abrasion marks and running creases grows. This undesired side effect can be controlled by blending lyocell with synthetic fibres, such as polyester or polyamide that do not or only slightly swell. For 100% lyocell fabrics an open-width alkaline treatment is recommended. Lyocell is much different than viscose in many properties. One of the main characteristics is the fibrillation of these fibres in wet state under mechanical stress. This property can be used for achieving special effects or can be removed from the surface of the material by cross linking the cellulose chains (see Fig. 5.15).

For a peach-skin woven fabric with lyocell after singeing and desizing in open width, the fabric is treated open width with alkali. The fabric swells and the alkali is washed off but the swelling remains. This helps in the subsequent processes being flexible without forming creases and during subsequent primary fibrillation which is done in a jet, which ensures uniform fibrillation, any fibrils that are not fully bonded are brought to the surface of the fabric. Since lyocell fabrics are prone to wet rigidity, temperatures <60°C should generally be avoided and a suitable crease inhibitor should be added to all baths to avoid crease marks. All precautions has to be taken to avoid any uneven fibrillation as it cannot be corrected by any process. Further, an enzyme treatment is given to remove any linty fibrils loosened during primary fibrillation. Exhaust dyeings should preferably be carried out on the same type of jets preferably by migration processes (migration at 80–90°C and dyeing at 60°C) improve levelness and trigger secondary fibrillation because elevated temperatures are used. If cold pad-batch (CPB) dyeing is used for reactive dyeing, the fibrillation has to be done before dyeing or after dyeing as the CPB dyeing does not cause any fibrillation due to the absence of mechanical forces. After CPB dyeing, tumble drying is a generally accepted process to raise the fibrils and thus achieving desired peach-skin effect. To prevent continued fibrillation, frosting and formation of crease marks in domestic laundering, a resin finish may be given.

5.3.2.9   Applications

Lyocell fibre cut into staple (usually 1.4–1.7 dtex fibre to 38 mm) and can be spun into yarn using spinning machinery for handling cotton fibres that are similar in dtex and length to lyocell. While spinning lyocell, the following characteristics of lyocell has to be considered – it is a cellulosic fibre and absorbs moisture readily, it has a smooth surface with a round cross section, it possesses a nondurable crimp, it has a high modulus, it has a high tensile

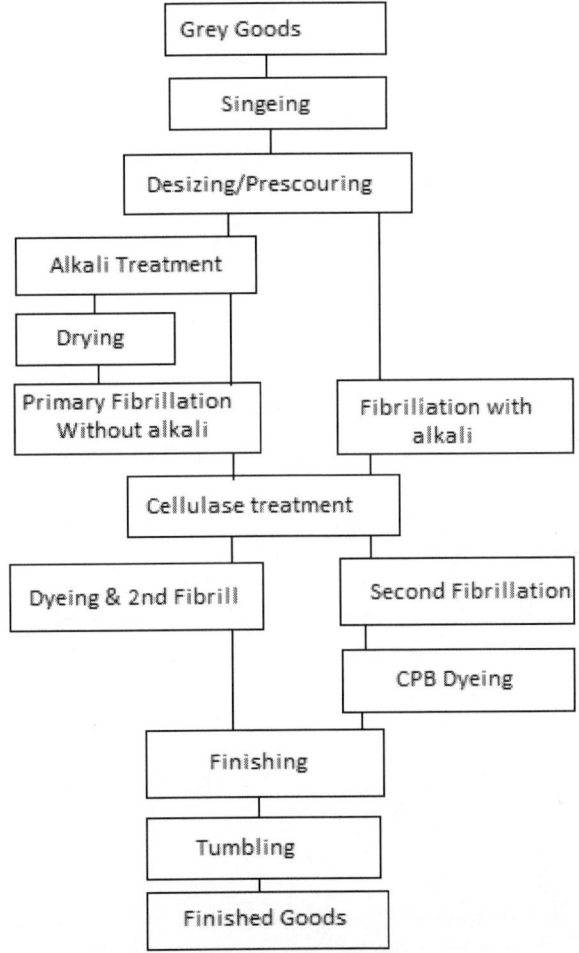

Figure 5.15. Sample process route for peach-skin lyocell fabric

strength, the fibre is very open and there is little fibre entanglement, the fibre supplied is usually lower in moisture content than cotton or viscose, etc. The adjustments and setting of the spinning machinery has to done taking these subjects into consideration. While spinning along with other fibres like cotton and viscose, care has to be taken to avoid contamination, as lyocell tends to dye slightly darker than viscose and cotton.

5.3.2.10   Fabric manufacture and processing

Lycocell can be woven on normal looms used for cotton/viscose weaving. While designing the weaving structure the abnormal swelling behaviour of lyocell, which is much higher than cotton or viscose, in presence of water has to be taken into consideration. Swelling easily causes fibres and yarns to move and can cause creasing. The swelling behaviour (increase in the diameter of the fibre/yarn) also causes increase the crimp, which will affect the fabric to contract (shrink) and become thicker. Thickness is due to an increase in the yarn crimp amplitude. The total amount of wet fabric contraction depends directly on the number of yarns in warp and weft and the size of those yarns. A predominance of warp yarn means that fabric length contraction will tend to be greater than that in the width and vice versa.

Thus while wet processing, the construction of the fabric plays an important part to achieve the desired result. The following factors can affect the wet processing of the fabric.

Processing tensions: For example, greater warp tensions will give either (a) more weft yarn crimp or (b) greater residual warp shrinkage (or both!).

Frequency and diameter of yarns 'yarn density': Where yarns are more numerous and/or larger in diameter then yarn crimp will develop only to the point at which adjacent yarns just touch. When such yarns touch one another, the fabric becomes 'jammed' or locked so that no more yarn crimp can develop without disrupting the alignment of yarns in that plane. 'Jamming' prevents further contraction in that plane but can readily force increased contraction (shrinkage) or fabric movement (creasing) elsewhere.

Fibre and yarn swelling in diameter: Swelling adds to the 'jamming' effect by thickening the yarns. Fabrics that are apparently quite loosely woven when dry can become stiff and firm when wet. Again, this limits the amount of yarn crimp that can develop within the plane of the fabric (i.e., without buckling).

The problems due to the structure of the fabric can be dealt with to some extent by setting of the fabric in wet processing. But the best way is to take care while deciding the constructions.

Knitting can be done with the lyocell yarn on the normal knitting machines used for cotton. It is better to select shorter stitch length to minimize the risk of work up and pilling during dyeing and subsequent use.

5.3.2.11   Dyeing

It is better to set the fabric by causticization the fabric to minimize many problems, which can arise due to the factors, explained above. It has been proved that the lyocell fabrics are best processed by conventional open-width techniques. Processing in rope form by conventional jet dyeing machine in problematical, because the fabric rope does not move sufficiently in running on the machine and uncontrolled localized fibrillation can develop. Air jet technology is more suited for lyocell fabric since it help in rope reorientation and hence fibrillation control, together with the introduction of new processing routes to reduce the high cost of lyocell piece finishing.

The swelling and fibrillation behaviour of lyocell can give following problems in dyeing – the swelling in water causes fabrics to stiffen in cold water, giving rise to creases which can in turn form abrasion and hence fibrillation. These localized fibrillation on the creases gives rise to white lines and damage marks in the finished goods. Correct machinery selection and the use of fabric causticization help to minimize these effects.

5.3.2.12   Causticization

Causticization can be done in open width with 10–12% caustic soda with width wise tension. The fibre swells in caustic soda in diameter but does not shrink in length. Because of this causticization is done with width wise tension to avoid the fabric to shrink. Otherwise caustic treatment will cause to increase the bulk and handle of the fabric significantly which is undesired. Instead of causticization, normal cotton mercerization alsocan be done on lyocell fabric.

The new set fabric with greater bulk and flexibility and new configuration is subsequently washed. The wet stiffness of the fabric is much reduced and this helps in further processing since it is less prone to crease damage marks during the dyeing processing. The caustic treatment also gives more rapid fibrillation removal in processing, changes the light reflectance behaviour and gives the 'bloom' or 'frosted' look on a peach-touch lyocell fabric.

5.3.2.13   Basic process for standard lyocell

The mostly followed lyocell processing have three stages. First, fibrillation is deliberately induced and then removed by a treatment in a cellulase enzyme. Secondary fibrillation is then created by further mechanical action, which can be simultaneous with dyeing if the fabric is not already predyed. This three stage process is time consuming and hence costly. Final finishing on a rope tumbling machine is required to fully generate the fibrillated effect. Lyocell can be successfully processed in garment dye or garment wash systems (including indigo). The high level of garment movement and mechanical aggression in the machine makes for rapid processing through the fibrillation, which can give a casual peach-touch result, or through enzyme processing steps an even surface effects.

Sometimes the lyocell fabric is treated with resin prior to dyeing to avoid the fibrillation in jet processing/dyeing. But the resin has to be selected since all resins are not suitable. The normal glyoxal-based easy-care resins significantly impair the dyeability of cellulosic fibres so they cannot be used as a treatment prior to dyeing. TAHT (triacroyl hexahydrotriazine; Fig. 5.16) will cross link lyocell under alkaline conditions. Pretreatment of the fibre using this chemical during the manufacturing process has proved successful (see Tencel A100).

5.3.2.14   Use of air jets

Air jets can be conveniently used for processing lyocell in rope form. The machine causes the fabric to 'explode' from the jet, giving excellent reorientation and high first quality results. The three-stage process can be followed to give a peach-touch result, or the order can be changed to give a clear fabric surface, but in this instance, resin finishing needs to be applied to prevent secondary fibrillation being generated in domestic washing. (Precausticization of the fabric can help as it imparts wet softness to the fabric, which allows better running on the machine and decreases the risk of creasing and damage.)

5.3.2.15   Special process routes

1. *TOP+C processing*: It is a special processing (tencel oxidative preparation plus caustic – TOP+C) involving a peroxide bleach (preferably pad steam) prior to causticization. By this method, the fabric requires much less mechanical action to generate the peach touch and the three-stage process is no longer required. In jet processing, the action on the fabric of a reactive dyeing process followed by rope tumbling is sufficient. This can halve the time needed on the jet and eliminate the cost of the enzymes. It is also applicable for garment processing,

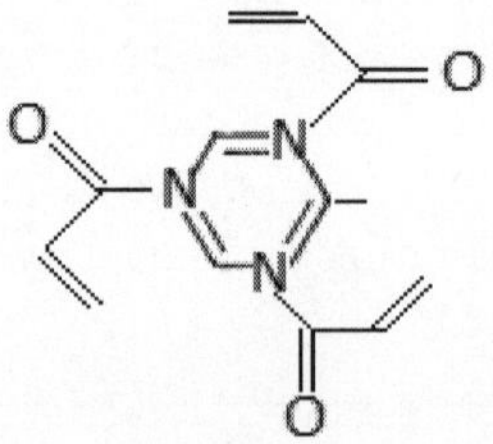

Figure 5.16. Triacroyl hexahydrotriazine (TAHT) resin

again showing significant time savings and chemical cost reductions, although enzymes can still have positive effects on finished fabric hand.

2. *MAGIC processing*: It is a simple short process, requiring only minimum basic machines. All pretreatment and dyeing is done on a padding mangle and thus doing away with prefibrillation and enzyme can be omitted from the jet process thus making the process much shorter and cost effective.

In open-width processing, one can get a flat classical fabric since the mechanical action on the fabric is virtually eliminated. No fibrillation occurs on lyocell fabrics that have been open-width processed, but they still retain a soft handle with an underlying bounce and resilience. Since the lyocell is having better strength, it can be processed by normal open-width processing routes. The processed fabric can be finished as per requirements of the customer like easy-care, etc., extracting both performance and value from the fibre.

5.3.2.16   Special finishes

Natural stretch lyocell – A limited stretch fabric can be made with lyocell as follows: The fabric is woven with enough space between the yarns to accommodation of yarn crimp. The fabric is causticized and allowed to shrink freely and impart a memory which together with the fibre's high modulus and resilience will allow stretch with high degrees of recovery, equal to that achievable with elastane containing fabrics. This finish can give only a limited stretch or a 'comfort' stretch rather than the 'power' stretch associated with elastomerics.

Easy-care lyocell: Easy-care finish in open-width fabric has two purposes: (1) to prevent fibrillation in usage and washing and (2) the easy-care finish. During finishing, it should be born in mind that if the chemical add-on is less then the fibrillation can occur in subsequent washing and if the chemical add-on is more it can affect the handle and strength of the fabric. Hence, suitable balance has to be made along with addition of softeners to balance the deterioration in handle. The general add-on of ~2–3% omf (on mass of fibre) fixed resin appears to be optimal for easy-care properties, if only fibrillation has to be avoided and easy-care properties are not necessary a 2% add-on will be sufficient. For softening effect, polyethylene emulsion or silicone softener (microemulsion is better than macroemulsions) may be employed. It is also worth remembering that caustic soda or liquid ammonia treatment in preparation will help to increase the easy-care rating of lyocell fabrics.

Because of its softness and strength lyocell/tencel is mainly used for making shirts. It is also used in slacks, dresses, coats and jeans.

5.3.2.17   Modified tencel – Tencel A100

As mentioned earlier, TAHT chemical is applied to lyocell to avoid fibrillation in subsequent processing, including jet dyeing. Tencel A100 is a specially developed by applying TAHT resin to the washed tow in the fibre production line. It is then dried, crimped, cut and baled in the normal way. The A100 fibre has comparable strength, elongation and modulus properties to the standard fibre and has an enhanced dye uptake that gives more economical dyestuff costs and strong, deep colouration. Its resistance to fibrillation means it can be processed on most dyeing machines and the fabrics produced have a good performance in subsequent washing.

This lyocell fibre is more attractive due to many special properties acquired due the special treatment. It can be processed in any normal processing machines including jet machines. The dye uptake of this fibre is higher than normal other viscose or cotton (40% higher than modal and cotton) and gives brighter shades. The drape and fluidity will generate more effectively than standard lyocell processed in an open-width system and resination is an option rather than a necessity. The exhaustion and fixation are also higher, hence the fastness properties are superior and same time washing is easier (higher dye fixation means less unfixed dye to be washed off) with less water consumption.

5.3.3   Cuprammonium rayon

Cuprammonium fibre is manufactured in small amounts by the regeneration of cellulose from cuprammonium solutions. This is also a cellulose hydrate fibre, but it differs from ordinary viscose fibre in that it is finer, its structure is more uniform (the skin is absent), it has a higher strength in the dry state and the loss of strength in the wet state is somewhat lower.

5.3.3.1   Manufacture

The main raw materials used are cotton linters and wood pulp. Cotton linters are more pure cellulose and hence it is preferred. If cotton linters are used, they are purified by kier boiling with dilute caustic soda at about 150°C, followed by bleaching with sodium hypochlorite or peroxide. If wood pulp is used it is selected and purified to yield a material of high alpha cellulose content (above 96%). (Alpha cellulose is that which does not dissolve in 17.5–18.0% caustic soda solution after 30 min at 20°C. It consists of cellulose, which has undergone a minimum of degradation, and it is the most satisfactory cellulose for use in fibre manufacture.)

In order to make a spinning dope the basic copper sulphate is dissolved in ammonia, giving a solution of cupritetrammino hydroxide and sulphate in the molecular ratio of 3:1 and purified linters are added. Caustic soda is then added to convert the sulphate to corresponding hydroxide and filtered by passing through nickel filter screens. The cellulose content of this solution is about 10%. The chemical reactions are:

$$C_6H_7O_2(OH)_3 + Cu(NH_4OH) \rightarrow C_6H_7O_2(OH) : Cu(NH_4OH) \rightarrow C_6H_7O_2(OH)_3 \, H_2SO_4$$

This solution is stable and may be stored for considerable periods without appreciable deterioration; in this respect, it contrasts strongly with viscose solution. The spinning solution is deaerated and pumped through the spinnerets into a funnel through which soft water is running. The movement of water stretches the newly formed filament. The fibres then move to spinning machines, where they are washed, put through a mild acid bath to remove any adhering solution, rinsed and twisted into yarns (Fig. 5.17).

5.3.3.2   Spinning

There are two types spinning practiced:

5.3.3.2.1   Continuous spinning

As explained above, the solution is pumped through the spinnerets into a funnel and from the funnel the filaments are passed through an enclosed bath of hot dilute acid called the pretreatment pan. This continues the coagulation of the cellulose, reducing the filaments to about one-third of their original

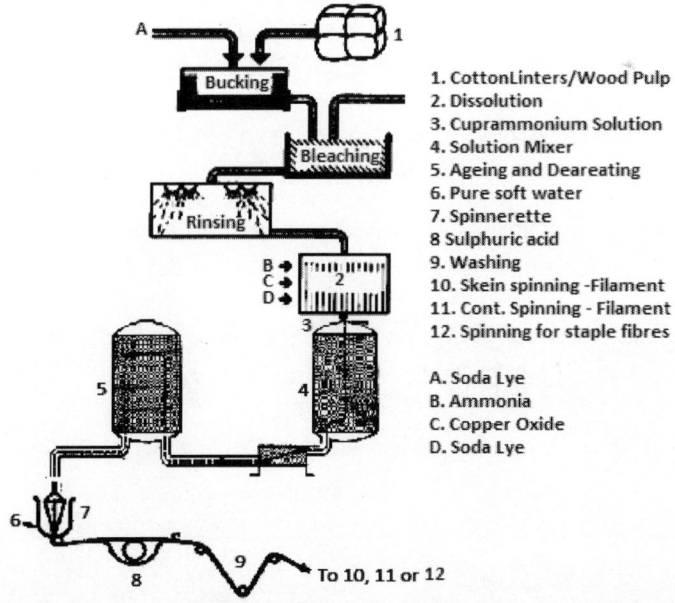

1. Cotton Linters/Wood Pulp
2. Dissolution
3. Cuprammonium Solution
4. Solution Mixer
5. Ageing and Deareating
6. Pure soft water
7. Spinnerette
8 Sulphuric acid
9. Washing
10. Skein spinning -Filament
11. Cont. Spinning - Filament
12. Spinning for staple fibres

A. Soda Lye
B. Ammonia
C. Copper Oxide
D. Soda Lye

Figure 5.17. Manufacturing process of cuprammonium rayon fibre

diameter. The oriented filaments of cellulose are sheathed in a film of una-ligned cellulose, and this is washed away in the pretreatment pan. If left, the unaligned cellulose would act as a glue, holding the filaments together.

Next, the thread of filaments passes through an acid trough where re-maining copper is removed as copper sulphate. The acid is washed away, as the thread moves through a water trough, lubricants, sizes, etc., are added as required by passing the thread over a preparation roll. During the whole pro-cess, the filaments are never handled and hence imperfections are thus held at a minimum. The filaments (usually 15–300 den/17–330 dtex) are of uniform structure and dimensions, and the properties are excellent. The thread passes through a succession of driers and over a roll, which applies coning oil before being wound onto flangeless spools.

5.3.3.2.2   Pot (batchwise) spinning

The ready spinning dope is pumped to a nickel spinneret, and extruded through holes of 0.8 mm in diameter. The jets of solution emerging from the spinneret holes flow into a glass funnel, where they meet a stream of pure water which is flowing down through the funnel. The water dissolves most of the ammonia and about one-third of the copper from the jets, bringing about coagulation of the cellulose to form plastic filaments. The filaments are carried along by the stream of water and are stretched continuously to form filaments of usually about 1.4 dtex (1.3 den). The loose thread of filaments emerging from the bot-tom of the funnel is carried round a guide rod, most of the water being flung off. The thread then passes round a roller which rotates in a trough of sulphu-ric acid; the remaining copper and ammonia are removed as copper sulphate and ammonium sulphate, respectively. The filaments are then wound either into skeins (reel spinning), or into cakes in a topham box (pot spinning). The skeins or cakes are washed to remove acid and any remaining copper sulphate or am-monium sulphate, softened by adding lubricants and dried. The yarn is com-monly given a second wash in soap and oil emulsion, or (if it is to be twisted later) in a soaking bath. It is then dried again (Tables 5.3–5.5).

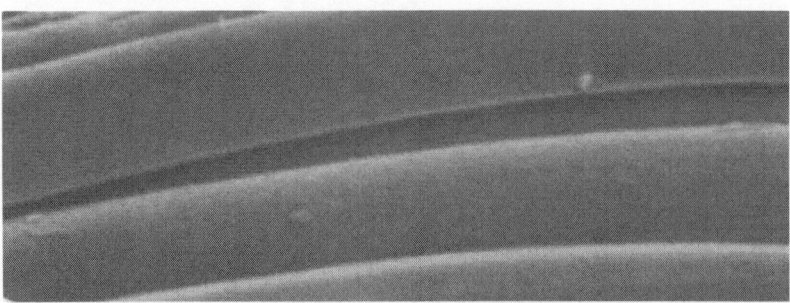

Figure 5.18. SEM view of cuprammonium rayon fibre (×1100)

Table 5.3. Comparison of different varieties of viscose (properties)

Characteristics	Regular viscose	Super high tenacity viscose rayon	High tenacity viscose rayons	Polynosic/modal (standard)	Cupro	Saponified cellulose ester
	Dry	Dry	Dry	Dry	Dry	Dry
Tenacity	18–23 cN/tex (2.0–2.6 g/den)	35–45 cN/tex (4–5 g/den)	26–44 cN/tex (3–5 g/den)	28–35 cN/tex (3.2–4.0 g/den)	15–20 cN/tex (1.7–2.3 g/den)	53–62 cN/tex (6–7 g/den)
Initial wet modulus	40–50	40–50		100–180	30–50	
Elongation, %	17–25	11–12	9.5–11.5	8–12	10–17	6
Elastic property	2% stretch recovers 67%	2% stretch recovers 70–100		Higher than cotton	20–75%	2% stretch recovers 60–70%
Specific gravity	1.5–1.52	1.52–1.54	1.52–1.54		1.54	1.5
Moisture regain, %, at standard conditions	13	Less than polynosic	Less than polynosic	11.5–12	12.5	10.7
Effect of sunlight (prolonged)	Discolouration, gradual loss in strength	Some degradation and loss of strength	Some degradation and loss of strength	Some degradation and loss of strength	Some degradation and loss of strength	Similar to cotton
Effect of acids	Similar to cotton. Disintegrates in hot dil. and cold conc. acids	Similar to cotton. Disintegrates in hot dil. and cold conc. acids	Similar to cotton. Disintegrates in hot dil. and cold conc. acids	Disintegrates in hot dil. and cold conc. acids	Disintegrates in hot dil. and cold conc. acids	Similar to cotton
Effect of alkali	Strong alkali swells and degrades	Strong alkali swells and degrades	Strong alkali swells and degrades	Withstand mercerizing conditions	Strong alkali swells and degrades	Similar to cotton
Weak oxidizing agents	Bleaches	Bleaches	Bleaches	Behaves as Cotton	Bleaches	Similar to cotton
Strong oxidizing agents	Degrades	Degrades	Degrades	Behaves as Cotton	Degrades	Similar to cotton
Organic solvents	Generally insoluble	Generally insoluble	Generally insoluble	Behaves as Cotton	Generally insoluble	Similar to cotton

Table 5.4. Comparison of various viscose fibres and cotton

Polymer	Approx. no. of cellobiose units	Approx. polymer length (mm)	Approx. polymer thickness (nm)	Approx. degree of polymerization
Viscose	175	180	0.8	175
Polynosic	300	310	0.8	300
Cuprammonium	250	260	0.8	250
Cotton	5000	5000	0.8	5000

Table 5.5. Microscopic views of different types of viscose

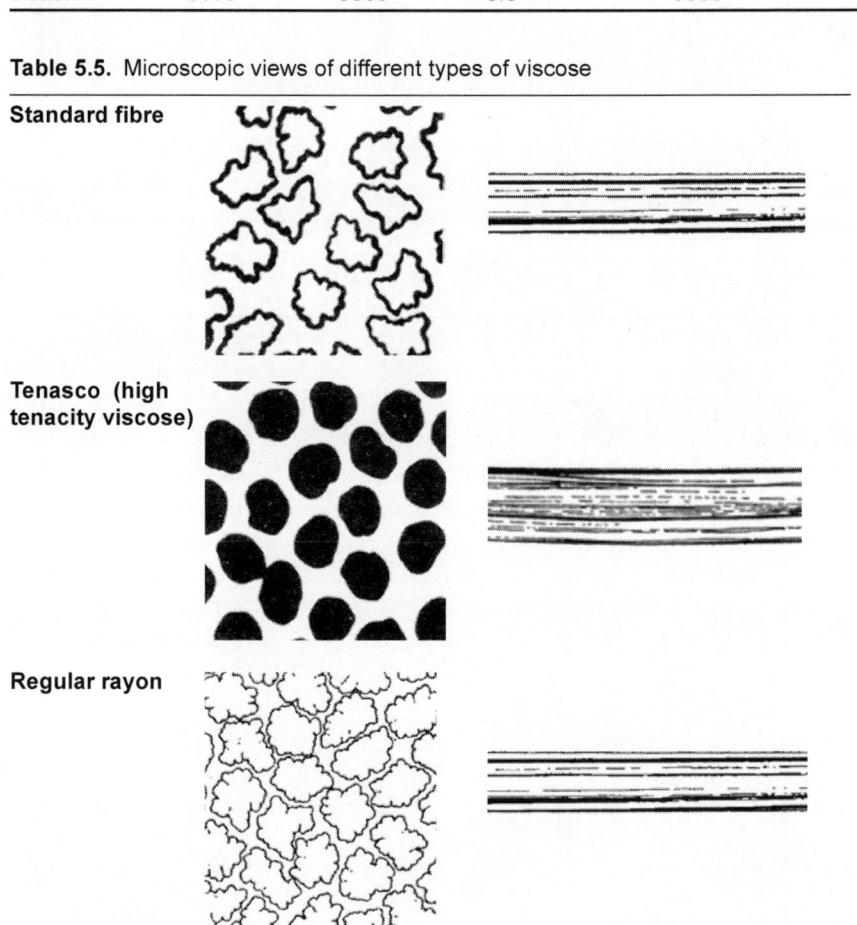

Standard fibre

Tenasco (high tenacity viscose)

Regular rayon

Cupro

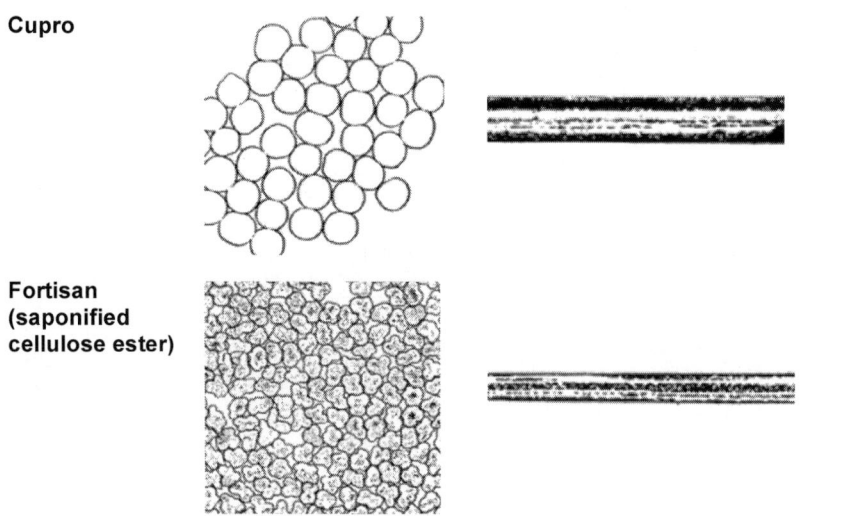

**Fortisan
(saponified
cellulose ester)**

5.3.4   Saponified cellulose ester

The poor tenacity was always a negative point of viscose rayons, even though its silk-like appearance and sheen were the plus point of the fibres. Researches were onto find out the possibilities of increasing the tenacity of these fibres. The main difficulty was apparently the problem in the difficulty of maintaining the extruded filament in a plastic condition after it entered the coagulating bath, and so providing an opportunity of stretching the filament to orientate the cellulose molecules. At the same time, in the effort of producing high tenacity of cellulose acetate, it was found that it was possible to bring to plastic condition after it has been made, as far as cellulose acetate was concerned. However, the problem was, there was a difference in the behaviour of cellulose and cellulose acetate towards solvents and heat are a consequence of the differences in their molecular structure. The hydroxyl groups of cellulose do not encourage solution of the molecule in organic solvents, whereas the acetyl groups of the cellulose acetate molecule do. Also, cellulose molecules are able to pack closely together and develop powerful forces of attraction associated with high crystallinity; cellulose acetate molecules, with their large pendant groups, do not permit of the close packing that results in high crystallinity, and cellulose acetate is softened by heat. Filaments of cellulose acetate, which have been softened by solvent or by heat, may be stretched to many times their original length, the long molecules of cellulose acetate sliding readily over one another as they are drawn into alignment. The stretching of solvent-plasticized cellulose acetate resulted in the production of yarns with tenacities in the region of 44–53 cN/tex (5–6 g/den). These yarns retained the essential characteristics of cellulose acetate, but they could be converted into cellulose by saponification with caustic soda solution, providing highly oriented filaments of regenerated cellulose. The molecules in these saponified cellulose acetate filaments were in a more highly oriented and crystalline condition than could be obtained by stretching filaments produced during coagulation of viscose rayon.

5.3.4.1   Manufacture

Saponified cellulose acetate fibres are made by heating cellulose acetate filament yarns in steam at about 2.1 kg/cm^2 (30 lb/in^2), and stretching the softened yarn by 4–10 times its original length. The stretched yarn is wound onto perforated bobbins and saponified by treatment with caustic soda solution. The yarn is then washed, oiled, dried and rewound. Very fine filaments of regenerated highly oriented cellulose may be produced in this way.

5.3.5   Acetate fibres

5.3.5.1   General

In contrast to the above described fibres, which are composed of cellulose hydrate, *acetate fibre* is an acetic acid ester of cellulose.

A manufactured fibre in which the fibre forming substance is cellulose acetate. Acetate is derived from cellulose by reacting purified cellulose from wood pulp with acetic acid and acetic anhydride in the presence of sulphuric acid.

Acetate fibre characteristics are luxurious feel and appearance, wide range of colours and lusters, excellent drapability and softness, relatively fast drying, shrink, moth and mildew resistant. Special dyes have been developed for acetate, since it does not accept dyes ordinarily used for cotton and rayon.

5.3.5.2   Milestones of cellulose acetate manufacture

1905 – Camille and Henri Dreyfus develop commercial process to manufacture cellulose acetate.

1910 – Cellulose acetate commercially used to make films, toilet articles and moulded articles. Product is also sold to aircraft industry to coat fabrics on airplane wings and fuselage.

1913 – Camille and Henri Dreyfus start producing acetate fibres in their laboratories.

1921 – Dreyfus Brothers develop dry spinning technique that allows fibre to be knit and woven for apparel.

1924 – First commercial acetate filament spun in the United States at Cumberland, Maryland.

1925 – Cellulose acetate fabrics are sold in silk constructions, such as taffeta, satin, ninon, voile and crepe.

1925 – Camille Dreyfus introduces modem fibre merchandizing to major department stores. Cellulose acetate is introduced as 'artificial silk'.

1930 – Acetate introduced for apparel linings.

1928–48 – Cellulose acetate manufacturing facilities open in Drummondville, Canada; Narrows, Virginia; Ocotlan, Mexico and Rock Hill, South Carolina, etc.

1994 – Antimicrobially treated acetate introduced.

5.3.6   Cellulose acetate fibres

5.3.6.1   Manufacture

Acetate fibres are produced from cotton linters or purified wood pulp, which are acetylated at temperature up to 50°C with acetic anhydride in presence of glacial acetic acid and concentrated sulphuric acid. It is then aged or ripened in the presence of water and hydrolysis occurs during the ripening and results in the formation of secondary acetate. The flakes are then dissolved in acetone containing 4% water as the solvent to form the spinning dope, which is filtered and then forced through the spinnerets into a warm-air chamber and the method of spinning is called dry spinning.

5.3.6.2   Method

Steeping: As in the case of viscose rayon manufacture cotton linters are purified by kier boiling using either soda ash or caustic soda and bleached, washed and dried. If wood pulp is used it is purified by suitable method to increase the percentage of cellulose. Next, it is steeped in glacial acetic acid to swell the fibres and increase their chemical reactivity. Modern processes use vapour-phase activation based on acetic-water mixtures.

Acetylation: Steeped cellulose is transferred to a closed reaction vessel containing a mixture of acetic acid and anhydride. The mixture has the following weight ratio:

- Purified linters – 1 part
- Acetic anhydride – 3 parts
- Glacial acetic acid – 5 parts

The reactants are mixed, and a small quantity (0.1 part) of sulphuric acid dissolved in glacial acetic acid is added. The acetylation of the cellulose now proceeds. The reaction is exothermic, and the vessel is cooled to maintain a predetermined temperature profile. After a period the temperature is allowed to rise and maintained at a higher temperature for a further period.

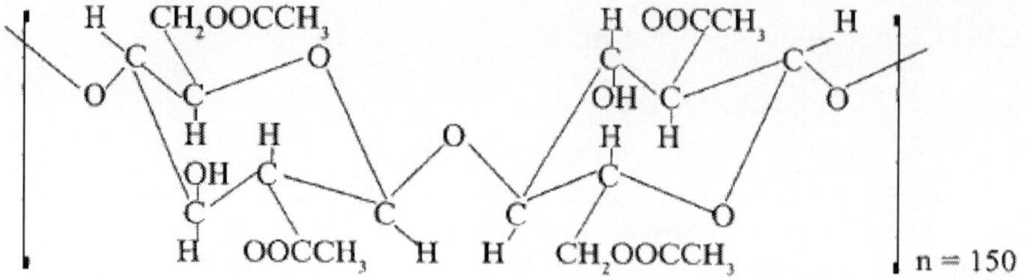

Figure 5.19. Polymetric structure of primary acetate

The D-glucose units in the cellulose polymer contain three hydroxyl groups, of which one is primary and two are secondary. By acetylating all of the hydroxyl group in cellulose, a triacetate is obtained. This is called primary acetate with a polymetric structure as represented in Fig. 5.19.:

5.3.6.3   Ripening (hydrolysis)

In the ripening process, the above solution is transferred to another vessel, and mixed with dilute acetic acid. The residual acetic anhydride in the solution reacts with the water to form acetic acid and this, together with the residual acetic acid from the acetylation mixture, forms a solution of acetic acid in water. Next, the cellulose triacetate is allowed to stand in this solution of acetic acid in water for up to 20 h. During this time, partial hydrolysis of the cellulose triacetate takes place, some of the acetyl groups being removed and hydroxyl groups formed again. This hydrolysis stage is, in effect, a partial reversal of the acetylation process, the completely acetylated cellulose triacetate being converted into a partially acetylated cellulose acetate. The restoration of some of the hydroxyl groups on the cellulose acetate molecule changes the solubility characteristics of the acetate. Cellulose triacetate is soluble in chloroform, but insoluble in acetone. The partially acetylated cellulose acetate is insoluble in chloroform, but soluble in acetone (Fig. 5.20).

The partially acetylated cellulose acetate made in this way is often called cellulose diacetate, signifying that one-third of the total acetyl groups have been removed from the cellulose triacetate, so that each glucose unit now has two of its three hydroxyl groups acetylated. But, in fact, the secondary acetate used in spinning acetate fibre does not correspond precisely with cellulose diacetate. The hydrolysis of the triacetate is allowed to proceed until each glucose unit in the cellulose molecule has, on average, about 24 of its hydroxyl groups acetylated The secondary acetate structure lies part way between that of triacetate and diacetate. When the reaction has

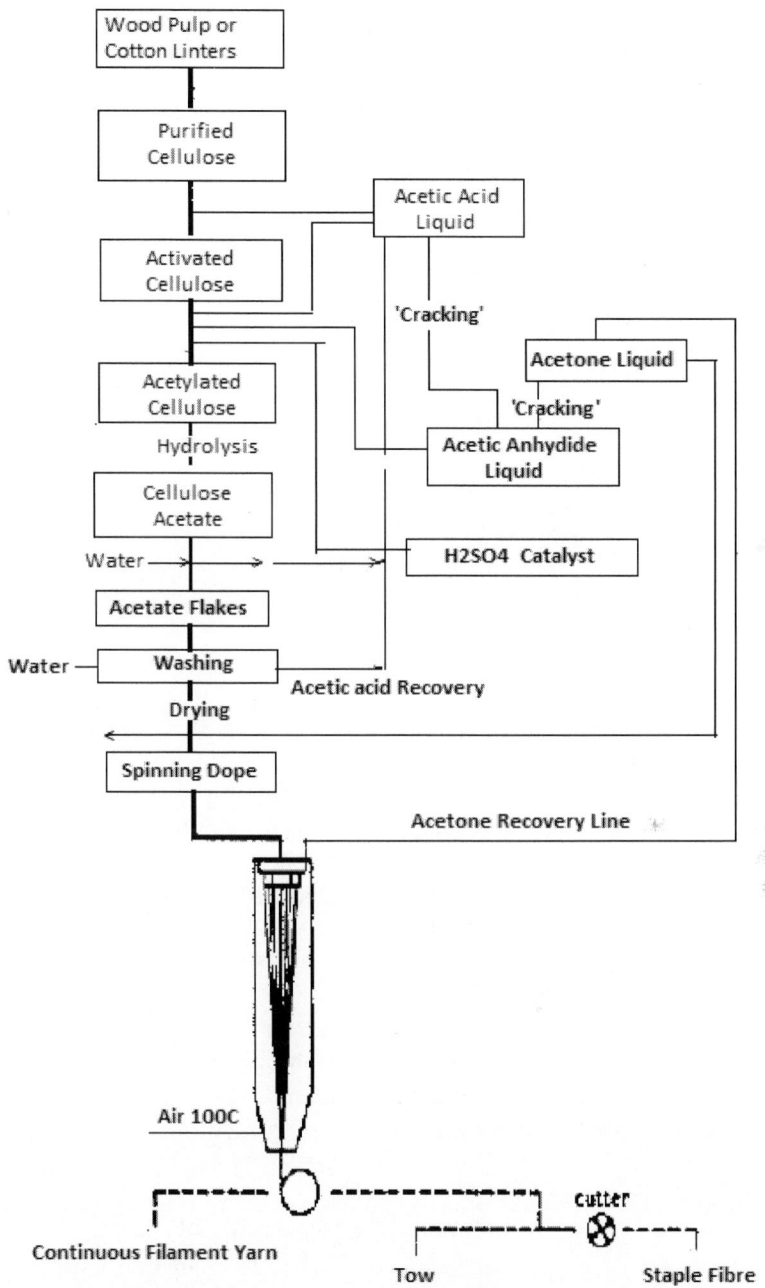

Figure 5.20. Flow chart of cellulose acetate manufacture

reached to the predetermined point, this whole mass is poured into excess water when the secondary acetate is precipitated as white flakes which is separated washed and dried. The acetic acid is recovered and reused.

Secondary acetate made is converted into spinning dope and spun by different methods.

5.3.6.4   Spinning

5.3.6.4.1   Dry spinning

The secondary acetate is dissolved in acetone containing 5% water in it. This spinning dope which is about 20–30% secondary acetate solution is filtered deaerated and pumped into the spinnerets after a second filtration. Spinnerets having hole size of 25–75 μ extrudes the fibre into a spinning tube. Through this enclosed tube hot air at 100°C is passed through and the acetone id evaporated leaving the filaments of cellulose acetate. This filament of cellulose acetate is stretched slightly while still plastic, to align the long molecules and increase the strength of the filament. As they come out from the tube, it is lubricated and wound as untwisted filaments onto a cylindrical tube, or insert twist and then wound the twisted yarn onto bobbins.

5.3.6.4.2   Wet spinning

In the dry spinning method after cellulose acetate is produced, it is precipitated by adding to excess water and then again made into solution in acetone. But when the cellulose acetate is formed, it is still in solution and hence the fibre can be spun from this solution. This is wet spinning principle. The solution of acetate resulting from hydrolysis could be extruded through a spinneret, and the cellulose acetate precipitated in the form of filaments in an aqueous coagulating bath.

5.3.6.4.3   Melt spinning

The cellulose acetate being a thermoplastic fibre it can be melt and spun. The acetate flakes are melt at 235°C and spun through spinnerets. But these fibres differ in the properties compared to dry spun fibres. When the dry spun fibre loose its lustre when boiled in water, melt spun fibres do not.

5.3.6.5   Properties

In acetate fibre, most of the hydroxyl groups in cellulose are replaced by acetyl groups, owing to which acetate fibres, as compared with viscose fibres, are less hygroscopic, their swelling in water is smaller, as is. The loss of strength in the wet state (40–45%). At the same time, due to the presence

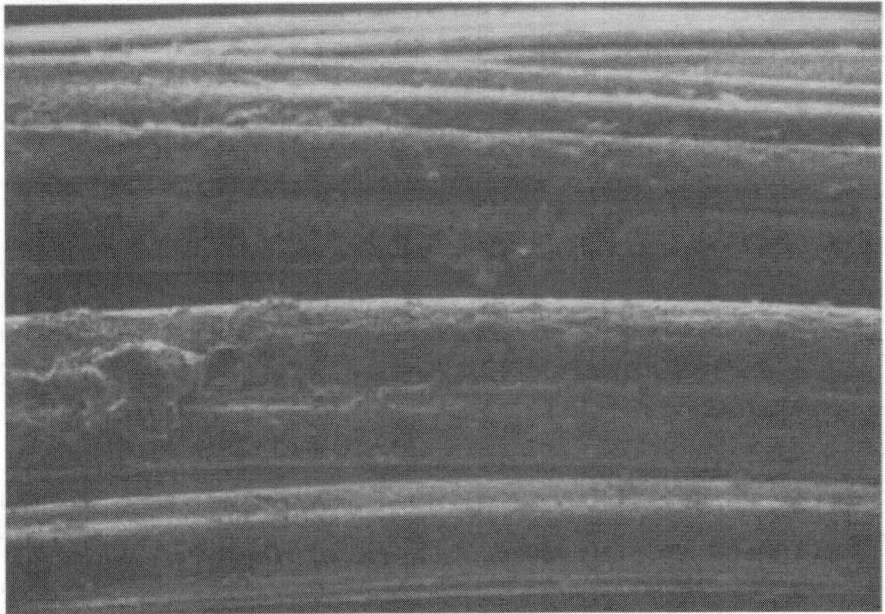

Figure 5.21. SEM view of acetate fibre (×1200)

of acetyl groups in the acetate fibre, it swells and dissolves in such organic solvents as acetone, methyl-ethyl ketone, methyl acetate, dioxane. Cuprammonium fibre neither dissolves nor swells in ethers (Fig. 5.21).

In contrast to viscose fibre, acetate fibre is characterized by a higher elastic elongation and a lower density (1.32 as compared with 1.52) but its abrasion resistance is lower than that of viscose fibre. Acetate fibre is also characterized by thermal plasticity and at a temperature above 140–150°C softening and deformation is observed, while at 230°C fusion with decomposition ensues. Therefore, acetate fabrics may be damaged if ironing is done with a hot iron.

Acids may cause hydrolytic degradation of cellulose acetate in the same way as that of cellulose, but cellulose acetate is somewhat more resistant. The resistance of acetate fibre to alkali is very low, as being an ester it is easily saponified by them and thus loses its valuable properties (conversion to cellulose hydrate). Acetate fibre has a high resistance to biological attack: it is not damaged by moths, mould fungus and microorganisms. Acetate fibre is a good insulator, is easily electrified and tends to accumulate charges of static electricity, so that in some cases it is necessary to treat acetate fibre or fabric with antistatic agents.

5.3.7   Cellulose triacetate

When cellulose is subjected to the action of acetic anhydride in the presence of glacial acetic acid and sulphuric acid (catalyzer), cellulose triacetate is obtained (Fig. 5.22):

$$[C_6H_7O_2(OH)_3]_n + 3n(CH_3CO)_2O \rightarrow [C_6H_7O_2(OCOCH_3)_3]_n + 3nCH_3COOH$$

Further, to achieve partial saponification, a small calculated amount of water is added to an acidic solution of the cellulose triacetate:

$$[C_6H_7O_2(OCOCH_3)_3]_n + nH_2O \rightarrow [C_6H_7OH(OCOCH_3)_2]_n + nCH_3COOH$$

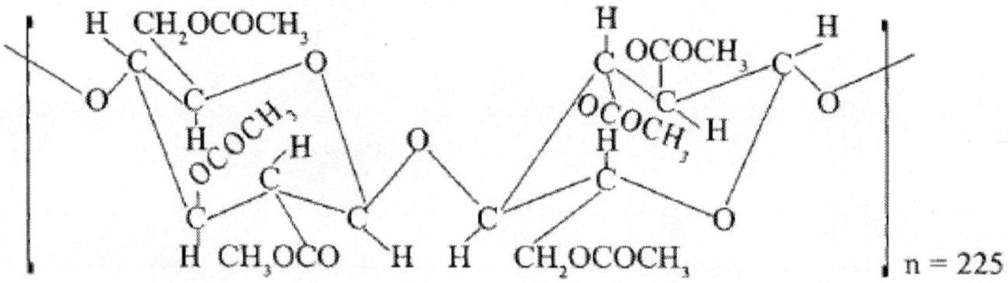

Figure 5.22. Triacetate fibre structure

5.3.7.1   *Manufacture*

Triacetate is manufactured from the same raw materials as secondary acetate, but the ripening stage in which hydrolysis occurs is omitted in triacetate production.

The cellulose is pretreated with acetic acid/water vapour and acetylated. The acetylation is done in such way as cellulose triacetate either goes into solution as it is formed or retains the structure of the original cellulose. There are two methods of acetylation.

5.3.7.1.1   *Solution process*

The cellulose is acetylated by treatment with acetic anhydride and sulphuric acid in the presence of acetic acid. As the acetylation proceeds, cellulose triacetate is formed, and it dissolves. Any sulphate group on the cellulose molecule has to be replaced and it is done by addition of magnesium acetate and water. After the ripening process is over, the mass is added to water where the cellulose triacetate is precipitated as in the case of secondary acetate. The same process can be done replacing acetic acid solvent with methylene chloride and small amount of sulphuric acid as catalyst. The earlier process produce triacetate of acetyl value of 61.5% but the latter method achieve acetyl value of 62% or more. The precipitated triacetate is washed free of acid and dried.

5.3.7.1.2   Nonsolution process

In this process, the original shape of cellulose is retained, as it is not dissolved. The pretreated cellulose is swollen using a solvent (e.g., benzene) and directly acetylated using acetic anhydride and perchloric acid (or other acid) catalyst. Since there is no solvent (only swelling agent) available, the cellulose is acetylated in the original form. It is then washed acid free and dried.

5.3.7.2   Spinning

To produce spinning solution, dried triacetate flake is dissolved in methylene chloride, containing a little alcohol and dry-spun into a warm-air chamber, where the methylene chloride and alcohol evaporates. The filament is the lubricated and further processed as in the case of secondary acetate explained above.

Melt spinning and wet spinning also can be done as explained above in secondary acetate.

5.3.7.3   Properties

As a result, cellulose acetate with DP 240–260 is obtained containing 53.5–56% of acetic acid, which is dissolved in a mixture of acetone and alcohol and then extruded through spinnerets. During spinning, the solvent is evaporated and the filament becomes hard (dry spinning).

The triacetate is soluble in chloroform, and a fibre known as *Lustron* was spun from chloroform solution in early small-scale American manufacture (1914–1924). If the triacetate is partially hydrolyzed to give a mixture with an average of 2 ½ acetyl groups per glucose residue the product loses solubility in chloroform, but becomes soluble in acetone. A different product, which is insoluble in acetone, is obtained by direct introduction of 2 ½-acetyl groups; presumably the less accessible hydroxyl groups are the last to be acetylated and the last to be reformed on hydrolysis. During the First World War, incompletely acetylated cellulose was produced on a large scale for use as a dope for aircraft fabric.

Ready triacetate fibre which is a cellulose acetate with DP 290–300 differs from the ordinary acetate fibre by its lower hygroscopicity (it absorbs 4.5% of moisture at 65% relative humidity), lower swelling in water and smaller loss of strength in the wet state (30%). The fibre swells and may be dissolved in many organic solvents: methylene chloride, chloroform, concentrated formic acid and glacial acetic acid (Fig. 5.23).

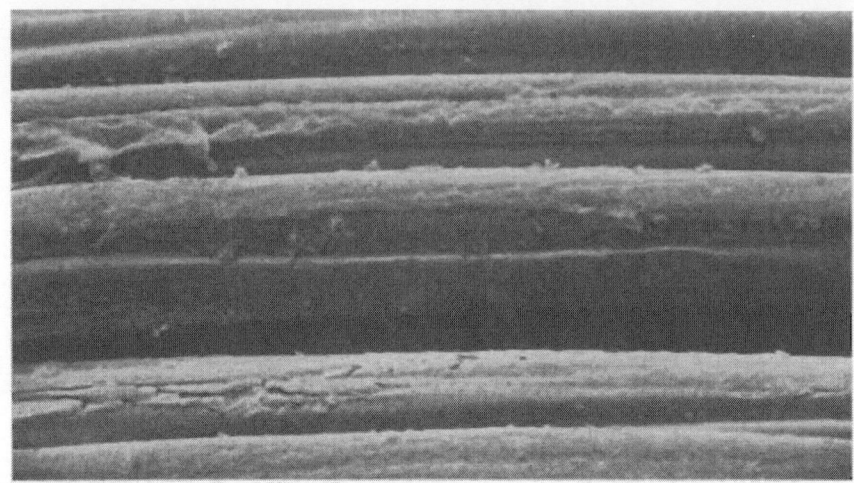

Figure 5.23. SEM view of triacetayte fibres

Triacetate fibre, as compared with acetate fibre, is more resistant to the action of acids, alkalies, oxidizing agents; in particular, triacetate fibre shows a high resistance to sodium hypochlorite, hydrogen peroxide, peracetic acid and sodium chlorite under the conditions usually used in bleaching. It is characterized by exceptionally high light and electrical resistance. The electric insulation properties of triacetate fibre are five times higher than those of fibre obtained from partially saponified cellulose acetate. Triacetate fibre, just as ordinary acetate fibre, is characterized by high thermoplasticity, but it melts with decomposition at a considerably higher temperature (290–300°C). Triacetate fibre is highly resistant to biological attack (Table 5.6).

Table 5.6. Comparison of physical properties of secondary acetate and triacetate

Properties	Secondary acetate	Triacetate
Tenacity, dry	9.7–11.5 cN/tex (1.1–1.3 g/den)	10.6–12.4 cN/tex (1.2–1.4 g/den)
Tensile strength	1260–1540 kg/cm^2	Dry: 10–13, wet: 52–80 (cN/tex)
Recovery at 4% elongation, %	45–65	50–65
Elasticity	48–65% recovery at 4% stretch	
Specific gravity	1.3	1.32
Moisture regain	6.5% at standard conditions	2.5–4.5% at standard conditions
Melting point	232°C, softening at 202°C	300°C, Softens at 225°C

Properties	Secondary acetate	Triacetate
Effect of sunlight	Deteriorates on prolonged exposure, no discolouration	Highly resistant, no yellowing
Effect of acid	Weak acid – no effect. Strong acid – decomposes	Resistant to dilute acids, affected by conc. acids
Effect of alkali	No effect up to pH 9.5, but strong alkali saponifies acetate	Resistant to dil. alkalis, attacked and hydrolyzed by hot strong alkalis
Organic solvents	White spirit or petrol (gasoline), ethyl ether, benzene, toluene, perchloroethylene, trichloroethylene, carbon tetrachloride, cyclohexanol, xylene	Dissolves in methylene chloride, chloroform, formic acid, acetic acid, dioxan and m-cresol
Cross section		
Magnified view		

5.3.7.4   Applications and uses of acetate fibres

Fabrics of acetate are soft and drapeable. They have a very luxurious and supple hand. Acetate has performance attributes of a natural fibre such as comfort, absorbency and breathability. However, unlike natural fibres, acetate will not pill or hold static electricity. Acetate dyes beautifully to produce richly coloured fabrics and outstanding prints. It blends with rayon, cotton, wool, silk, linen, nylon, acrylic, polyester and spandex to create a full range of fabric types – from comfortable slinky knits for contemporary sportswear, as well as woven crepes and gabardines for career dressing – to elegant satins, taffetas and velvets for evening and bridal wear. It is also used in a broad selection of luxurious home fashions and draperies. Acetate, a cellulosic fibre made of wood pulp (from reforested trees) and acetic acid, is biodegradable.

Acetate is preferred for linings by today's top designers of women and men's fashions since they are inherently breathable yet absorbent, without pilling or static cling and are with soft, drapable hand and in colours that suit every fashion in every season. Most important, acetate linings deliver outstanding sewing performance and available in a variety of different fabric weights and constructions. There is a perfect lining for every kind of garment, from casual sportswear to tailored career and eveningwear. It is made in remarkable versatility in satin, taffeta, twill and sheath constructions. No other lining feels quite like an acetate lining: luxuriously soft and supple to the hand, beautifully drapable and wonderfully cool and comfortable to wear. Acetate fibres dye into a marvellous spectrum of true rich colours for linings. Acetate is a cellulosic fibre that is environmentally friendly. It is made from acetic acid and wood pulp, a renewable resource.

5.3.8   Regenerated Protein Fibres

The raw materials used for the preparation of regenerated protein fibres may be milk, soya beans, peanuts and zein. Sometimes alkaline solutions of gelatin, albumin and other raw materials like waste wool, silk and feathers may be used. Fibrolane (Great Britain) and Merineva (Italy) are made by dissolving casein in sodium hydroxide, and then by extrusion into an acid/salt bath. The fibres formed in this way tow and partially stabilized by treatment with formaldehyde.

The general properties of regenerated protein fibres include the wool-like attributes of resilience, warmth and soft handle. The strength is lower than wool. These fibres do not contain cystine linkage, which results in more open structure. These fibres are far less successful than hoped and never seriously challenged wool.

$$R\text{-}NH_2 + HCHO + H_2N\text{-}R \rightarrow RNHCH_2NHR$$
$$\text{Casein}$$
$$+ \qquad\qquad +$$
$$[-NH + CH_2O + NH-] \quad à \quad [-HN-CH_2O-NH-]$$
$$\text{Vicara}$$

5.3.9   Casein fibres

5.3.9.1   Manufacture

Raw material for the production of casein fibre is skimmed milk. The skimmed milk is treated with acid when the casein is obtained (coagulates) as curd. It is then washed dried and powdered. Thirty-five litres can give one kilogram of casein (Fig. 5.24).

5.3.9.2   Spinning

The casein made as explained above is dissolved in caustic soda solution and allowed to stand to reach a predetermined viscosity (ripening). This solution is pumped to spinnerets and under pressure, the filaments are extruded into a coagulating solution.

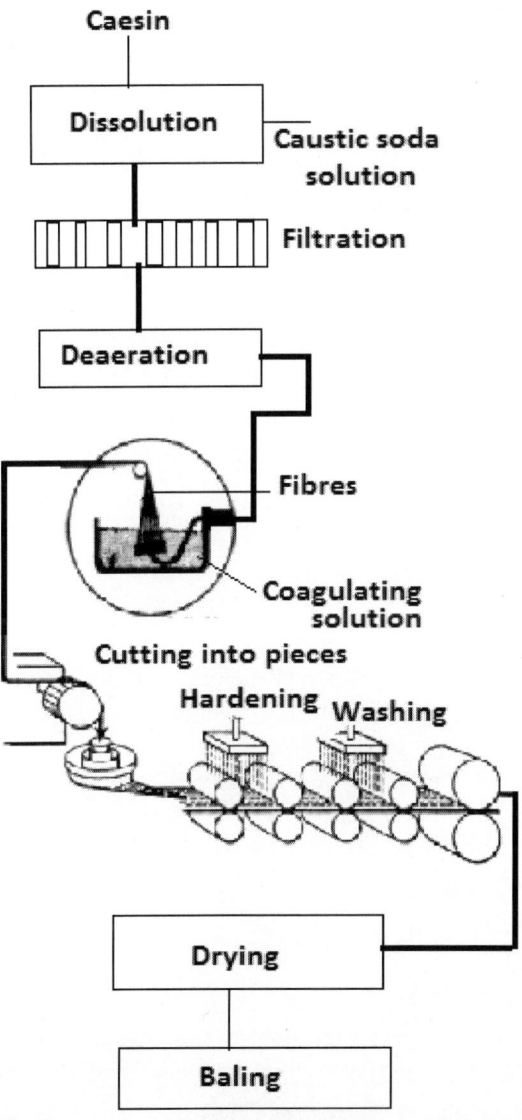

Figure 5.24. Flow chart of casein fibre manufacture

Example of coagulating solution (recipe):

2 parts	Sulphuric acid
5 parts	Formaldehyde
20 parts	Glucose
100 parts	Water

During coagulation, the filaments are stretched to some extent. During the spinning operation, the molecules are aligned to some extent but are not crystallized hence they are weak and soft, if handled will break and also water can enter the fibre and push the molecules apart hence it has got little use in textiles. To make it stronger, the casein is cross linked which will hold the molecule together. Cross-linked casein acquires an increased resistance to the effect of water, retaining a higher degree of tensile strength and resistance to swelling.

5.3.9.3   Properties

The fibre is naturally white, smooth surfaced, with faint striations. Cross section is bean shaped to round with dappled shape due to pitting. It can be spun into extremely fine filaments of 20–30 μ in diameter.

Physical/chemical properties of casein summarized in Table 5.7.

Table 5.7. Physical and chemical properties of casein

Properties	Casein
Tenacity, dry	9.7–8.0 cN/tex (1.1–0.9 g/den)
Elongation	60–70%
Specific gravity	1.3
Effect of heat	Turns yellow if heated at 100°C itself for long time, at 150°C it decomposes
Moisture regain	14%
Effect of sunlight	Very little, like wool
Effect of acid	Weak acid – no effect. Strong mineral acid – decomposes
Effect of alkali	Sensitive to alkali. Sodium bicarbonate and disodium hydrogen phosphate have little effect. Strong alkali disintegrates
Organic solvents	Stable to dry-cleaning solvents

Cross section

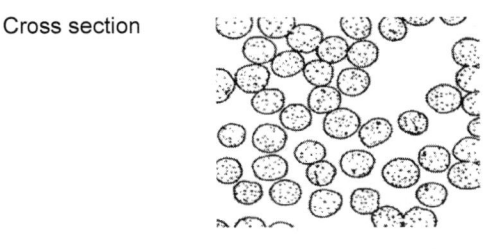

Magnified view

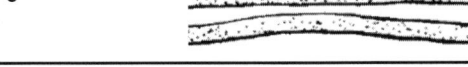

5.3.10   Soya bean fibre

5.3.10.1   Manufacture

As the name suggests, the raw material for this fibre is soya bean. This bean has a protein content of about 35% and is abundantly grown in the United States and other countries. After crushing the beans, the oil is extracted using a solvent like hexane. The protein, which is available in the remaining material is dissolved out using sodium sulphite solution. The protein is father recovered from this solution by acidification (Fig. 5.25).

5.3.10.2   Spinning

The spinning process is almost like the casein fibre. The recovered soya bean protein is dissolved in caustic soda solution, filtered, ripened and pumped to spinnerets. The jets emerge into an acid coagulating bath, and the filaments are stretched, hardened, washed, dried and cut into staple.

5.3.11   Zein fibre (vicara)

Zein fibre is made from the protein extracted from the raw material, maize. The procedure, process, etc., are same as above.

5.3.12   Groundnut protein fibre (ardil)

5.3.12.1   Manufacture

For this fibre, the protein in groundnut is used as the raw material. Other than extracting the protein from the groundnuts, remaining process almost same as the other protein fibres. Groundnuts are shelled or decorticated. The red skins are then removed from the shelled nuts, together with foreign matters. The nuts contain about 50% of oil. They are crushed and pressed to remove

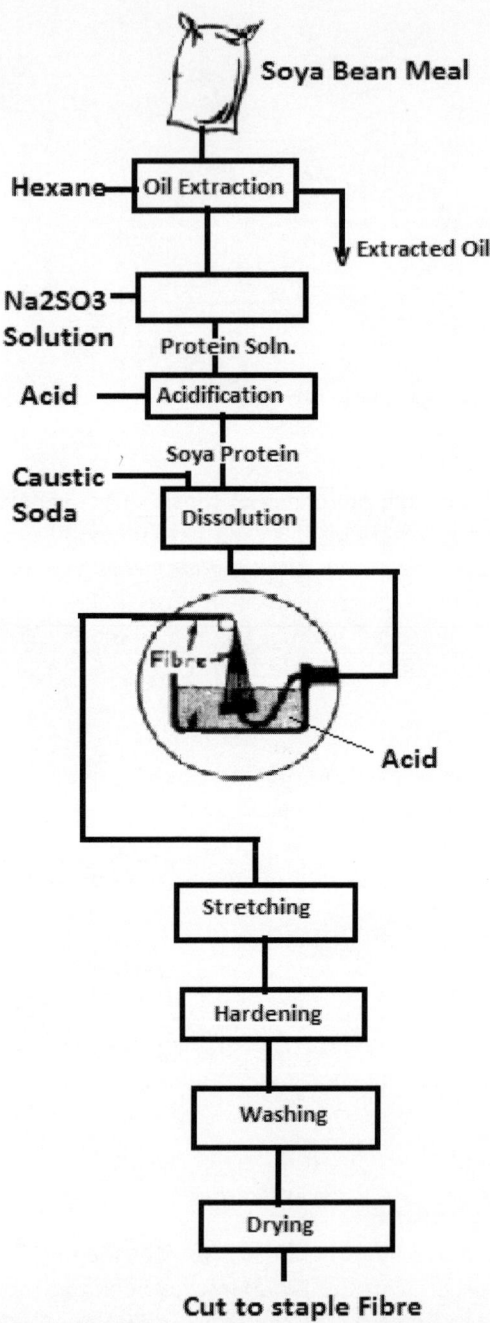

Figure 5.25. Flow chart of the manufacture of soya bean fibre

80% of the total oil content. The resultant meal is reduced in breaker rolls and passes through flaking roll which converts the meal into flakes. These thin flakes pass via a series of buckets on an endless chain into an extraction plant. As they pass through the plant, the buckets of meal are subjected to a thorough washing with solvent (hexane) which removes the remainder of the arachis oil. The extracted meal is heated under low pressure in steam jacketed pans to remove residual solvent. It is then cooled, screened, weighed and bagged. The meals produced this way will contain about 50% protein. The protein is extracted using caustic soda solution and then acidified when the pure protein is precipitated.

5.3.12.2   Spinning

The protein produced as above is again dissolved to 10–12% solid content solution using dilute caustic soda solution. Other solvents for this protein are urea, ammonia and solutions of detergents such as alkyl benzene sulphonate. The protein solution is extruded through the spinneret having holes of 0.07–0.1 mm size. The spun fibres pass through a coagulating liquor containing sulphuric acid, sodium sulphate and other auxiliaries at a temperature of 12–40°C. Further the filament is stretched slightly increase its strength and hardened as in the case of casein fibres.

Natural Non-cellulosic Man-Made Fibres

6.1 Rubber fibre

Rubber is an elastic hydrocarbon polymer that naturally occurs colloidal suspension, or latex, in the sap of some plants like rubber tree, *Hevea brasiliensis* (Fig. 6.1).

Figure 6.1. Rubber tree

6.1.1 Manufacturing process

6.1.1.1 Cut filaments

The latex is collected from rubber tree. The rubber latex is coagulated, and the raw rubber is mixed with vulcanizing agents and other ingredients on a mill. It is then passed through a calender, which produces a thin sheets of 0.3–1.3 mm thickness (which can be very accurately controlled), depending on the final size of the rubber thread desired. Multiple sheets are layered, heat treated to vulcanize and then slit into threads for textile uses. Individual threads have either square or rectangular cross sections (Fig. 6.2).

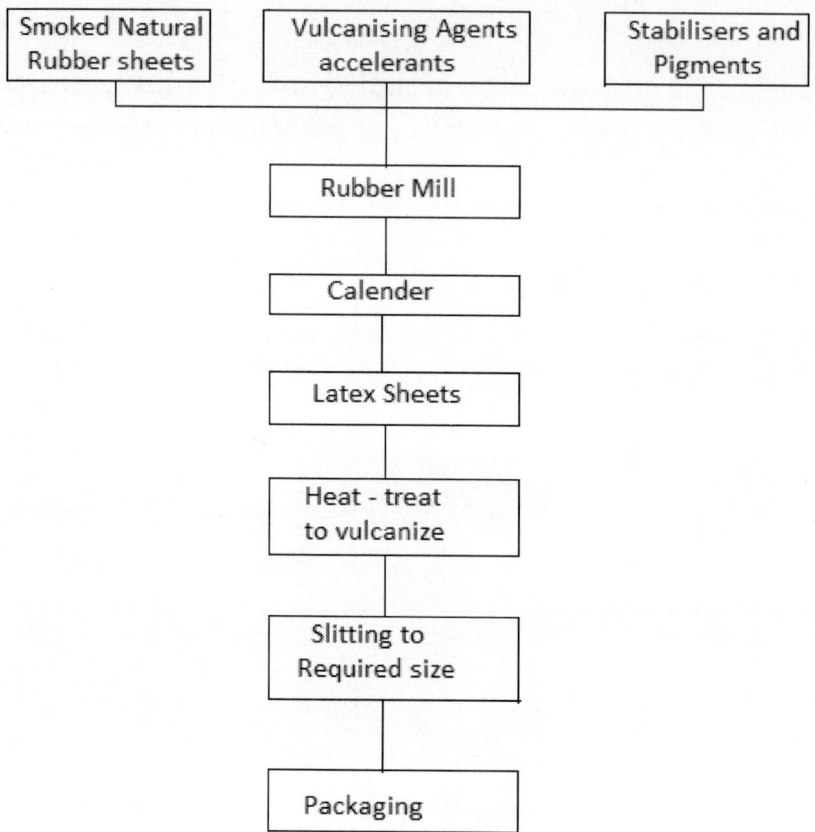

Figure 6.2. Flow chart of cut filament manufacture

6.1.1.2 Extruded filaments

This manufacturing process consists of extruding the natural rubber latex into a coagulating bath to form filament; then, the material is cross linked to obtain fibres which exhibit high stretch. It can be synthesized. The cut filaments are not round (square cross section) in shape, and the next method can produce filaments can produce filaments of round cross section. In this method also, the rubber latex is the starting point. It is mixed with vulcanizing agents, accelerators, antioxidants, pigments and other materials, and this compounded latex is held under controlled temperature conditions until partial vulcanization occurs. This has the effect of increasing wet strength and thus the processability of the extruded threads. The matured latex is extruded at constant pressure through precision-bore glass spinnerets into a coagulating bath of 15–55% acetic acid bath where coagulation into thread form occurs. The filaments are washed free of excess acid with water and conducted through a dryer, after which a silicone oil-based finish is applied and the threads are formed into multi-end ribbons. The ribbons are then vulcanized by multiple passes on a conveyer belt through an oven that can increase curing temperature in stages up to about 150°C. The filaments are thus converted into fine, highly elastic threads, which are dusted with talc if necessary, to provide a smooth surface which facilitates processing. Latex thread production rates vary with thread size and equipment but owing to hydrodynamic drag and the weak nature of the coagulating thread, maximum line take-up speeds are about 30 m/min. They are strong and uniform and can be spun to very fine diameter and in almost any continuous length. Using this technique, it is possible to produce filaments to counts as fine as 160s (Fig. 6.3).

The molecular weight of the polymer between cross links is of the order of 3000–10,000. The cis-1,4-polyisoprene chains are flexible and highly mobile because of the ease of rotation of the four-carbon unit in the polymer molecules. The chains are randomly oriented in the relaxed state at room temperature. The force required to stretch the fibre initially is very low, but at higher extensions, the chains begin to crystallize, increasing the modulus and breaking strength of the fibre. Stretch-induced crystallization is an important property for elastomeric materials so that they may survive textile processing and wear without breaking. If the material is stretched to an extension below its breaking point, elastomeric fibres have the ability to return to their original unstretched dimension. After the stretching force is removed the covalent cross links in the rubber thread, and the mobility of the polymer chains are responsible for returning the stretched fibre to its original dimension.

Rubber threads are susceptible to oxidative degradation, and high concentrations of antioxidants are added to the mixture to make the high surface area fibres more resistant. Pigments, such as titanium dioxide, are used as fillers or to impart whiteness to the thread. Carbon black reinforcement is used in coloured rubber materials. Other agents include accelerators and

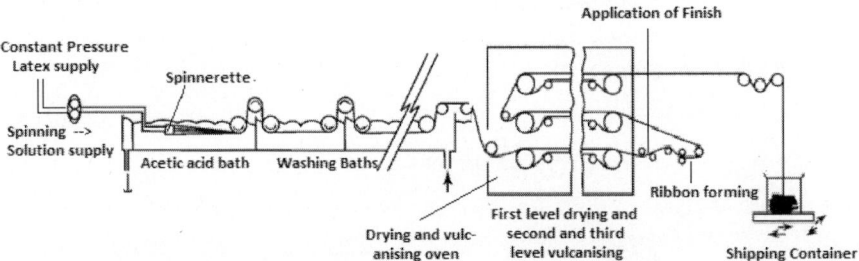

Figure 6.3. Schematic diagram of latex thread manufacture

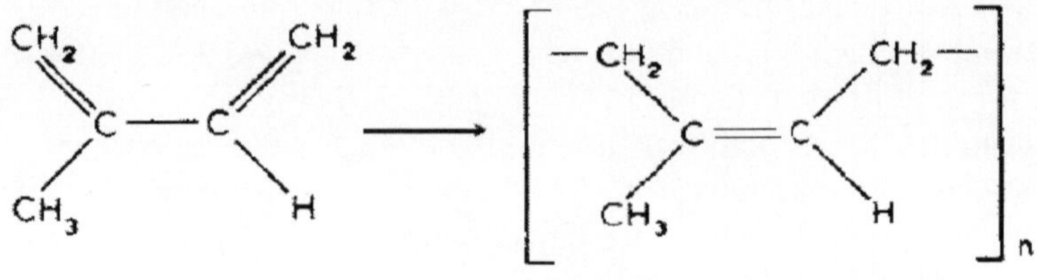

Isoprene   Polyisoprene (natural rubber)

Figure 6.4. Chemical structure of natural rubber

activators to promote the vulcanization process. With all the additives, a typical high-grade rubber thread contains less than 85% elastomer. Synthetic rubber threads can also be produced, but do not have the same elastomeric properties as natural rubber because the polymer chains are less mobile and do not crystallize at high elongations. They are used in applications, which require solvent resistance.

Natural rubber is essentially a polymer of isoprene units, a hydrocarbon diene monomer. Synthetic rubber can be made as a polymer of neooprene or various other monomers. The rubber molecule has a DP of about 10,000, i.e., there are some 10,000 isoprene units in the chain. The material properties of natural rubber make it an elastomer (Fig. 6.4).

Natural rubber, cis-1,4-polyisoprene, itself is not elastomeric but is converted into an elastomer for elastomeric fibres by blending the polymer with sulphur, and curing the material at an elevated temperature to promote vulcanization.

Rubber exhibits unique physical and chemical properties. Rubber's stress-strain behaviour exhibits the Mullins effect, the Payne effect and is often modelled hyper elastic. The 'rubber-like' behaviour of rubber is due to the unusual form of its molecules, which are highly folded. When a piece of rubber is pulled, the molecules tend to straighten out, and if the stretching force is sustained, the molecules may begin to slide over one another and take up new positions. When the stretching force is released, the molecules will return towards their folded state, remaining in their new positions with respect to one another (Table 6.1).

Table 6.1. Properties of natural rubber

Property	Rubber fibre	Spandex
Tensile strength	490–700 kg/cm^2	385 kg/cm^2
Tenacity	6.2 cN/tex (0.7 g/den)	4.0 cN/tex (0.45 g/den)
Elongation	700–800%	700–900%
Modulus	0.013–0.045 N/tex	0.004–0.005 N/tex
Sizes available	16–610 tex	1.1–250 tex
Specific gravity	0.96–1.066	
Stress decay, stress loss, %	Rubber thread 60 gauge Percent	Spandex fibre, 622 dtex (560 den) Percent
After 5 min	23.4	46.6
After 10 min	25.0	50.2
After 20 min	27.1	53.2
After 40 min	28.0	55.2
After 80 min	28.9	86.0
Stability		
UV light	Fair	Good
Ozone	Poor	Good
NO$_x$	Poor	Fair, yellows
Active Cl	Poor	Fair, yellows
Body oils	Poor	Fair
Cosmetics	Fair	Good
Dyeability	Not dyeable	Dyeable
Abrasion resistance	Poor	Very good
Effect of moisture	Negligible	
Effect of temperature	Near 140°C hardens due to further vulcanization. At 350°C becomes hard and brittle, avoid above 95°C	
Effect of sunlight	Discolouration and deterioration	
Effect of acid	Resistant to most inorganic acids. Concentric sulphuric and nitric acids may affect	
Effect of alkali	Resistant to alkali	
Effect of organic solvents	Avoid dry-cleaning solvents, attacked by hydrocarbons, oils and fats. Swells in hydrocarbons, vegetable oils	

6.2   Alginate fibre

Minor fibre made of a jelly-like calcium alginate derived from certain forms of seaweed used as scaffolding in such fabrics as surgical dressings which can be dissolved away.

The most important source of commercial alginates is brown algae. Alginate is found in the cell wall and intercellular regions of brown algae. However, only three types of brown algae are sufficiently abundant or suitable for commercial extraction of the alginic acid. In order of abundance, they are laminaria (British Isles, Norway, France, N. America, Japan), microcystis (USA) and ascophyllum (British Isles). The high viscosity alginate in commercial use has a molecular weight of about 150,000 and a DP of about 750 but the average molecular weight of ordinary alginate is 15,000. Harvesting is easy because most brown seaweed grow in shallow water.

6.2.1   Manufacturing process

Raw material is sea weed which contains alginic acid which is a polymer of d-mannuronic acid of molecular weight in excess of 15,000, which accounts of the one-third its weight of many dried sea weeds. The dried seaweed is powdered and (stored) is treated with a solution of sodium carbonate and caustic soda where by the alginic acid is converted to sodium alginate. Resultant solution allowed to stand and all undissolved matters are allowed to settle and removed. The viscous solution of sodium alginate is purified by sedimentation then bleached and sterilized by the addition of sodium hypochlorite. The alginic acid is then precipitated by acidification, which is later washed and reconverted to the pure sodium salt. The sodium alginate salt is made into a thick paste, dried and milled to make sodium alginate powder (Fig. 6.5).

A dilute solution of sodium alginate is made, filtered, then spun by the viscose spinning method into a coagulation bath (wet spinning) containing

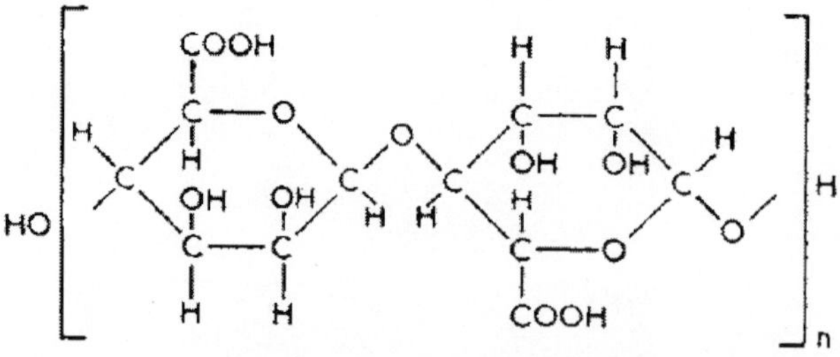

Figure 6.5. Alginic acid

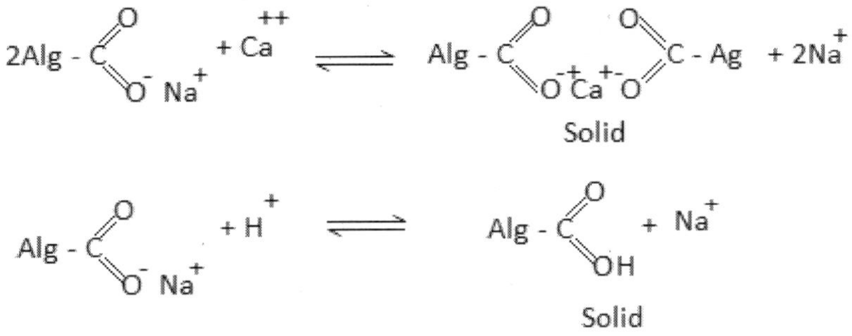

Figure 6.6. Reactions during alginate fibre manufacture

certain polyvalent cation salts (Ca++, Al+++, etc.) or inorganic acid solution; about 0.02 N hydrochloric acid, emulsified oil and a small quantity of a cationic surface agent. The water soluble sodium alginate is thus precipitated in filament form as an alginic acid metal salt, e.g., calcium alginate or alginic acid. The filaments are drawn together, washed, lubricated, dried and wound. The alginic acid is again converted to sodium alginate by neutralization with soda ash (Fig. 6.6).

This solution is extruded through spinneret into coagulating bath containing hydrochloric acid, calcium chloride and a little surface active agent. Streams emerging coagulate into calcium alginate fibres which are washed, oiled, dried and wound onto bobbins or cut (sometimes stretched to break the fibres by varying the relative feed rates which control the degree of stretch-breaking effect) to the required staple length suitable for non-woven products. As calcium or hydrogen ions are exchanged with sodium ion, the reaction proceeds until the sodium alginate is converted to calcium alginate or alginic acid.

6.2.2   Spinning

Alginate can be wet spun in apparatus similar to those used for spinning regenerated cellulose fibres. The precipitation (coagulation) bath can be situated horizontally or vertically (fibre moving upwards or downwards). Further manufacturing operations (drawing, washing, drying, etc.) can be realized continuously or periodically as separate operations. A typical spinning process may be as follows: a 6.4% by weight aqueous sodium alginate solution is extruded through a jet containing 20 holes into a bath containing 5% salt

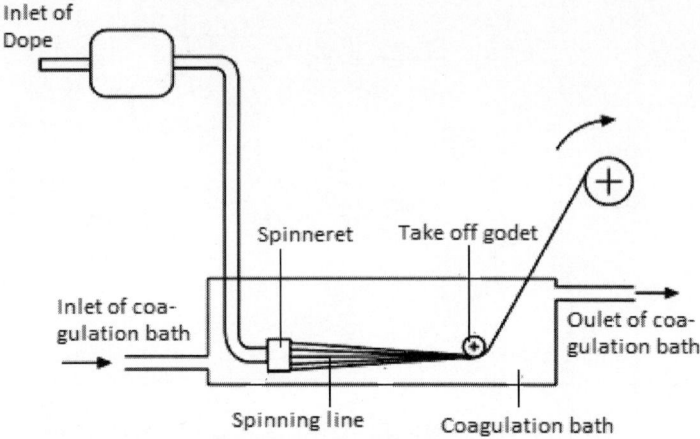

Figure 6.7. Schematic diagram of alginate fibre spinning process

of calcium chloride, a 0.2% acetic acid and 0.05% cetylpyridinium chloride (cation active compound) at 40°C. By this process, it is possible to obtain alginate fibres which do not adhere to one another without addition of emulsified oil to the bath. The threads produced are 37% stretched by passing them over godets and reeled into skeins. They are then washed in a 0.1% solution of calcium chloride at 80°C and then dried at room temperature and conditioned (Fig. 6.7).

It has got good dry strength but the strength is lost when wet. It is non-flammable. Used as scaffolding to support other yarns in the manufacture of lightweight, sheer and lacy fabrics. Medical application for dressing. Flameproof characteristics has also fibre suitable for limited applications.

A major use of alginate yarns are as removable linkages, e.g., in collar bindings in the production of hosiery. They may be removed from the fabrics by washing in dilute solutions of sodium carbonate or sequestering agents such as Calgon. Goods should have sufficient freedom of movement to ensure that the dissolving solution reaches the alginate threads. The following treatments are suggested to remove the alginate yarn from the corresponding mixture fibres given.

1. Alginate along with cotton, viscose rayon, linen or nylon

Quantity	Unit	Additions
2.5	g/l	Soda ash
5	g/l	Common salt

Treat 20–40 min with the above solution at 50–60°C.

2. Alginate along with wool, cellulose acetate or other alkali-sensitive fibres

Quantity	Unit	Additions
2	g/l	Detergent
3	g/l	Calgon
5	g/l	Common salt

Treat the material with above solution for 20–40 min at a temperature of at least 50°C.

6.2.3   Properties

The main properties of alginate fibres are given in Table 6.2.

Table 6.2. Properties of alginate fibres

Property	Alginate fibre
Tenacity	14–18 cN/tex (1.6–2.0 g/den) dry; 4.4 cN/tex (0.5 g/den) wet
Elongation at break	10–14% under normal conditions; 25% wet
Moisture regain	20–35
Specific gravity	1.779
Effect of moisture	Calcium alginate is not soluble in water
Effect of alkali	Calcium alginate readily soluble in dilute alkali
Effect of organic solvents	Not soluble
1. Cross section 2. Lengthwise enlarged	

7
Inorganic Fibres

Inorganic fibres are classified into following categories: glass (e.g., silica sand, lime stone and other minerals), ceramic (e.g., alumina, silica and graphite [carbon] fibres) and metallic fibres (e.g., aluminium, silver, gold and stainless steel).

7.1   Glass fibre

Glass is an inorganic nonmetallic material. Generally, the glass state is defined as the frozen state of a supercooled and thus a solidified liquid. It results from the suppression of the crystallization of a melt. It is also known as fibreglass that is a material made from extremely fine fibres of glass. Glass fibre is formed when thin strands of silica-based or other formulation glass are extruded into many fibres with small diameters textile processing.

The glass is basically an alkali calcium silicate. The common glass is lime-soda glass which is produced by the fusion of silicon dioxide (silica), calcium oxide (lime) and sodium carbonate (soda ash). Soda ash can be replaced by potash, and the glass produced this way is called potash glass (Fig. 7.1).

Silicate glass is composed of a network of the two components SiO_2 and silicate. Depending on the composition, the diameter of the glass and the implementation of additional components, A-glass, C-glass, D-glass, E-glass, ECR-glass, R-glass and S-glass and some other special types are distinguished. A network modifier is a technical term for an atom that modifies the glass network and changes the glass properties (Fig. 7.2).

It has a high degree of viscosity. The basis of textile grade glass fibres is silica, SiO_2. In its pure form, it exists as a polymer, $(SiO_2)n$. To induce crystallization, it must be heated to higher temperatures.

7.1.1   Manufacture

The fibre is manufactured by drawing molten glass. At present, of all processes, drag spinning is most often used, with a market share of about 90% for all glass fibres (E-fibres 97%). The glass melt is kept at a temperature of about 1250–1350°C. Due to gravitational forces, the glass exits through the spinneret, which is perforated with holes of about 1–2 mm in diameter. The

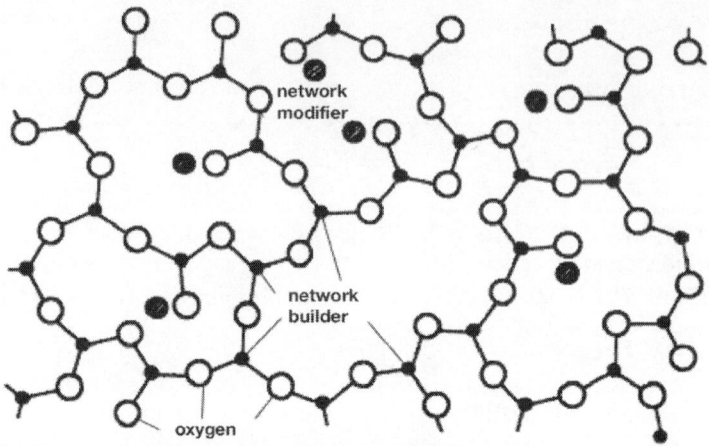

Figure 7.1. Inner structure of glass

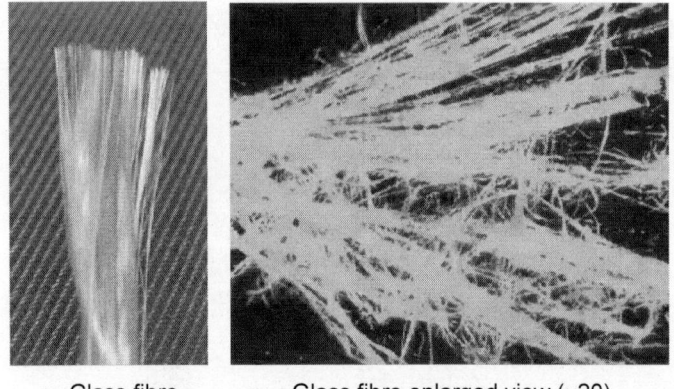

Glass fibre   Glass fibre enlarged view (×20)
Figure 7.2. Morphology of glass fibre

number of capillaries varies between 400 and 2400. After exiting the spinneret, the glass filaments are drafted mechanically and continuously at high speed. The extremely high pull-off velocity makes an additional cooling device necessary. Therefore, cooling fins are arranged between the filaments on the lower side of the jet. After the cooling phase, an aqueous liquid or oil is applied to the glass filaments. This step is called 'sizing'. Then the filaments are wound onto a cone.

The fibre is also spun from manufactured glass pellets. Pellets are melted in a spinning bath and drawn through spinnerets (1200–2000) at a speed of 2000–3000 m/min or more. The drawn filaments are immediately cooled by suitable method, and the fibre attains a solid state. Further drawing is not

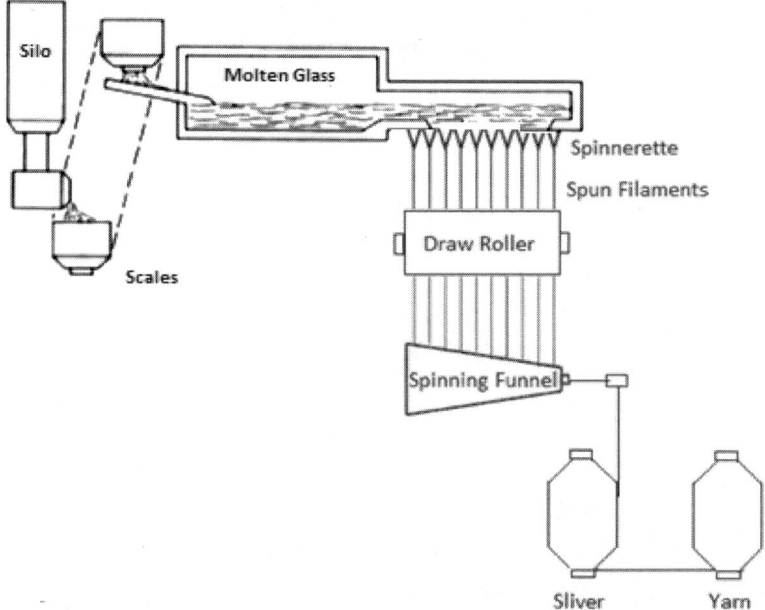

Figure 7.3. Schematic diagram of manufacturing of glass fibre

possible. Finer the fibre, faster the cooling and formation of the solid fibre. A frozen – in induced orientation – birefringence exists in glass fibres which is, however, difficult to determine. Structural changes in the fibre are caused by ageing or reheating. Structural changes take place, and at the same time density is increased along with contraction. The exceptional structure of the fibre is one reason for the higher strength of the fibre when compared to solid glass (Fig. 7.3).

Table 7.1. Properties of glass fibre

	D-glass	E-glass	R-glass	AR-glass
Density (g/cm³)	2.14	2.6	2.53	2.68
Filament tenacity (MPa)	2500	3400	4400	3000
Filament tenacity in composite (MPa)	1650	2400	3600	1800
Elongation at break (%)	4.5	4.5	5.2	4.3
E-modulus (MPa)	55,000	73,000	86,000	73,000
Softening temperature (°C)	775	846	985	773

7.1.2 Uses

The first type of glass used for fibre was soda-lime glass or a glass, which was by trapping air within them, blocks of glass fibre make is used as a reinforcing agent for many polymer products. It has a good thermal insulation, with a thermal conductivity of 0.05 W/m. Because glass has an amorphous structure, its properties are the same along the length. Humidity is an important factor in the tensile strength (Table 7.1). If adsorbed, humidity can worsen microscopic cracks and defects, and lessen tenacity. It has no effect on exposure to sunlight even after extended periods. In the commercial use, the fibre diameters are important criteria along with the cost. For a general reinforcing purpose, fibre diameters of 9–11 μm are required but due to the cost, diameters of 13–15 μm are also sometimes used. Glass fibre find application in chemistry, chemical technology, insulation of machines, pipes and containers, in sound insulation, electronics, boat making, etc. Fabric made with glass fibres find use in interior furnishings wherever high demands of safety requirements like in ships, hotels, cinemas, wall coverings, etc.

7.2   Silica fibres

French scientist M. Gauding discovered that the quartz, a form of silica, could be drawn into filaments. This filament found use as spring which has perfect elasticity and does not undergo deterioration due to from fatigue or corrosion, in torsion balances, and it is being used even today. The main attractive property of silica is its high melting point. Quartz, for example, softens at about 1500°C and melts at 1710–1755°C. Another welcome property is that it is chemically extremely stable, resisting the attack of almost all common chemicals. Silica occurs as agate, amethyst, chalcedony, cristobalite, flint, jasper, onyx, opal, quartz, rock crystal, sand and tridymite. Nowadays silica fibre is used for purposes, where the special properties of silica are useful. The different types of silica fibres made are

1. Quartz fibres
2. Silica from glass fibres (silica (G) fibres)
3. Silica (viscose process) fibres (silica (V) fibres)

The main hurdle of manufacturing silica fibres is the manufacturing units to with stand the high softening or melting point of the raw materials like quartz. Quartz fibres of 5–10 micron diameter are made by softening quartz rods in an oxy-hydrogen flame, and drawing the rods out into filaments of about 0.1 mm diameter. These are then passed through a series of oxy-hydrogen jets, and the quartz being blown forward by the jets to form fine fibres has been collected on a rotating drum. Continuous filaments may be made by softening quartz rods in an oxy-hydrogen flame, and then drawing out the fused quartz into filaments of diameter 0.7 micron and less (Table 7.2).

Table 7.2. General properties of silica fibre

Property	Silica fibre
Tensile strength	650 kg/cm^2
Tenacity	
Elastic recovery	Its elastic recovery is almost 100%
Specific gravity	2.6
Effect of moisture	Negligible
Effect of temperature	Softening point: 1500°C Melting point: 1700–1756°C
Effect of high temperature	Can be heated prolonged periods at temperatures up to 1400–1600°C without undergoing deterioration
Flammability	Nonflammable
Thermal conductivity	Low
Effect of sunlight	Negligible
Effect of acid	Highly resistant to all acid except hydrofluoric acid
Effect of alkali	Highly resistant to most alkalis
General	Highly resistant to all common chemicals

Quartz fibres are widely used as filtration materials, especially where resistance to corrosive substances and/or high temperature is essential. These fibres are also used as insulation materials, serving at temperatures above those where mineral silicate fibres are normally used. Examples of applications include rockets and missiles, jet aircraft, nuclear power plants and industrial furnaces.

7.3   Metallic fibres

Metallic fibres are manufactured fibres composed of metal, plastic-coated metal, metal-coated plastic or a core completely covered by metal. Gold and silver have been used since ancient yarns for fabric decoration. More recently, aluminium yarns and aluminized plastic nylon yarns have replaced gold. The metal generally used is aluminium, and the plastic used is cellulose acetate butyrate or mylar. They are made through laminating process. Coated metallic filaments help to minimize tarnishing. It is available only in filament form.

7.3.1   Properties of metallic yarns

1. Bright and appealing.
2. The yarns are flexibility and some extensibility.
3. Good strength and hence can be used as warp or weft yarn unsupported.
4. All metallic yarns are moth proof.
5. Good chemical and biological resistance.

Because of its brightness and appeal in apparels, the metal yarns are mainly used for decorative purposes in women's dresses, blouses and skirts (Fig. 7.4).

The earliest production of metal fibres dates back to around 3000 BC when metal wires were used to decorate textiles. During the industrial revolution many metal-working businesses were established, and wire drawing developed from a craft into an industry. A thin endless metal object

Figure 7.4. Uses and enlarged view of metallic yarns

with a diameter below 100 μm is called 'filament'; above that it is considered to be a 'wire'. Apart from pure metals (such as copper), often alloys of different metals are processed (widespread for aluminium- and steel-based fibres).

There is a range of production processes available:

1. Wire drawing (coarse fibres)
2. Bundle drawing (medium to fine fibres)
3. Cutting (staple fibres)
4. Taylor (very fine fibres)
5. Melt drag (coarse fibres)

Fig. 7.5 depicts the principle design of a wire-drawing machine. It is one of the oldest production processes. The wire is pulled through subsequent drawing dies, which cause it to become thinner and longer. An annealing process stabilizes the inner structure of the drawn fibres.

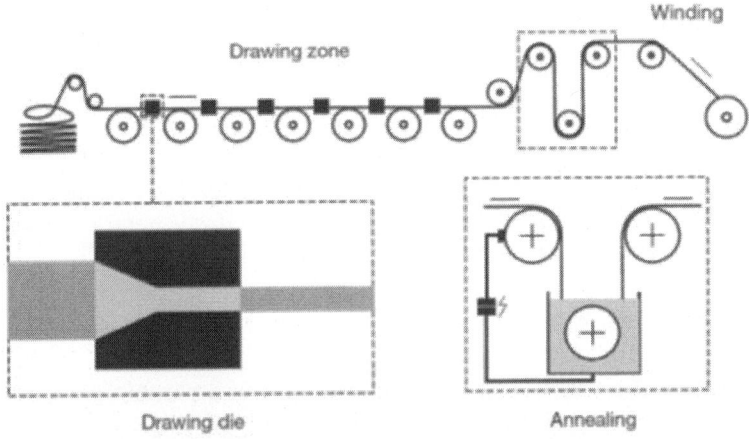

Figure 7.5. Wire drawing process

In the Taylor process, a metal rod is molten inside a glass tube with a slightly higher melting point. The electric current in the coil heats up both materials, the wire melts, and the glass tube softens. Then the glass tube is

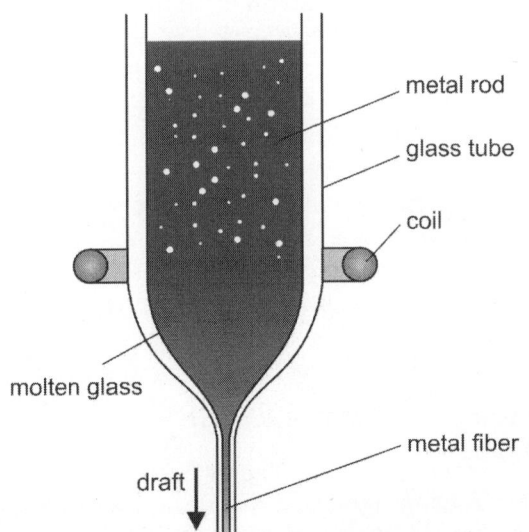

Figure 7.6. Manufacturing of metallic fibre

drawn, causing the metal to become very thin. After cooling, the fine metal fibres are taken up whereas the glass tube is crushed, using ultrasound. Fibre diameters can be lower than 50 μm, but production speed is low and the process can only be used for specific metals that do not fuse with the glass tube (Fig. 7.6).

Typical properties of metal fibres include high density (steel: 8 g/cm^3), high tensile strength (steel: 200–1600 MPa), high elongation at break, high E-modulus (steel: 210,000 MPa), high heat and electrical conductivity, resistance to corrosion and many chemicals, and flame resistance.

Metal fibres are mainly used for woven and knitted fabrics. Typical applications of metal fibres are filters (as in polymer melt spinning), antistatic applications (e.g., filters, protective clothing), reinforcement structures (tire cord), and also sensors (smart textiles) and architecture (metal fabric façade). When suitable adhesives and films are used, they are not affected by salt water, chlorinated water in swimming pools or climatic conditions. If possible anything made with metallic fibres should be dry-cleaned. Ironing can be problematic because the heat from the iron, especially at high temperatures, can melt the fibres. They are used mainly for decorative purposes.

7.4   Asbestos

Asbestos is the only natural mineral fibre obtained from varieties of rocks. It is fibrous form of silicate of magnesium and calcium containing iron and aluminium and other minerals. It is acid proof, flame proof and rust proof. Its particles are carcinogenic and hence its use is restricted. General name that applies to several types of fibrous silicate minerals, existing in nature in metamorphic or altered basic and ultra basic igneous rocks. Six minerals defined as 'asbestos' are chrysotile, amosite, crocidolite, tremolite, anthophyllite and actinolite. Asbestos can be subdivided into two major classifications of minerals: amphiboles and serpentines. All but one form, chrysotile, is amphiboles. Chrysotile is a serpentine. It is nonflammable, insoluble in water and organic solvents, resistant to acids and has high strength. Used in fireproofing, low-density insulation board and ceiling tiles, thermal and chemical insulation, mud and texture coats, clutch plates.

7.5   Rock wool (stone wool)

Rock wool is an amorphous silicate manufactured from rock. Stone wool is a furnace product of molten stone, at a temperature of about 1600°C, through which is blown a stream of air or steam. More high-tech production techniques are based on spinning molten rock (lava).

Man-Made Synthetic Fibre

8.1 General

8.1.1 Classification of synthetic fibres

The polymers used for synthetic fibres are produced by lining up single atoms or atom groups, so-called monomers. For the production of polymers for synthetic fibres, three different reaction mechanisms have been used (Fig. 8.1).

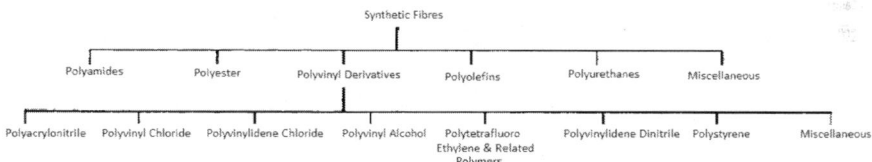

Figure 8.1. Classification of synthetic fibres

1. Polymerization: Identical monomers are connected by covalent bonds between the single monomers to form a long-chain molecule. To initiate chain growth, double bonds within the monomers must first be broken. Polymerization reaction can be represented as follows shown in Fig. 8.2.

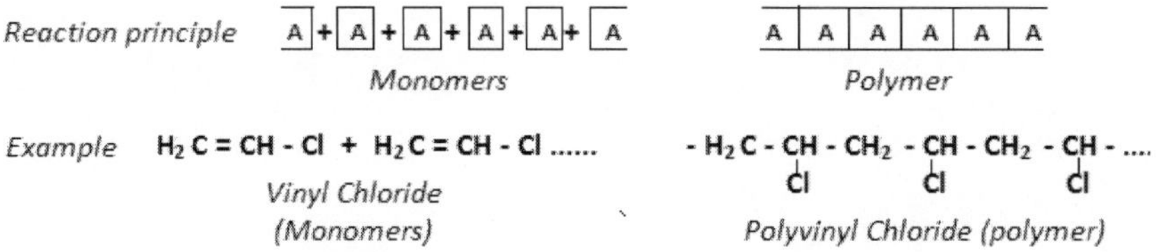

Reaction principle   A + A + A + A + A + A   A A A A A A
                         Monomers   *Polymer*

Example $H_2C = CH - Cl + H_2C = CH - Cl$ ...... $- H_2C - CH - CH_2 - CH - CH_2 - CH -$ ....
                                                                      $\overset{|}{Cl}$ $\overset{|}{Cl}$ $\overset{|}{Cl}$
            Vinyl Chloride   *Polyvinyl Chloride (polymer)*
            (Monomers)

Figure 8.2. Polymerization reaction

2. Another mechanism to produce synthetic polymers is polycondensation. With polycondensation, different or identical molecules bind together while by-products split off (e.g., water, short-chain alcohols). This reaction requires that each monomer has at least two reactive groups. Polyester and polyamide are produced by polycondensation (Fig. 8.3).

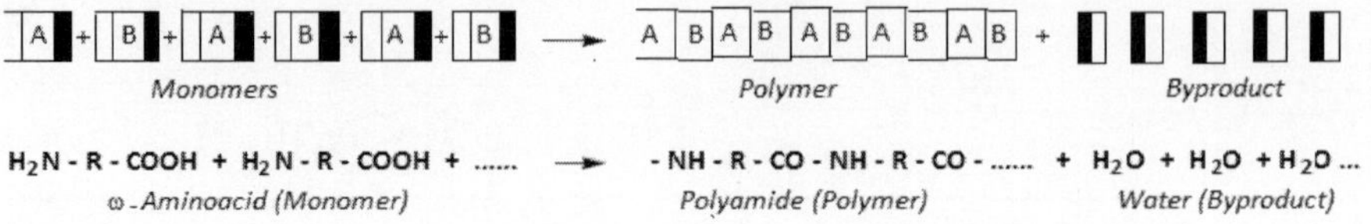

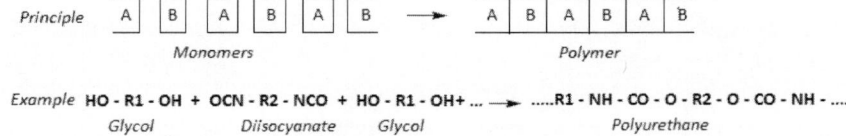

$H_2N - R - COOH + H_2N - R - COOH + ...$ → $- NH - R - CO - NH - R - CO - ...... + H_2O + H_2O + H_2O ...$
ω-Aminoacid (Monomer)   Polyamide (Polymer)   Water (Byproduct)

Figure 8.3. Principle and example of polycondensation

3. Polyaddition, the third mechanism to produce synthetic polymers, is used mainly for polyurethane elastomers. Hydrogen atoms are exchanged between the different monomers, which have at least two reactive groups each. The results are long-chain macromolecules with no by-products.

Principle | A | B | A | B | A | B | → | A | B | A | B | A | B |
Monomers   Polymer

Example $HO - R1 - OH + OCN - R2 - NCO + HO - R1 - OH + ...$ → $.....R1 - NH - CO - O - R2 - O - CO - NH - ....$
          Glycol   Diisocyanate   Glycol   Polyurethane

Figure 8.4. Polyaddition

8.1.2   Overview of the important synthetic fibres according to above polymerization mechanisms

Table 8.1. Processes involved in the production of polymers

Fibre	Structural formula	Specific properties	Application
Polyamide (PA)	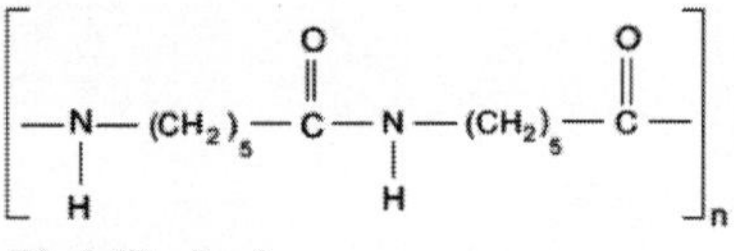 PA 6 (Perlon) 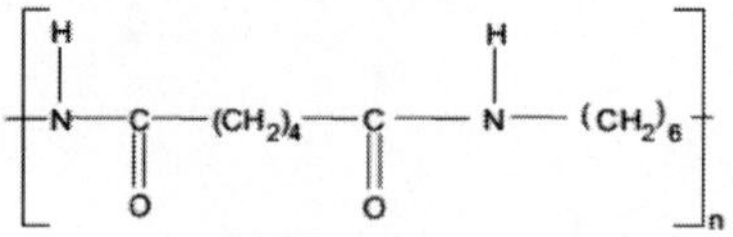 PA 6.6 (Nylon)	*Outstanding tensile properties *High tenacity *Reduction in tenacity under action of light	*Carpet tuft yarns *Tights, balloon silk *Substrates, tricot materials *Linings, Oxford yarns (soft luggage and outdoor leisure

Table 8.1. Processes involved in the production of polymers

Fibre	Structural formula	Specific properties	Application
Aramid (AR)	PMI (Poly-m-Phenylene Isophtalamide) *meta*-Aramid PPTA (Poly-p-Phenylene Terephthalamide)	*Good temperature stability and resistance to chemicals *Outstanding high tenacity and E-modulus	*Asbestos substitute *Safety gear and preservatives *Cover and decorative textiles *Electrical isolations *Filtrations of hot gases *Friction linings, seals *Tire reinforcements *Fibre-reinforced composites *Technical wovens *Ropes, cables, nets
Polyester (PES)	PES (Polyester) n-Polyethylene-terephthalate (PET)	*High tear strength, abrasion resistance *High tenacity *Good light fastness	*Apparel *Home and furnishing textiles *Sewing threads *Nonwovens, needled felts *Textile floor coverings *Technical textiles *Leather cloths

2. Polymerization

Fibre	Structural formula	Specific properties	Application
Polyvinyl alcohol (PVAL)	CH_2 — CH — OH	*Solubility in boiling water *Low elasticity	*Binder fibre nonwovens

2. Polymerization

Fibre	Structural formula	Specific properties	Application
Polyvinyl chloride (PVC)	$-CH_2-CH-$ with Cl substituent, bracketed with subscript p	*Noninflammability *Very high elasticity *Low abrasion resistance	*Home and furnishing textiles, nonwovens, filter *Safety gear
Polyacrylonitrile (PAN)	$-CH_2-CH-$ with $C\equiv N$ substituent, bracketed with subscript p	*Very low moisture absorbency *High light fastness *High bulk elasticity	*Stuffing *Decoration and furniture fabrics *Tilts (cover of truck beds)
Polytetrafluoroethylene (PTFE)	$-CF_2-CF_2-$ bracketed with subscript p	*Outstanding resistance to chemicals *High temperature fastness *Resistance against UV radiation	*Filter material for liquids and gases *Electrical isolations *Sealing material
Polyolefin: polyethylene (PE), polypropylene (PP)	$-CH_2-CH_2-$ bracketed with subscript p (PE) Polyethylene $-CH_2-CH-$ with CH_3 substituent, bracketed with subscript p (PP) Polypropylene	*Lowest density of all fibre materials *Very high specific tenacity *Very high elasticity *Lowest moisture absorption of all fibre materials	*Needle-punched carpets *Carpet floor wovens *Safety gear *Athletic apparel *Foils

3. Polyaddition

Fibre	Structural formula	Specific properties	Application
Elastane (EL)	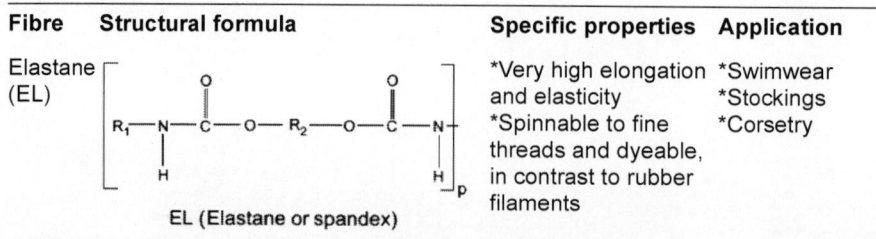 EL (Elastane or spandex)	*Very high elongation and elasticity *Spinnable to fine threads and dyeable, in contrast to rubber filaments	*Swimwear *Stockings *Corsetry

8.2   Acrylic (polyacrylonitrile) fibres

The basic material for acrylic fibre production is acrylonitrile, which can be manufactured by various methods.

Before 1960, acrylonitrile was commercially produced by adding hydrogen cyanide to acetylene and dehydrating the resulting ethylene cyanohydrins to acrylonitrile. This process is only of secondary significance today.

$$CH_2-O-CH_2 + HCN \rightarrow HOCH_2CH_2CN \rightarrow (-H_2O) \rightarrow CH_2=CH-CN \rightarrow$$
$$-CH_2-CH(CN)-CH(CN)-CH-$$
Polyacrylonitrile

Another process is based on acetylene and hydrocyanic acid:

$$HCCH + HCN \rightarrow CH_2=CH-CN$$

The modern process is called the Sohio (Standard Oil Ohio) process. The starting compound is the very cheap propylene which is aminated in the gaseous phase with ammonia react at high temperature in the presence of suitable catalysts such as bismuth phosphomolybdate and then dehydrogenated to acrylonitrile. This process has appreciably reduced production costs because of the lower cost of raw materials and the single step nature of the process.

Commercial acrylonitrile is synthetic polymeride fibres, the chain of which is made up of at least 85% of acrylonitrile. The composition of the polyacrylonitrile fibres varies due to the polymer addition of affinity increasing foreign substances (monomers such as vinyl acetate, vinyl chloride, vinylidene chloride, vinyl pyridine, methacrylate, methacrylamide, etc.). 'Pure' polyacrylonitrile fibres are subdivided into (a) fibres consisting solely of polyacrylonitrile (nondyeable) and (b) so-called modified polyacrylonitrile fibres(Modacrylic fibres) with up to 15% foreign substances (good dyeing properties), whereas if there is (c) a high foreign substance content, we talk of polyacrylonitrile copolymer fibres.

8.2.1   Manufacture of acrylic fibres

8.2.1.1   *Starting from ethylene cyanhydrin*

Ethylene cyanhydrin is made either by treatment of ethylene oxide with hydrogen cyanide or by reaction of ethylene chlorohydrin with alkali cyanides. Ethylene cyanhydrin is hydrated in the presence of a catalyst either in liquid phase or vapour phase to get acrylonitrile.

$$H_2C - CH + HCN \longrightarrow HOCH_2 - CH_2CN \xrightarrow[\text{Hydration}]{\text{Catalyst}} CH_2 = CH\ CN$$

Ethylene   Hydrogen   Ethylene   Acrylonitrile
Oxide   Cyanide   Cyanhydrin

$$HO\ CH_2 - CH_2\ CN + KCN \longrightarrow HOCH_2 - CH_2\ CN \xrightarrow[\text{Hydration}]{\text{Catalyst}} CH_2 = CH\ CN$$

Ethylene   Potassium Ethylene   Acrylonitrile
chlohydrin   Cyanide   Cyanhydrin

8.2.1.2   *Starting from acetylene and hydrogen cyanide*

Acetylene on reaction with hydrogen cyanide can directly give acrylonitrile.

$$HC \equiv CH + HCN \longrightarrow H2C = CH\ CN$$

Acetylene   Hydrogen   Acrylonitrile
               Cyanide

8.2.1.3   *Starting from propylene*

Propylene can be oxidized to get acrolein, which on reaction with ammonia will give a hydroxyl amino compound. This compound on dehydration and dehydrogenation will form acrylonitrile.

$$H_2C = CH\ CH_3 \longrightarrow H_2C = CH\ CHO \longrightarrow H\ C = CH \overset{NH}{\underset{OH}{\Big\langle}} \longrightarrow H_2C = CH.CN$$

Propylene   Acrolein   Acrylonirile

8.2.1.4   *Starting from acetaldehyde*

Acetaldehyde on reaction with hydrogen cyanide to form the cyanhydrin which is dehydrated to acrylonitrile.

$$H3C\ CHO + HCN \longrightarrow CH3\ CH \overset{OH}{\underset{CN}{}} \xrightarrow{\text{Dehydration}} H_2C = CH.CN$$

Acetaldehyde Hydrogen   Cyanhydrin   Acrylonitrile
              Cyanide

8.2.2   Polymerization

8.2.2.1   *Aqueous dispersion polymerization*

By far the most widely used method of polymerization in the acrylic fibres industry is aqueous dispersion (also called suspension). Polymerization of acrylonitrile is usually achieved in the presence of peroxy disulphate and sulphite or thiosulphate as activators (radical generators) in an aqueous solution of the monomer. Reaction between peroxy disulphate and sulphite or thiosulphate involves the formation of intermediate radicals that are extremely short-lived and immediately undergo further reaction. The formation of intermediate radicals is encouraged by adding so-called promoters, e.g., copper II or iron III salts, to the reaction solution. The formation of sulphite and sulphate radicals constitutes the intermediate stages of this reaction, and the following electron formulae can be described as:

$$
\left[\; . \; \overset{\overline{|O|}}{\underset{|O|}{S}} - \overline{O]} \; \right]^{-}
\qquad
\left[\; . \; \overline{O} - \overset{\overline{|O|}}{\underset{|O|}{S}} - \overline{O|} \; \right]^{-}
$$

The so-formulated radicals differ from sulphite and sulphate ions in that they have only one single negative charge and furthermore possess a free electron, represented by a dot. This free electron is responsible for the instability and high reactivity. Both the sulphite and sulphate radicals act as initiators of the polymerization of acrylonitrile. They attach themselves to an acrylonitrile molecule, so that this in turn becomes a C radical carrier:

$$
^{-}O_3S \; . + C = \underset{\underset{CN}{|}}{C} \longrightarrow \; ^{-}O_3S - C - \underset{\underset{CN}{|}}{C} \; .
$$

Other acrylonitrile molecules are added and the radical character of the growing molecular chain is always maintained. The polymerization reaction eventually produces a giant polyacrylonitrile molecule with a molecular weight of over 40,000, its radical position being finally saturated by the addition of a sulphite or sulphate radical:

$$
^{-}O_3S \left(C - \underset{\underset{CN}{|}}{C} \right)_{n-1} C - \underset{\underset{CN}{|}}{C} \; . + . SO_3^{-} \longrightarrow \; ^{-}O_3S \left(C - \underset{\underset{CN}{|}}{C} \right)_{n} SO_3^{-}
$$

The polyacrylonitrile produced by this or another similar process thus contains sulphite or sulphate groups at the ends of the giant molecule. These are incorporated as initiating and terminating radicals in the polymerization process. For this reason, polyacrylonitrile is anionic.

In commercial practice, polymerization is effected in a continuous-stirred tank reactor, a system in which all components are fed continuously and mixed, and the product is continuously discharged. For start-up, the reactor is charged with a certain amount of pH-adjusted water or the reactor is filled with overflow from another reactor already operating at steady state. The polymerization can be carried out by mixing one part of acrylonitrile with 10 parts of water and adding an excess of promoters (Cu^{++} or Fe^{+++}) to the reaction solution. The peroxy-sulphate and sulphite or thiosulphate activators are then allowed to flow slowly and continuously into the reaction mixture. The reactor feeds are metered in at a constant rate for the entire course of the production run, which normally continues until equipment cleaning or maintenance is needed. The reaction vessel is normally an aluminium alloy; this minimizes scale build-up as the wall provides a sacrificial surface. The reactor is jacketed; steam may be introduced to heat the contents for start-up, but once the polymerization is initiated, water is circulated in the jacket to remove the heat of polymerization and maintain a constant temperature, usually 50–60°C. The availability of activators has a very important bearing on the polymerization because the amount is responsible for the molecular weight achieved. As polymerization proceeds, the polymer (which is insoluble in water) is precipitated to form a slurry. A steady state is established by taking an overflow of slurry stream at the same mass flow rate as the combined feed streams. This is filtered, and the polymer is washed and dried.

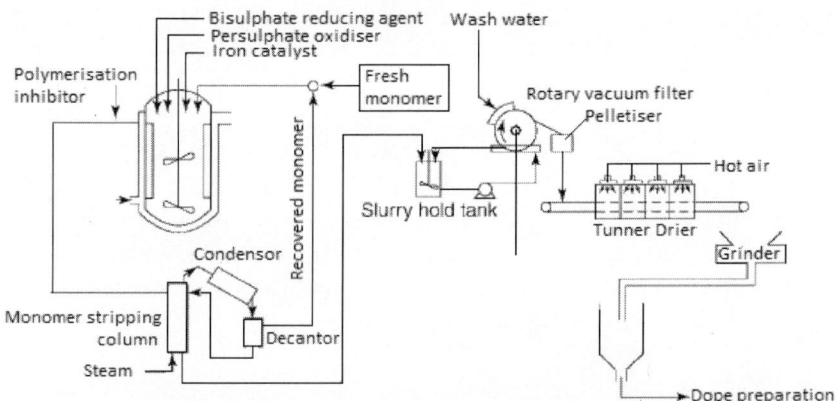

Figure 8.5. CSTR dispersion polymerization process

The extent of polymerization can be checked during the reaction by measuring the molecular weight. The process may be carried out as a batch process, or on a continuous basis. In the latter case, monomer, water and other materials are fed into a reaction vessel, and slurry is withdrawn continuously. Unreacted monomer is recovered and returned to the polymerization.

A monomer mixture composed of acrylonitrile and up to 10% of a neutral comonomer, such as methyl acrylate or vinyl acetate, is fed continuously (see Fig. 8.5). Polymerization is initiated by feeding aqueous solutions of potassium persulphate (oxidizer), sulphur dioxide (reducing agent), ferrous iron (promoter), and sodium bicarbonate (buffering agent). Alternately the system may employ a sodium bisulphite/sulphur dioxide or a sodium bisulphite/sulphuric acid buffer. The aqueous and monomer feed streams are fed at rates that give a reactor dwell time of 40–120 min, and a feed ratio of water to monomer in the range 2–5. The reactor overflow, an aqueous slurry of polymer particles, is mixed with an iron chelating agent, or the pH is raised to stop the polymerization. The slurry is then fed to the top section of a baffled monomer-separation column. The separation of unreacted monomer is effected by contacting the slurry with a countercurrent flow of steam introduced at the bottom of the column. Monomer plus water is condensed from the overheads stream and the monomer separated using a decanter, the water phase being returned to the column. The stripped slurry is taken from the column bottoms stream, and the polymer separated using a continuous vacuum filter. After filtration and washing, the polymer is pelletized, dried, ground and then stored for later spinning.

8.2.2.2   Solution polymerization

The reaction is carried out in a homogeneous medium by using a solvent for the polymer. Suitable solvents are aqueous sodium thiocyanate (NaSCN) used by some manufacturers and dimethyl sulphoxide (DMSO) by some other. The homogeneous solution polymerization of acrylonitrile follows the conventional kinetic scheme developed for vinyl monomers. Thermally activated initiators such as azobisisobutyronitrile (AIBN), ammonium persulphate or benzoyl peroxide can be used in solution polymerization, but these initiators are slow acting at temperatures required for fibre-grade polymer processes. Half-lives for this type of initiator are in the range of 10–20 h at 50–60°C. Therefore, these initiators are used mainly in batch processes where the reaction is carried out over an extended time. Redox initiators, such as the ammonium persulphate/sodium bisulphite/copper system, have much higher initiation rates and are reported to be employed in the NaSCN process. A typical continuous solution polymerization equipment is shown Fig. 8.6.

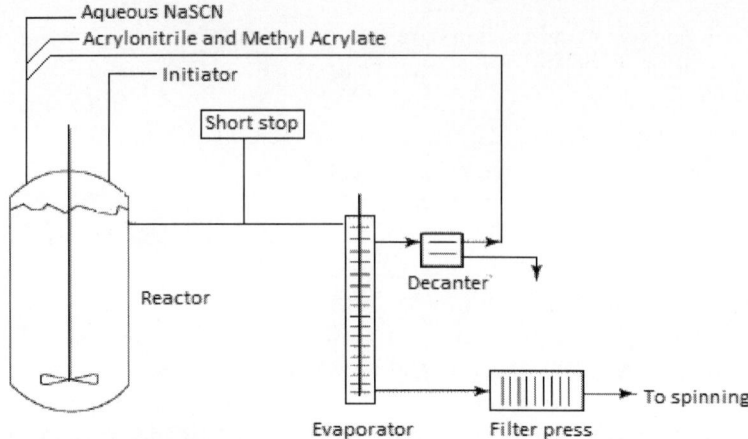

Figure 8.6. A typical solution polymerization equipment

Of the two common comonomers incorporated in textile-grade acrylics, methyl acrylate is the least active in chain transfer, whereas vinyl acetate is as active in chain transfer as dimethylformamide (DMF). Vinyl acetate is also known to participate in the chain transfer-to-polymer reaction. This occurs primarily at high conversion, where the concentration of polymer is high and monomer is scarce. The advantage of solution polymerization is that the polymer solution can be converted directly to spin dope by removing the unreacted monomer. Incorporation of nonvolatile monomers, such as the sulphonated monomers, can be a problem. The sulphonated monomers must be converted to a soluble form such as the amine salt. Nonvolatile monomers are difficult to recover or purge from the reaction medium. Monomer recovery systems based on carbon adsorption have been developed. However, the usual practice is to maximize the single-pass conversion of these monomers. Subsequent to the polymer reactor, acrylonitrile and volatile comonomers are removed in a thin-film evaporator. Additives such as pigments or stabilizers may be incorporated using a static or active mixer before the dope is transferred to the spinning area.

8.2.2.3   Bulk polymerization
The bulk polymerization is attractive, since the polymer would not require water removal and the process would not have the low propagation rates and high chain transfer rates of solution processes. However, bulk polymerization of acrylonitrile is complex and is not clearly understood. When initiator is first added, the reaction medium remains clear while particles 10–20

nm in diameter are formed. As the polymerization proceeds, the particle size increases, giving the reaction medium a white milky appearance. When a thermal initiator, such as AIBN or benzoyl peroxide, is used the reaction is autocatalytic. This contrasts sharply with normal homogeneous polymerizations in which the rate of polymerization decreases monotonically with time. With acrylonitrile bulk polymerization, three propagation reactions occur simultaneously accounting for the anomalous autoacceleration. These are chain growth in the continuous monomer phase, chain growth of radicals that have precipitated from solution onto the particle surface, and chain growth of radicals within the polymer particles. Bulk polymerization is not used commercially because the autocatalytic nature of the reaction makes control difficult.

8.2.2.4   Emulsion polymerization

This process is limited to the manufacture of modacrylic compositions. It was shown that the emulsifier disperses a small portion of the monomer in aggregates of 50–100 molecules approximately 5 nm in diameter called micelles. The majority of the monomer stays suspended in droplet form. These droplets are typically 1000 nm in diameter, much larger than the micelles. Since a water soluble radical initiator is used, polymerization begins in the aqueous phase. The micelle concentration is normally so high that the aqueous radicals are rapidly captured. The micelle is essentially a tiny reservoir of monomer; therefore, polymerization proceeds rapidly, converting the micelle to a polymer particle nucleus. Since the halogen-containing monomers have little water solubility, the micelle promotes their ability to react. The ability of emulsion polymerization to segregate radicals from one another is of great importance commercially. The effect is to minimize the rate of radical recombination, allowing high rates of polymerization to be achieved along with high molecular weight. This is important in modacrylic polymerizations where chain-transfer constants of the halogen monomers are high.

8.2.3   Copolymerization

Even though, acrylonitrile has already been polymerized as early as 1929, the potential exploitation of polyacrylonitrile in the manufacture of fibres was immediately realized, but the first acrylic fibre did not arrive on the market until 1950. The reason for taking 21 years to manufacture an acrylic fibre is because the melting point of polyacrylonitrile is near the decomposition point which does not allow a melt spinning process such as that used in the manufacture of polyamide and polyester. It was also not possible to utilize the method of spinning polyacrylonitrile from a solution because no suitable solvent was available. For a long time, polyacrylonitrile was considered worthless because of processing difficulties. It gained some significance in the form of a copolymer with butadiene which became known as synthetic rubber under the designation Buna N and was used in industrial application. Only after Rein discovered in 1942 that polyacrylonitrile was soluble in DMF and that fibres could be spun from this solution, the major obstacle to further development was overcome.

Nowadays the acrylonitrile monomer is then combined with other monomers to produce co-polymers or polymerized alone to form homopolymer acrylic. Special properties of the fibre, however, are influenced far more by the type and amount of modifying monomers. Vinyl compounds, e.g., vinyl chloride, vinyl acetate, methylacrylate, vinylpyridine and vinylpyrazine, are suitable monomers of this type. Methacrylic acid and styrene sulphonic acid are also used. To qualify for the description acrylic, the final polymer must contain at least 85% by weight of acrylonitrile units. Acrylonitrile is an addition polymer, the monomers adding or joining end-to-end without liberating any by-product.

8.2.4   Spinning

In 1950, the first acrylic fibre became finally available under the trade name Orion. Later on, systematic research showed that solvents other than dimethyl formamide are suitable for the spinning of polyacrylonitrile fibres. The solvents involved were mainly those which exhibit a high dipolar moment; these include dinitrile malonic acid, dinitrile succinic acid, dinitrile adipic acid, nitrile glycolic acid, DMSO, ethylene carbonate, propylene carbonate, butyrolactone, butyrolactarn, dimethyloxamide, nitrophenols, dialkylcyanamide and oxypyrrolidon, furthermore concentrated nitric acid and aqueous solutions of inorganic salts, such as ammonium thiocyanate or zinc chloride, which have been found suitable for spinning polyacrylonitrile. Acrylic fibres are spun either by the wet or by the dry spinning method. In both processes, the polyacrylonitrile is dissolved first and the solution pressed through spinnerets. The two processes differ from each other only in the way the solvents are removed while the filament is forming. In the dry spinning method, the filament is passed through a heating stack in which the first acrylic fibres such as Orion 41, Orion 81 and PAN were pure polymers which contained only acrylonitrile as monomer. They exhibited excellent textile properties but had the serious disadvantage of being very difficult to dye. The reason for this was that the glass transition temperature of the pure polymer lies above 100°C. This glass transition temperature, which is also known as second order transition temperature, is a specific temperature for a certain type of fibre at which the glass-like amorphous ranges of the fibre begin to soften. This temperature can be determined by measuring a physical property such as specific heat, refraction index or the specific volume at various temperatures and plotting it against the temperature. A bend in the curve indicates the glass transition temperature. The glass transition temperature is a deciding factor in the dyeability of a fibre. Below this temperature, all areas of the fibre are frozen and thus the prerequisite for a dye pick up by the fibre is not given. The conventional dyeing plants did not allow the temperature of the bath to be raised above 100°C. It was therefore necessary to reduce the glass transition temperature to at least 85°C by modifying the acrylic fibre. This could be accomplished by incorporation of comonomers. The pure polymers such as Orion 41, Orion 81, PAN and Dralon T are no longer of significance in the textile field but the readily dyeable copolymers have gained a considerable share of the market. The acrylic fibres that are available in many commercial forms often differ from one another in their technological properties and dyeing behaviour. This is mainly a result of the different solvents, which are used for spinning and the different spinning methods, i.e., the wet spinning and the dry spinning methods.

8.2.4.1   Solution spinning

As the acrylic fibre industry has matured, the wide range of spin solvents that were commercialized in the 1950–1960s has narrowed. DMSO and zinc chloride are each limited to one producer; no processes based on ethylene carbonate solvent remain in operation. Most new plants are based on either Dam or Nascent wet spinning. Table 8.2 shows commercial solvents and the dissolved polymer concentration range for a spin solution. For dry polymer, the dope-making process may use chilled solvent to form a slurry and wet out the polymer particles before they begin to dissolve, or may use hot solvent so that the solutioning process occurs immediately. Additives such as

Table 8.2. Polymer concentrations suitable for solution spinning

Solvent	Polymer concentration, %
Dimethylformamide (DMF)	20–32
Dimethylacetamide (DMAc)	20–27
Dimethyl sulphoxide (DMSO)	20–30
Ethylene carbonate (EC)	15–18
Sodium thiocyanate (NaSCN), 45–55% in water	10–15
Zinc chloride ($ZnCl_2$) 55–65% in water	8–12
Nitric acid (HNO_3) 65–75% in water	8–12

thermal stabilizers and delusterant (TiO_2) are added at this time. In both cases, active mixing is required. Subsequently, the suspension is pumped through a shell tube heat exchanger to complete dissolution. The resulting dope is degassed and filtered (plate and frame) before being pumped to the spin area.

8.2.4.2   Dry spinning

For acrylic fibres, the only dry spinning solvent used commercially is DMF. The DMF spin dope coming from the dope preparation unit is filtered and then heated to approximately 140°C. It is pumped through spinnerets of up to 2800 holes placed at the top of a solvent removal tower. The DMF is evaporated by circulating an inert gas through the tower at 300–350°C. The tower walls are also heated to prevent any solvent condensation. With such a high boiling solvent (boiling point 153°C), it is not possible (or desirable) to remove all solvent in the tower. Consequently, the fibre from the bottom of the tower contains 10–25% solvent (Fig. 8.7).

In discontinuous processes, the fibre exiting the tower is wet with water and combined with the product from other threadlines into a rope; the rope is

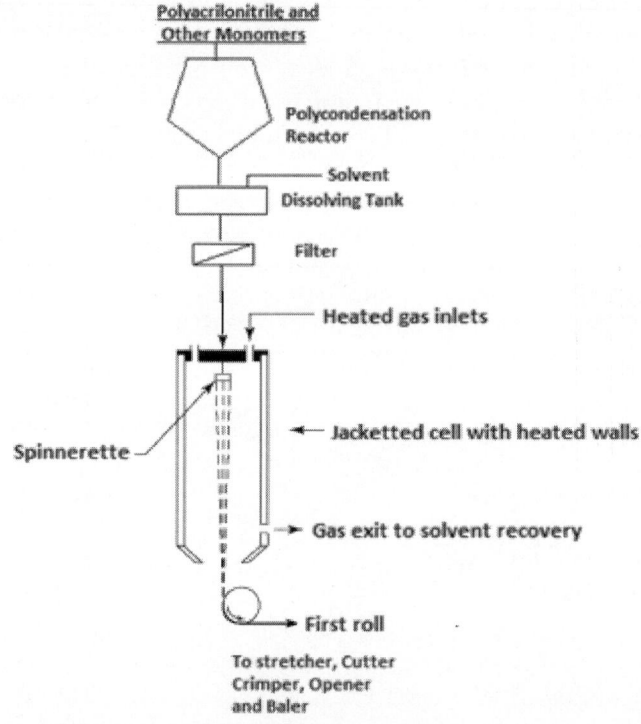

Figure 8.7. Dry spinning

Figure 8.8. Schematic representation of dry spinning

plaited into a can. The residual DMF is removed in a second step by passing the rope via roll sets through a series of hot water baths. A more modern process, introduced by Bayer, washes the fibre by sprays while passing on a belt. The as-spun fibre has little orientation, and so it is stretched three- to six-fold either before or concurrent with the washing step. The fibre is crimped to improve bulk and textile processing, then dried by heated air on a moving belt. During drying, the fibre structure collapses to the same density as solid polymer and the length decreases as the structure relaxes. A 'finish' comprising an antistatic agent and a lubricant are applied by spray or kiss rolls, and the product is either cut to staple or packaged directly as tow. The process steps are shown in Fig. 8.8.

8.2.4.3   Wet spinning

In the wet spinning process, the filament is passed through a bath in which the solvent is washed out of the fibrous material. The solvent trapped in the coagulated filaments is first removed and then the filaments stretched to many times their original length at temperatures of 140 and 180°C. This causes orientation of the crystalline areas of the fibre and improves the strength properties of the fibre markedly. Wet spinning differs from dry spinning primarily in the way solvent is removed from the extruded filaments. The solvent is the same as the dope solvent and the nonsolvent is usually water. Filament fusion is less of a problem in wet spinning, and so the number of capillaries in wet spinning spinnerets is much larger than in dry spinning. The spinnerets in commercial processes may have anywhere from 3000 to 100,000+ capillaries, which may range in diameter from 0.05 to 0.25 mm; it is common to use multiple spinnerets in a single spinbath. The critical part of this process is the transition from a liquid to a solid phase within the filaments. Precipitation is favoured when the solvent is organic and the nonsolvent is water, as the solubility of polymer decreases abruptly with water concentrations of only a few per cent.

The fibre emerging from the spin bath is a highly swollen gel containing both solvent and nonsolvent from the spin bath. The fibres are essentially unrented except at the fibre skin. The microstructure consists of a febrile network. The spaces between fibrils are called micro voids. Depending on the conditions of coagulation, the filaments may also contain large voids radiating out from the centre of the fibre. The best combination of tensile properties, abrasion resistance and fatigue life is realized when the coagulated fibre has a homogeneous, dense structure with small

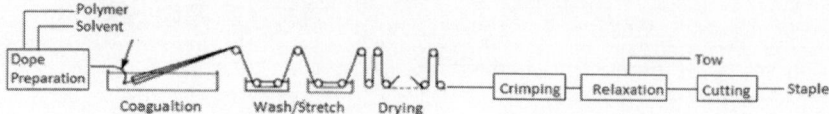

Figure 8.9. Organic solvent wet spinning process

fibrils and no macro voids. Fibre cross-sectional shape is determined by the coagulation conditions.

After the spin-bath or spin-tower step, the tow processing is similar for both wet and dry spun yarns. Wet-spun tows, however, may contain 100–300% of solvent/no solvent. Therefore, the initial washing steps differ in their details. The key wet-spinning steps are washing, stretching, finish application, collapse, drying, crimping and relaxing. The washing step consists of several countercurrent stages, with the effluent being recycled to a solvent recovery process. The washing step may be followed by additional stretching. The porous febrile structure of wet-spun fibres increases in density with stretching. After the fibre is washed, stretched and optionally dyed, finish may be applied using a bath or similar device. If drying is accomplished on heated rolls (see Fig. 8.9), a prettying finish is required to prevent fibre fusion. In other processes (see Fig. 8.10), finish application may be postponed until the fibre is dried and collapsed. After drying/collapsing, the tow is relaxed. Relaxation is essential because it reduces the tendency for fibrillation and increases the dimensional stability of the fibre. Fibre shrinkage during relaxation ranges from 10% to 40% depending on the temperature, the polymer composition and the amount of prior orientation. Fibre crimping using a stuffer box device may be done before in-line relaxation or before autoclaving. The relaxation process tends to 'set' the crimp. Process speeds for wet spinning vary from 55 to 260 m/min. The limitations are the speed at which the fibre can move through the spin bath without filament breakage and the equipment line length required to complete the washing and drying processes. A single machine may have up to 48 spinnerets (six rows of eight) with a total productivity of 50 ton/day.

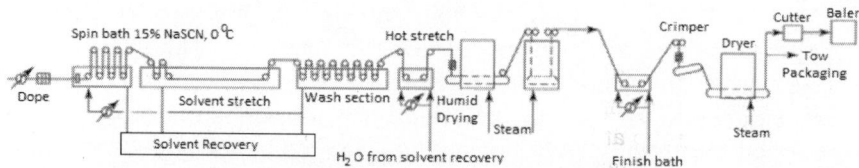

Figure 8.10. NaSCN salt wet spinning process

8.2.4.4   Air gap spinning

Air gap spinning or dry jet spinning is mainly used to produce filament yarn for textile use or to convert to carbon fibres. It is suitable for producing the small bundles required for these end uses because the filament has been drawn before it enters the bath, and so drag forces are less likely to cause breakage; thus, much higher line speeds can be achieved. In theory, any acrylic solvent like DMAc, NaSCN and DMSO can be used. To achieve quick gelation on extrusion, normally the polymer percentage is kept higher. The spinneret is positioned a short distance above the bath, which is a solvent/nonsolvent mixture, typically at low temperature. The fibre is spun vertically into the bath, and then rerouted out via a tube or pulley as shown in Fig. 8.11. Spinnerets may be less than 1000 holes for a textile product or as many as 4000 for a carbon fibre precursor. The remainder of the process resembles the wet-spinning process and the final product is taken up on

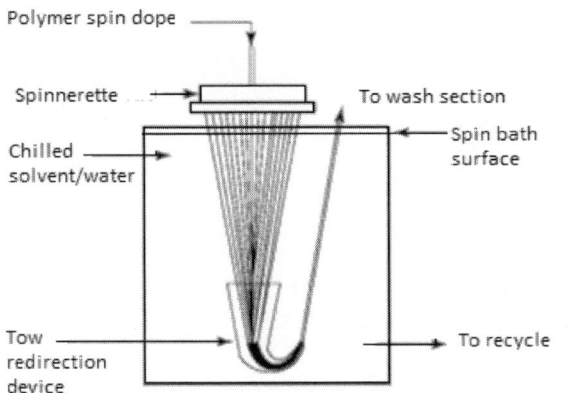

Figure 8.11. Air-gap spinning

bobbins. Final line speeds may be up to 500 m/min, but because of the small bundle size, machine productivity may be only 5 ton/day.

8.2.4.5   Melt spinning

As explained earlier, melt spinning was a problem with PAN fibres. But the lengthy process, solvent recovery and cost has forced manufacturers to standardize a melt spinning process. A true melt spinning process has been developed by a group at Standard Oil Co. Their approach was to make a polymer containing substantial comonomer content by a process, which minimized 'blocking' of acrylonitrile groups. The resultant polymer was melt processible without degradation. Possible limitations of this approach are that the high comonomer content leads to high relaxation shrinkage as well as lower softening and sticking temperatures. These are disadvantages in modern textile processes. Not yet commercially exploited.

8.2.5   Properties of polyacrylonitrile

Acrylic fibre density is lower than other synthetic fibres <1.13 (modified polyacrylonitrile fibres possibly higher: up to 1.37). Looked at as replacement for wool acrylic fibre has full handle, absorbency (below 0.5%), swelling (below 1%), highly resistant against microorganisms and insects including termites and physiologically harmless. Excellent light and weather resistance; heat resistance relatively good (boiling fast, ironing fast up to 150°C), however, shrinks at higher temperatures (approx. 4% with 100% hot air). Low hot-wet modulus gives rise to the danger of slight irreversible distortion or deformation particularly in knitted goods during hot-wet treatments above 60°C. Softening range 190–230°C (modified fibres possibly 150–160°C), becoming sticky at 300–320°C (does not melt, but decomposes). High breaking strength; lower abrasion resistance than polyamide and polyester. Crease reforming good (similar to polyester); breaking strength in bending high (similar to polyamide); elastic stretching and bulk elasticity good. Chemical resistance against acids, oxidants and solvents are good but alkali sensitive. They are washable and quick drying. Dye affinity varies (dyeing of acrylic fibres). Acrylic can be electrostatically charged. Application: sport and work clothing, swimming costumes, rainwear, imitation fur, clothing material, fleece, quilt fillings, curtains, decorative and furniture materials, tarpaulins, technical fabrics (acid protective clothing, filter cloths, leader and back grey cloths). Blended fibre with wool, polyester, polyamide, viscose, copper and acetate fibres are quite common (Table 8.3).

Table 8.3. Modification of acrylic fibres

Modification	Means of modification	Fibre obtained
Chemical modification	By comonomers	Enhanced dissolution and dyeing, Flame resistant, antistatic, improved hydrophilicity. Dyeing simultaneously with acid and basic colours
	By polymer mixture	Graft polymerization, bicomponent fibres, fibres from polymer mixture
	Incorporation of additives	Enhanced whiteness and fibre stability, flame retardant, antisoiling, antistatic, antipilling

Modification	Means of modification	Fibre obtained
Physical modification	At polymerization stage, dope preparation and spinning condition.	High bulk fibres, hollow fibres, change of surface structure, extreme denier fibre

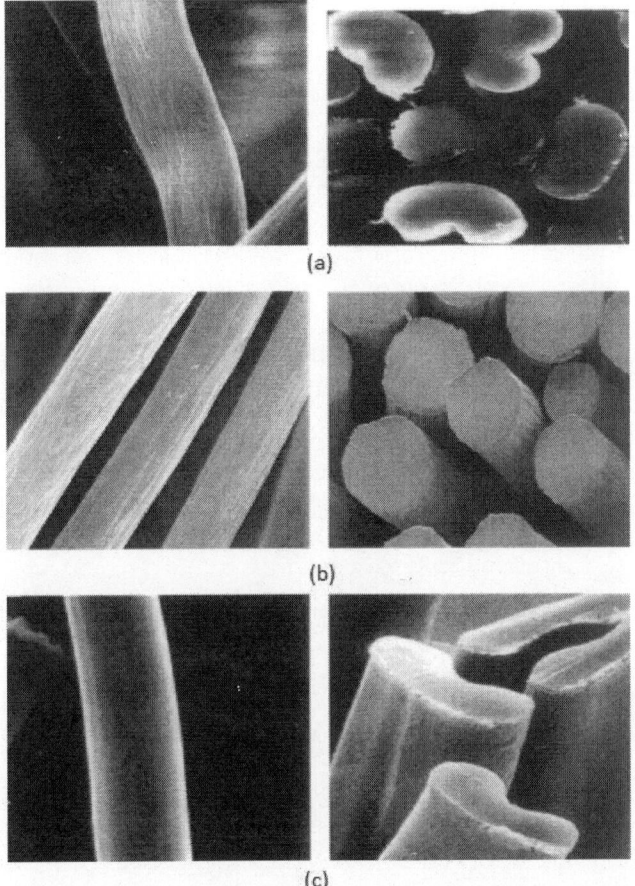

Figure 8.12. Acrylic fibre microscopy comparisons (scale 1 mm = 0.5 μm).
(a) Organic solvent wet spun. (b) Salt solvent set spun.
(c) Organic solvent dry spun

Fig. 8.12 shows lobed (dog-bone shaped) cross section of dry-spun acrylic fibre (LHS) and lobed (kidney shaped) cross section of wet-spun acrylic fibre (RHS).

Nitrogen-free nonionic monomers are used to lower the glass transition temperature and ionic monomers for a further modification, which mainly manifests itself in the uptake of dye. Anionic vinyl compounds, e.g., methacrylic acid or styrene sulphonic acid, raise the saturation values for cationic dyes and thus enable the production of deep shades. The dyeing behaviour of acrylic fibres is changed far more drastically by introduction of basic vinyl compounds such as vinylpyridine or vinylpyrazine. Acrylic fibres which contain basic monomers are also known as cationic fibres which can be dyed with anionic dyes. It is possible to differentiate between anionic- and cationic-modified acrylic fibres depending on the type of monomers contained. The anionic-modified fibres are the usual types on the market.

As it is possible to subdivide the acrylic fibres into two groups which differ from each other in their dyeing behaviour, it is also possible to do so according to the amount of modifying monomers. Differentiation is made between normal acrylic fibres and modacryl fibres. Normal acrylic fibres contain by definition less than 15% modifying monomers and modacryl fibres more than 15%.

Examples of acrylic fibres with anionic nature are Cashmilon, Courtelle, Dralon, Orlon, etc., and with cationic nature are Acrilan 41, Creslan 58, Orlon 28, 44, Vonnel. Examples of modacrylic fibres are Verel, Dynel, etc. (Table 8.4).

Table 8.4. Comparison of properties of acrylic and modacrylic fibres

Property	Polyacrylonitrile	Modacrylic (Verel)
Tensile strength, kg/cm²	Staple: 2100–3150 kg/cm² Filament: 3500–5250 kg/cm²	2940–3290 kg/cm²
Tenacity cN/tex staple or tow	Dry: 17.7–31.8 cN/tex (2.0–3.6 g/den) Wet: 14.1–23.9 cN/tex (1.6–2.7 g/den)	Dry: 15.9–22.1 cN/tex (1.8–2.5 g/den) Wet: 15.0–21.2 cN/tex (1.7–2.4 g/den)
Tenacity cN/tex filament	Dry: 35.3–36.2 cN/tex (4.0–4.1 g/den) Wet: 26.5–33.5 cN/tex (3.0–3.8 g/den)	
Elongation, %	Staple: 20–55 Filament: 30–36	35–40%, dry or wet
Elastic recovery	90–95% at 1% extension; 50–60% at 10% extension	88% at 4%; 55% at 10%
Initial modulus of elasticity, cN/tex	Staple: 353–441 cN/tex (40–50 g/den) Filament: 141–362 cN/tex (16–41 g/den)	
Average toughness	Staple: 0.40–0.70 g/den Filament: 0.22–0.49 g/den	4.33 cN/tex (0.49 g/den)

Property	Polyacrylonitrile	Modacrylic (Verel)
Specific gravity	1.16–1.18	1.37
Moisture regain	1.0–3.0%	3.0–3.5%
Melting point, °C	250	
Softening point	215–255°C	180–185°C
Sticking temperature	230–240°C	
Specific heat	0.27 cal/g °C at 20°C	Resistant dry cleaning agents
Heat setting, °C		200–210
Flammability	Acrylic fibres will burn, but they are not dangerously flammable fibres	Modacrylic fibre has a very good flame resistance
Effect of sunlight	Acrylic fibres have excellent resistance to sunlight	Good weathering characteristics
Effect of acid	Unaffected by dilute solutions of mineral and acids, but they are attacked by concentric acids	Excellent resistance even at high concentrations
Effect of alkali	Dilute alkalis have no effect, but strong alkalis attack the fibre	Under moderate conditions no effect on tenacity, but can cause discolouration
Solvents	Acrylic fibres generally have a good resistance to common organic solvents, including those normally used in dry cleaning	Fibre resists all dry cleaning solvents and most common organic solvents. It dissolves in warm acetone
Cross section (enlarged)		
Fibre (enlarged)		

The modacrylic fibres have similar properties to those of acrylics and are flame resistant. Mostly this fibre is based on a 60/40 or 50/50 copolymer of acrylonitrile with vinylidine chloride ($CH_2=CCl_2$) together with small proportion of ternary monomer to improve ionic dyeability or hydrophilicity. The better known modacrylic fibres have a ribbon-shaped or peanut-shaped cross section. One problem encountered with modacrylic fibre is loss of lustre at the boil. This fibre is used for apparel, home furnishing, wigs, etc.

Polyacrylonitrile is at present the only purely carbon chain polymer besides polyvinyl chloride, polyvinyl alcohol and polypropylene suitable for the production of fibres. It can be considered as a high-molecular hydrocarbon substituted with cyano groups. This explains the excellent stability of acrylic fibres to acids, caustic, bacteria and light and its rapid drying properties as a result of its hydrophobic nature (Table 8.5). Further properties exhibited by textiles of acrylic fibres are the wool-like, soft handle, the good heat-retaining properties, the reduced tendency to creasing and shrinking, the excellent resiliency and the low specific weight. Owing to these properties, acrylic fibres are particularly suitable for the production of pullovers, curtains, hangings, textiles for home use, canvas awnings, carpets, blankets, laundered goods and, in blends with wool, polyester and cellulosic fibres, for the production of shirting and ladies' and men's outerwear material (Fig. 8.13).

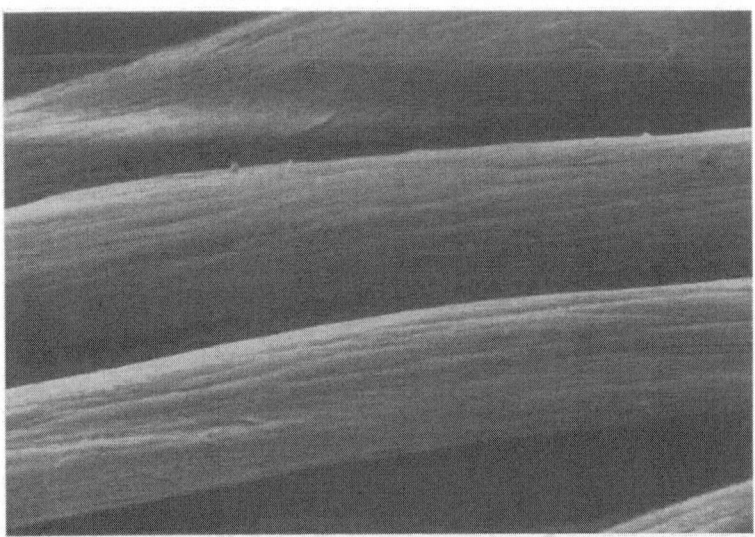

Figure 8.13. Electron microscopic view of acrylic fibre (×1300)

Table 8.5. Comparison of properties of staple acrylic fibres with other staple fibres

Property	Acrylic	Modacrylic	Nylon 66	Polyester	Polyolefin	Cotton	Wool
Specific gravity	1.14–1.19	1.28–1.37	1.14	1.38	0.9–1	1.54	1.28–1.32
Tenacity (N/tex)							
Dry	0.09–0.33	0.13–0.25	0.26–0.64	0.31–0.53	0.31–0.40	0.18–0.44	0.09–0.15
Wet	0.14–0.24	0.11–0.23	0.22–0.54	0.31–0.53	0.31–0.40	0.21–0.53	0.07–0.14
Average modulus N/tex dry elastic recovery % – 10% stretch		95	99	57–74	96		
Elec. resistance	High	High	Very high	High	High	Low	Low
Static build up	Moderate	Moderate	Very high	High	High	Low	Low
Flammability	Moderate	Low	Self extinguishing	Moderate	Moderate	Ignition at 360°C	Self extingushing
LOI	0.18	0.27	0.2	0.21		0.18	0.25
Char/melt	Melts	Melts	Melts, drips	Melts, drips	Melts	Chars	Chars
Resistance to sunlight	Excellent	Excellent	Poor; must be stabilized	Good	Poor; must be stabilized	Fair; degrades	Fair; degrades
Resistance to chemical attack	Excellent	Excellent	Good	Good	Excellent	Attacked by acids	Affected by alkalis, oxidizing reducing agents
Abrasion resistance	Moderate	Moderate	Very good	Very good	Excellent	Good	Moderate
Index of birefringence	0.1		0.6	0.16			0.01
Moisture regain 65% RH 21°C, %	1.5–2.5	1.5–3.5	4–5	0.1–0.2	0	7–8	13–15

8.3   Polyester

Polyester is one of the great man-made fibre discoveries of the forties and has been applied on many various scales all over the world. It has long played a dominant role in our life activities. The first polyester fibre was Terylene. In 1946, second polyester fibre prepared was Dacron. In 1958, another polyester fibre called Kodel was developed by Eastman. Today, polyester is still widely regarded as a 'cheap, uncomfortable, fibre, but even now this image is slowly beginning to change with the emergence of polyester luxury fibres such as polyester microfibre.

Polyesters are defined as polymers containing at least one ester linking group per repeating unit. The term 'polyester' is most commonly used to refer to polyethylene terephthalate (PET). Polyester fibre is a 'manufactured fibre' in which the fibre forming substance is any long chain synthetic polymer composed at least 85% by weight of an ester of a dihydric alcohol (HOROH) and terephthalic acid (p-HOOC–C_6H_4COOH).

Polyester (polyester terephthalate) is made by the condensation of terephthalic acid (or a derivative like dimethyl terephthalate – DMT) and ethylene glycol.

The DMT can be made by the reaction of terephthalic acid with methyl alcohol.

Based on their chemical structures polyester can be group in to three types:

1. PET polyester fibres
2. Poly-1,4-cyclohexylene-dimethylene terephthalate fibres (PCDT polyester fibres)
3. Other type of polyester fibres

8.3.1   Manufacture

As explained above, the raw material required for the manufacture of poly-ester are ethylene glycol, terephthalic acid or DMT. Ethylene glycol can be synthesized starting from ethylene which is a petroleum cracking product. Catalytic oxidation of ethylene gives ethylene oxide, and hydration of ethyl-ene oxide gives ethylene glycol. Terephthalic acid (PTA), produced directly from p-xylene with bromide-controlled oxidation (air or nitric acid). Tereph-thalic acid on reaction with methyl alcohol produces DMT. Also, DMT made in the early stages by esterification of terephthalic acid. However, nowadays a different process involving two oxidation and esterification is used for man-ufacturing most DMT.

$$CH_2 = CH_2 \xrightarrow[\text{Catalyst}]{\text{Oxidation}} CH_2 \overset{O}{\diagup\diagdown} CH_2 \xrightarrow{\text{Hydration}} CH_2OH - CH_2OH$$

Ethylene   Ethylene oxide   Ethylene glycol

P - Xylene $\xrightarrow{\text{Nitric acid}}$ Terephthalic acid $+ CH_3OH \longrightarrow$ Dimethyl Terephthalate

The general manufacturing procedure is as follows, but the process vary with the manufacturers.

8.3.2   Polymerization

The PET is a condensation polymer and is industrially produced by either terephthalic acid or DMT with ethylene glycol in two stages. First as a result of ester interchange, diglycol terephthalate is obtained:

$$CH_3 - O - CO - \langle C_6H_4 \rangle - CO - CH_3 + 2HO - (CH_2)_2 - OH \longrightarrow$$

$$\longrightarrow HO - (CH_2)_2 - O - CO - \langle C_6H_4 \rangle - CO - O - (CH_2)_2 - OH + 2CH_3OH$$

and then follows the polycondensation reaction:

$$n\,HO-(CH_2)_2-O-CO-\hspace{-0.5em}\left\langle\right\rangle\hspace{-0.5em}-CO-O-(CH_2)_2-OH \rightarrow$$

$$\rightarrow \left[-CO-\hspace{-0.5em}\left\langle\right\rangle\hspace{-0.5em}-CO-O-(CH_2)_2-O-\right]_n + (n-1)HO-(CH_2)_2-OH$$

The resulting polymer has a molecular weight of 15,000–20,000 and a DP of 85–120. It is characterized by considerable polydispersion, rigidity of its chain and its crystallinity. When the desired DP has been reached, the clear, colourless polyester is extruded through a slot on to a casting wheel. The polymer solidifies into an endless ribbon, which is fed to a cutter and cut into chips in the form of cubes with 3–6 mm (1/8–1/4 in) sides.

8.3.3   Spinning

Polyester is a 'melt spun' fibre, which means that it is heated, extruded through the spinnerets, and cools upon hitting the air. From there it is wound around cylinders. PET melts at about 260°C, and the molten polymer is stable so long as oxygen is rigorously excluded. Every care is taken during melt spinning, as in the polymerization process, to prevent air coming into contact with the molten polymer. The spinning process is almost like in case of nylon spinning which is explained in detail in this book. As the filaments emerge, they solidify and are wound into packages of undrawn yarn.

8.3.4   Drawing

To produce uniform PET, the drawing process is carried out at temperature above the glass transition temperature (80–90°C) with draw ratios (3:1–6:1) lengthwise on draw-twist machines, the stretching being carried out usually at elevated temperature. If high tenacity yarn is being made, the filaments are drawn to a higher degree than in the manufacture of regular tenacity yarn. As a rule polyester yarn has to be drawn hot, as this gives better results, however sometimes monofilaments are drawn cold.

For staple fibre, many filaments are spun together as a tow, which is drawn, crimped and set. Then it is cut into desired length and used for blending or to for spun yarn (Fig. 8.14, Table 8.6).

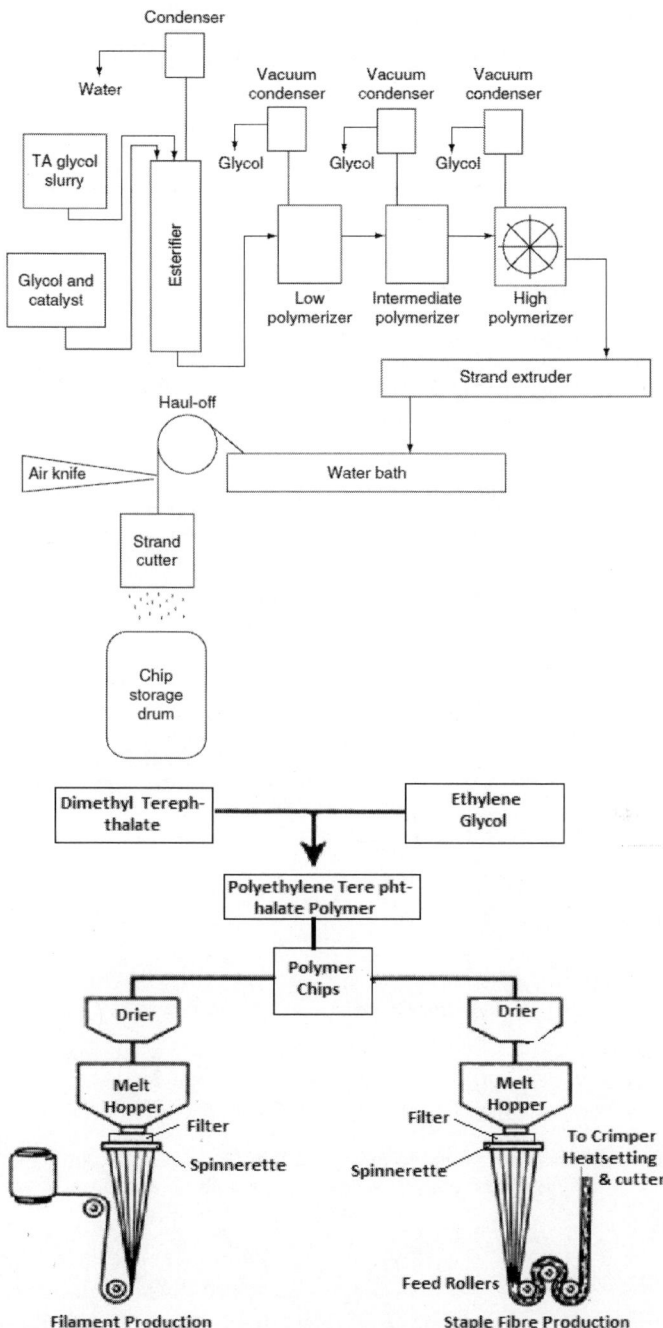

Figure 8.14. Flow diagram of polyester manufacturing

Table 8.6. Properties of polyester

Property	Polyester filament	Polyester staple
Tensile strength, kg/cm^2	High Tenacity: 7350–8750 Medium Tenacity: 4900–5950	High Tenacity: 5250–7350 Medium Tenacity: 4900–5950
Tenacity, cN/tex	High Tenacity dry: 56.5–70.6 Medium Tenacity dry: 35.3–44.1	High Tenacity dry: 48.6–57.4 Medium Tenacity dry: 35.3–44.1
Elongation, %	High Tenacity: 8–11 Medium Tenacity: 15–30	High Tenacity: 20–30 Medium Tenacity: 30–50
Elastic recovery	Good	Good
Initial modulus of elasticity, cN/tex	High Tenacity: 971–1148 Medium Tenacity: 883–1015	High Tenacity: 706 Medium Tenacity: 265–530
Average toughness g-cm/ denier-cm	High Tenacity: 0.325 Medium Tenacity: 0.500	High Tenacity: 0.610
Specific gravity	1.38	1.38
Moisture regain	0.4	0.4
Melting point, °C	265	265
Softening point	260°C (approx.)	260°C (approx.)
Sticking temperature	230–240°C	230–240°C
Specific heat	0.27 cal/g °C at 20°C.	Resistant dry cleaning agents
Heat setting, °C	200–210	200–210
Flammability	On ignition, it fuses and form beads	On ignition, it fuses and form beads
Thermal conductivity	2x 10^{-4} cal/g cm °C	
Effect of sunlight	PE has a high resistance to degradation by light	PE has a high resistance to degradation by light
Effect of acid	Good resistance to the majority of mineral and organic acids	Good resistance to the majority of mineral and organic acids
Effect of alkali	PE withstand the conditions encountered in mercerizing and in dyeing	PE withstand the conditions encountered in mercerizing and in dyeing

Property	Polyester filament	Polyester staple
Solvents	Solvents: mono-, di- and tri-chloroacetic acid, ortho-chlorophenol, tetrachloroethane (146°C), cyclohexanone (155°C), benzyl alcohol (205°C), nitrobenzene (210°C), naphthalene (218°C), diphenyl (254°C) and dimethyl phthalate (282°C)	Solvents: mono-, di- and tri-chloroacetic acid, ortho-chlorophenol, tetrachloroethane (146°C), cyclohexanone (155°C), benzyl alcohol (205°C), nitrobenzene (210°C), naphthalene (218°C), diphenyl (254°C) and dimethyl phthalate (282°C)

8.3.5   PCDT polyester

The PCDT polyester is made by condensation of terephathalic acid and 1,4-cyclohexanedimethanol.

1,4 - Cyclohexanedimethanol   Terephthalic Acid   Poly 1,4 - Cyclohexylene-dimethylene terephthalate

The two raw materials are mixed and heated to 200°C in the presence of a catalyst. As the condensation takes place, the methyl alcohol is evaporated and gradually the temperature is raised to 300°C and same time vacuum is applied. The reaction is allowed to continue till the polymer has achieved the desired molecular weight.

Spinning is done the same way as polyester. Molten polymer is pumped, and as it is passed through the spinneret, the filament meets the cold air where it solidifies. These filaments are further drawn to 4.5–5 time in length at 120°C.

8.3.6   Modifications on the basic polyester

The chemical modifications of polyester is mainly intended to overcome some of the inherent drawbacks of the fibre such as low dyeability, low moisture regain, static accumulation, soiling tendency, pilling property and flammability.

Fibres of different properties can be created by doing one or more of the following:

1. Adding a delustrant (dulling agent): Polyester is a naturally bright fibre, but can be made dull or semi-dull by the addition of a TiO_2.

2. Drawing out the fibre to five times its original length is normal, but polyester can be stretched even further. Drawing it out more than normal may also affect the strength, elasticity and dyeability.

3. Adding dye stuffs: In its natural state, polyester is a slightly transparent off-white. Adding dye stuffs at the manufacturing stage can create brilliant colours like electric blue and atomic red.

4. Crimping: When the fibre is drawn out it is long and smooth. Crimping can give the fibre more texture and bulk and can increase its insulation properties, as well as its elasticity. Another word for this is texturizing.

5. Spin finishes: Fibres need to be treated with surface finishes or lubricants to allow high-speed processing. The various processing steps such as drawing, bulking and textile processing would be impossible without these spin finishes because so many of them rely on specific frictional properties of the fibre (e.g., friction twisting). Spin finishes are often water emulsions of various surface-active agents and lubricant oils; their formulation is a complex process and sometimes more of an art as well as a science. Finish application is made early in the process, before the cooling thread line from the spinner hits the first godet. Earlier, finish was applied from a lick roll rotating slowly in a bath. As spinning speeds increased, the finish was applied directly via a special hollow ceramic yarn guide as a neat oil formulation and metered at precise levels via a metering pump. Staple fibre is sprayed with emulsified finish or the whole tow may be immersed in large baths of finish. Some staple processes use a draw stage in a hot bath of finish.

8.3.7   Anti-pill polyester

Pilling is a problem inherent to polyester staple, rather many of the synthetic staple fibre and it is common to all staple fibres, particularly if the level of yarn twist is low, so that the fibre has many loose ends. It has been found that if the pill in formed and it is fallen of it does not pose much problem. The falling of the balls formed is related to the strength of the fibre (tenacity of polyester is ca. 5 g/decitex). This was very apparent in the polyester wool fabrics which was a blend with instant success of the various blends introduced with polyester after the introduction of the polyester in 1950s. However, consumers soon noticed an annoying problem. It was the formation of small fuzzy balls (called 'pills') on the surface of fabrics. The pills rub off harmlessly with wool because wool is a weak fibre. However, PET is a strong fibre and therefore pills do not rub off; instead, they cling and have a negative impact on fabric aesthetics.

To reduce pilling, the fibre is made weaker by different methods. These do not pill so obviously because the pills break away. If the melt viscosity is made so low, the molten polymer becomes fluid so that the process became unstable. A method had to be found to raise the effective melt viscosity of the polymer while maintaining the low-pill properties to give an acceptable melt-spinning process. The method adopted was to introduce branching points into the polymer chain by adding a multifunctional component (either a polyacid or a polyhydric alcohol) so as to produce a star-branched polymer. Such polymers are known to have higher melt viscosities. The branching agent added (ca. 1 mol%) was usually pentaerythritol. Too much additive would lead to gel formation by forming cross-linked networks, but this is not a problem at low levels.

8.3.8   Multilobal cross-sectional polyester

As with usual synthetic fibres, polyester is also spun such a way as to get multilobal cross section there by increasing the surface area, etc., and as a result changes in the property of the fibre.

8.3.9   Chemical modifications

Fig 8.15 shows some of the chemical modification in the polymer to get better properties than the basic polyester. Kodel II is based on 1,4-dimethylcyclohexane terephthalate and gives increased stability to hydrolysis. A-Tell is polyethyleneoxybenzoate (PEB) and has a silk-like handle and drape, good wrinkle resistance and easy care properties. The self polymer of poly (pivalolactone) has better resistance to hydrolysis and a much higher melting point than the original aliphatic polyester. Polybutylene terephthalate (PBT) is a carrier free dyeable polyester with lower glass-transition temperature (Tg). Vycron (USA) is a modified polyester where terephthalic acid is partly reduced by isophthalic acid to open up the compact structure. Sulphonic acid group is introduced additionally into the isophthalic acid to dye with basic dye (CDPET). Similarly anionic dyeable polyester is also introduced containing nitrogenous compounds (polyamines) which offer basic group as sites for the adsorption of acid dyes.

8.3.10   Properties

8.3.10.1   *Physical properties*

Polyester is strong, resistant to stretching and shrinking, resistant to most chemicals, quick drying, crisp and resilient when wet or dry, wrinkle resistant, mildew resistant, abrasion resistant, able to retain heat-set pleats and creases and easily washed (Fig. 8.15).

Repeating Unit	Modified Fibre
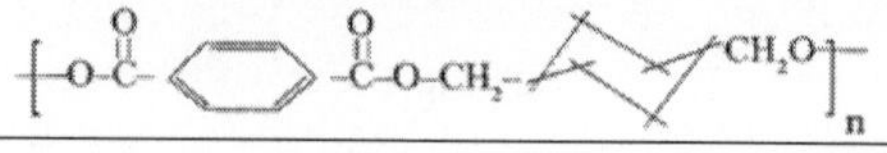	Kodel III (Kodak)
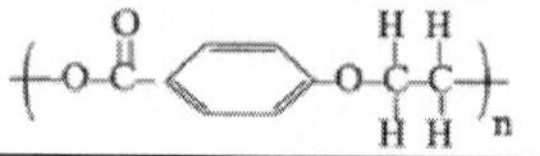	A-Tel (Japan)
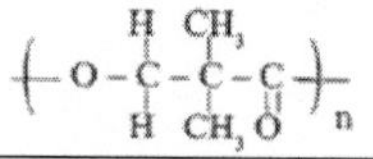	Poly(pivalolactone)
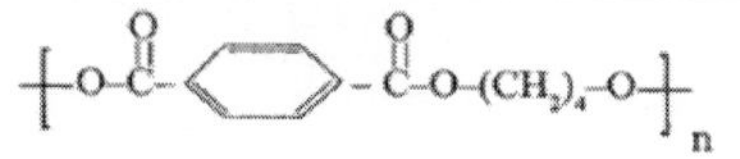	Polybutylene terephthalate
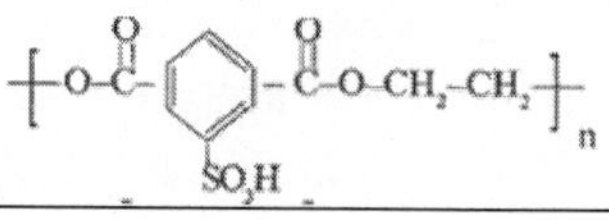	Cationic Dyeable Polyester (CDPET)
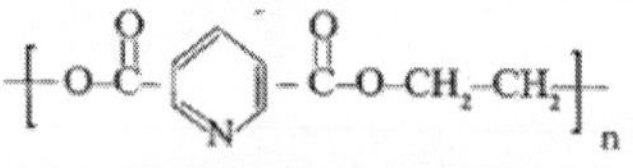	Anionic Dyeable Polyester

Figure 8.15. Repeating units of some modified polyester

8.3.10.2   Chemical properties

The fibre has a well-ordered internal structure; interaction between the greatly extended macromolecules is attributed to van der Waals' forces and, according to the assumptions of certain authors, to hydrogen bonds between the oxygen atoms of the ester or carboxyl groups and the hydrogen in the benzene ring. Due to its compact structure and to the absence of hydrophilic groups (since hydroxyl and carboxyl groups may be present in macromolecules only as terminal groups), the fibre acquires hydrophobic properties. The hygroscopicity of the fibre at a 65% relative humidity is 0.4% and at 100% relative humidity, it is 0.6–0.8%. Due to its hydrophobic properties, the fibre does not swell in aqueous media and this has an unfavourable effect on the dye ability of these fibres; moreover, strong static charges are formed on polyester fibre which creates difficulties during mechanical treatment of polyester and exploitation of goods containing this fibre.

Polyester fibres have good resistance to weak mineral acids, even at boiling temperature, and to most strong acids at room temperature. Hydrolysis is highly dependent on temperature. Thus, conventional PET fibres soaked in water at 70°C for several weeks do not show a measurable loss in strength, but after 1 week at 100°C, the strength is reduced by approximately 20%. Polyesters are highly sensitive to bases such as sodium hydroxide and methylamine, which serve as catalysts in the hydrolysis reaction. Alkaline attack is sometimes used to modify the fabric aesthetics during the finishing process. The porous structures produced on the fibre surface by this technique contribute to higher wettability and better wear properties. Polyester displays excellent resistance to oxidizing agents, such as conventional textile bleaches, and is resistant to cleaning solvents and surfactants. Concentrated solutions of benzoic acid and o-phenylphenol have a swelling effect. PET is both hydrophobic and oleophilic. The hydrophobic nature imparts water repellency and rapid drying. But because of the oleophilic property, removal of oil stains is difficult. Polyester fibres have a low moisture regain of around 0.4%, which contributes to good electrical insulating properties even at high temperatures. The tensile properties of the wet fibre are similar to those of dry fibre. The low moisture content, however, can lead to static problems that affect fabric processing and soiling. PET has optical characteristics of many thermoplastics, providing bright, shiny effects desirable for some end uses, such as silk-like apparel. Some polyester microfibre with a linear density of less than 1.0 denier per filament (dpf), achieves the feel and lustre of natural silk. Because of its rigid structure, well-developed crystallinity and lack of reactive dyesites, PET absorbs very little dye in conventional dye systems. Polyester fibres are therefore dyed almost exclusively with disperse dyes. Polymerizing a third monomer, such as dimethyl ester, has successfully produced a cationic dyeable polyester fibre into the macro-molecular chain. The third monomer makes the structure of cationic dyeable polyester less compact than that of normal PET fibres. The disturbed structure is good for the penetration of dyes into the fibre.

But disadvantage of adding a third monomer is the decrease of the tensile strength. Polyester fibres display good resistance to sunlight but long-term degradation appears to be initiated by UV radiation. Although PET is flammable, the fabric usually melts and drops away instead of spreading the flame. Polyester has good oxidative and thermal resistance. The resistance of polyester fibres to mildew, aging and abrasion is excellent. Moulds, mildew and fungus may grow on some of the lubricants or finishes, but do not attack the fibre. Dyes, they are thermoplastic, have good strength and are hydrophobic. The fibre has a rod-like shape with a smooth surface, it is lustrous and its hand is crisp. It has excellent resiliency and is the best wash and wear fabric.

8.3.11   Noncircular cross-section fibres

Changing the shape of the holes in the spinneret: the simplest and most common shape is a circle, but by changing the shape of the spinneret, square, oval and bean-shaped fibres can be formed. Fortunately, a melt-spun fibre lends itself NCCS (noncircular cross section) well to the production of (NCCS) fibres by varying the shape of spinneret orifice, provided the melt viscosity is high enough so that surface tension does not cause the filament to resume a circular shape. Since the holes had to be very small (about 0.015 in overall), machining a multiplicity of holes at a uniform size and shape was a major engineering problem, particularly in the hard metal alloys used for spinneret plates. Laser etching is one technique used. A hole shaped like a T gave tri-lobal filaments. In the pioneering days, much of this work was entirely heu-ristic, but gradually emerged some rules of thumb. Multilobed yarn cross sections (trilobal and octalobal) can give quite different appearances. Trilobal is glittery as the incident light reflects off the fibre surface, while octalobal gives an opaque matte effect, as the light is effectively absorbed by multi-ple reflections from the many acute angles. Sharp-edged filaments have the prized rustle and high frictional characteristics of pure silk, where it is called 'scroop'. Flat rectangular filaments give fabrics an unpleasant 'slimy' handle. Gradually, these principles were applied to commercial yarns, and many fil-ament yarns for the apparel and BCF carpet markets now use NCCS fibres.

One can even create a hollow fibre. The different shapes affect the hand and strength of the fibre. For example, the cross section of the regular pol-yester is round, of Dacron (Du point) is trilobal, of Fortel and Encron are T-shape, of Trevira (Hoechst) is pentalobal and of Kodel is trilateral. Changes that occur during the cross section produce fibres with different hand and appearance.

8.3.12   Antistatic and antisoiling polyester

Antisoiling and antistatic are related because the origins of the problems are interrelated. Synthetic fibres in general, and PET in particular, are hydropho-bic materials – PET has a moisture regain of 0.4% at 60% RH. PET fibres are difficult to wet and rapidly build up static electrical charges by friction because as water effectively leaks away, voltage is produced. Static charges also lead to attraction of dust and dirt. To avoid these problems, the moisture uptake of the polyester should be increased by combining it with hydrophilic materials that are wash fast. One additive that has been used repeatedly is polyethylene oxide (PEO), a stable, functional, highly hydrophilic, water sol-uble and humectant polymer:

$$HO–CH_2CH_2O(CH_2CH_2O–)_n–H$$

The molecular weight can be from a few hundred to many millions. Co-polymers of PET with PEO having a molecular weight of approximately 500–2000 Da were made, and it was possible to incorporate permanently enough PEG without drastic reduction of the PET properties to greatly improve the fibre moisture uptake, but at the expense of severe reduction in the light stability of dyed fibres. Other processes used a PET/PEO block copolymer in aqueous dispersion that was padded and baked onto the fibre as a textile finish. This relied on cocrystallization of the PET segments with the polymer to make the treatment wash fast. The most satisfactory technique is probably to make a bicomponent fibre with a thin coating of a PET/PEO copolymer on a PET core in a core–sheath configuration. This does not affect fibre properties and minimizes the light fastness issue.

8.3.12.1   Biodegradable polyester

Biodegradable polyester research has its origin from the development of surgical sutures, which slowly disappear in vivo and do not need subsequent surgical removal. The first commercial samples were introduced in the early 1970s by Ethicon Corporation. These sutures were monofil fibres spun from a copolymer of glycolic acid and d-lactic acid. Such aliphatic hydroxy acids are completely biocompatible and harmless. The properties of polyglycolide and stereochemically pure d- or l-polylactide polymers are quite good, and they form strong, highly crystalline fibres by melt spinning. Synthetic lactones such as e-caprolactone and 2-dioxanone have been copolymerized with glycolide and lactide. ICI began working on poly (3-hydroxybutyric acid) in the 1970s and later developed a copolymer with 3-hydroxyvaleric acid. Both polyhydroxyacids are stereochemically pure and give crystalline polymers, which can be processed into fibres and films. The interesting feature of these polymers is that they are made in very high molecular weight form by bacteria. Certain microorganisms, when cultivated and starved of nitrogen sources, synthesize aliphatic polyesters instead of proteins. The number average molecular weight of the as-harvested polymer can be several million daltons and it must be reduced to allow the polymer to be processed and fabricated. ICI (now Astra-Zeneca) first developed 'Biopol' as one product and although others have been introduced by different companies, little has been targeted towards fibre end-use. All the polyhydroxyacids are unstable and degrade on exposure or composting, but the degradation rate is very much governed by the ratio of hydrophobic/hydrophilic properties. While hydrolysis is important, catalyzed degradation by various lipases is also a factor.

8.3.13   Polyester ethers fibre (A-TELL)

Introduced by Nippon Rayon Corpn, Japan A-TELL is a polyester ether fibres. It is produced in filament form only. It has a very good lustre and low moisture retension (approx. 0.05%). Because of its high lustre, it finds use in ties and high end party wear.

8.4   Polyamides

Polyamides are polymers, which contain recurring amide groups as integral parts of the main polymer chains. Naturally occurring polyamides include the protein fibres, e.g., silk and wool. Synthetic polyamide fibres form one of the most important of all classes of textile fibre, which we know today as nylon. Nylon is one of the most common polymers used as a fibre. There are several forms of nylon depending upon chemical synthesis such as nylon 4, 6, 6.6, 6.10, 6.12, 8, 10 and 11.

1. Polycaprolactam fibres produced from polycaprolactam is generally known as nylon 6. Fibres of this group are produced according to the following equation:

$$nNH-(CH_2)_5-CO \xrightarrow{+H_2O} -+H[-NH-(CH_2)_5-CO-]_nOH$$

2. Polyhexamethylene–adipamide fibre obtained by polycondensation on of hexamethylenediamine and adipic acid, and each component having six carbon atoms, hence the name nylon 66.

The following polyamide fibres are also obtained by polycondensation:

(a) polyaminoenant fibre from aminoenant acid; called nylon 7:

$$n[H_2N-(CH_2)_6-COOH] \rightarrow H[-NH-(CH_2)_6-CO]_n OH + (n-1) H_2O$$

(b) polyhexamethylene sebacinamide fibre from hexamethylenediamine and sebacic acid; nylon 6, 10:

$$nH_2N-(CH_2)_6-NH_2 + nHOOC-(CH_2)_8COOH \rightarrow$$
$$H[-NH-(CH_2)_6-NH-CO-(CH_2)_8-CO]_nOH + (n-1)H_2O$$

(c) polyaminoundecanoic fibre from aminoundecanoic acid; called nylon 11:

$$H_2N(CH_2)_1COOH \rightarrow H[-NH-(CH_2)_1CO]OH + (n-1)H_2O$$

and some others.

Thus, the chemical structure of these fibres is characterized by the presence in the macromolecules of amido groups: –NH–CO–, which link together separate parts of chains consisting of methylene groups, of the general formula:

$$-NH-(CH_2)_x-CO-NH-(CH_2)_x-CO-NH-(CH_2)_x-CO- \ldots$$

or the formula

$$-NH(CH_2)_x-NH-CO-(CH_2)_y-CO-NH(CH_2)_x-CO-NH-(CH_2)_y-CO...$$

Polyamide fibres exhibit well-ordered crystalline regions separated by regions of disorder (crystallinity). Extended macromolecules are interconnected in the crystalline regions by intermolecular van der Waals' forces and hydrogen bonds. Hydrogen bonds are formed between the >NH and >C–O groups of adjacent chains and play a particularly important role in the stabilization of the supermolecular structure of polyamide fibres. The structure of polyamide fibres, like that of viscose fibres, may not be homogeneous, i.e., there may be a surface layer of higher orientation (an oriented shell). The physical and mechanical properties of polyamide fibres depend on their peculiar structure. These fibres exhibit a great breaking

Repeating Unit	Fibre
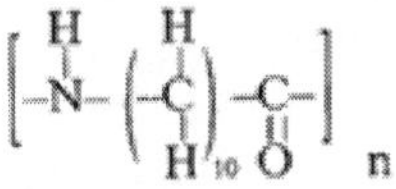	Nylon 11
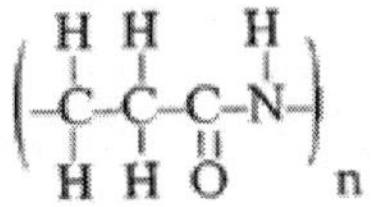	Nylon 3
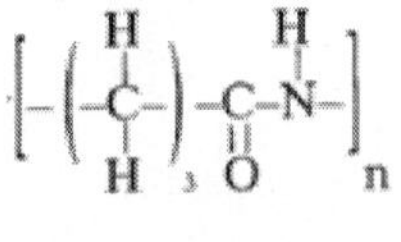	Nylon 4
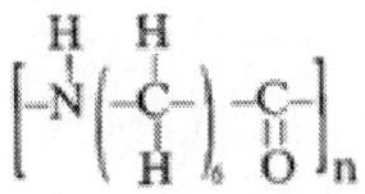	Nylon 7
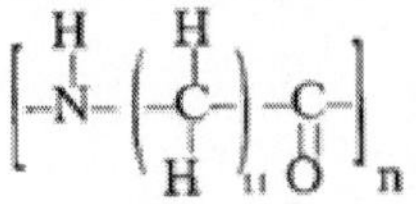	Nylon 12
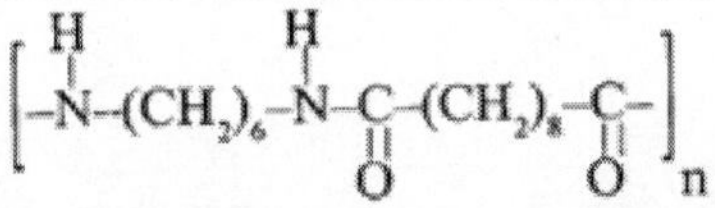	Nylon 6, 10

Figure 8.16. Different nylons and their structure

length (40–50 km), high elasticity (complete recovery amounts to 35–40% of the total elongation); the recovery of nylon 66 is equal to 100% after stretching it by 8%. The high abrasion resistance of polyamide fibres is an extremely valuable property in practical use. The abrasion resistance of polyamide fibres excels that of natural and other man-made fibres. The behaviour of polyamide fibres in water is somewhat specific. Swelling of polyamide fibres in water is not considerable and their strength in the wet state is very slightly reduced (by 5–10%). At a relative humidity of 65%, these fibres absorb 3.5–4% of moisture. This is probably due to the low content of hydrophilic groups. Polyamides belong to thermoplastic polymers and melt without decomposition: nylon 6 at 215°C, nylon 66 at 255°C nylon 7 at 225°C and nylon 11 at 186–187°C. As compared with nylon 6, nylon 66 melts at a considerably higher temperature, which is due to its greater crystallinity and to the larger number of hydrogen bonds between amido groups (Fig. 8.16).

Nylon is found in clothing all the time, but also in other places, in the form of a thermoplastic material. Nylons are also called polyamides, because of the characteristic amide groups in the backbone chain. These amide groups are very polar and are linked with each other with hydrogen bonds. Nylon is a regular and symmetrical fibre with crystalline regions and make very fibres. The fibre has a smooth rod-like shape with a smooth surface.

8.4.1   Nylon 6

Nylon 6 can be produced by the self-condensation of an amino acid, or a derivative such as a lactam:

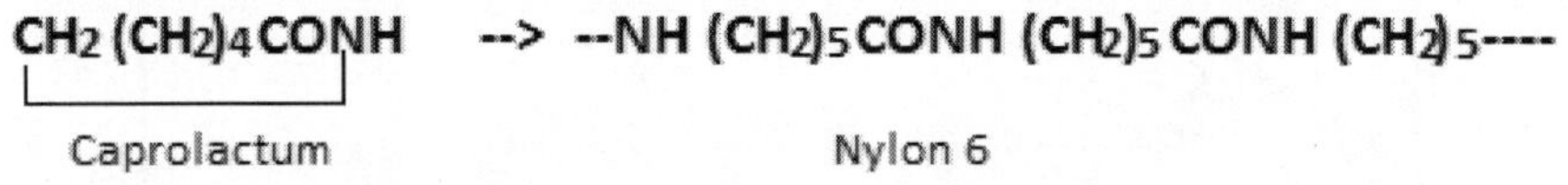

$$CH_2(CH_2)_4CONH \quad \rightarrow \quad --NH(CH_2)_5CONH(CH_2)_5CONH(CH_2)_5----$$

Caprolactum   Nylon 6

The average DP and the molecular weight for nylon 6 are 150–200 and 15,000–20,000, respectively, and elementary unit is $-NH(CH_2)_5CONH(CH_2)_5CO-$ as shown above.

8.4.2   Nylon 66

Nylon 66 is produced by the interaction of a diamine and a dibasic acid, e.g., hexamethylene diamine and adipic acid:

$NH_2(CH_2)_6NH_2 + HOOC(CH_2)_4COOH \longrightarrow --CO\ NH\ (CH_2)_6NH\ CO\ (CH_2)_4CO\ NH\ (CH_2)_6NH\ CO\ (CH_2)_4CO--$

Hexamethylene   Adipic Acid   Nylon 6.6 (Polyhexamethylene Adipamide)
diamine

The average DP and molecular weight of nylon 66 is 80–100 (assuming that the elementary unit is $-CO-(CH_2)_4-CO-NH-\ (CH_2)_6-NH-)$ and 12,000–20,000, respectively. It should be borne in mind, however, that polyamides present considerable polydispersion. Nylon 6 and nylon 66 together account for almost the entire production of polyamide fibres. The versatility of nylon 6 and nylon 66 has increased steadily over the years as research has introduced all manner of modifications and improvements to the basic types of fibre. The production of trilobal filaments and textured yarns in comparatively recent times, e.g., has opened up vast new fields of application for nylon fibres. The chemical structures of nylon 6 and nylon 66 are virtually identical, differing only in the arrangement of the atoms in the amide groups. The melting point of nylon 6 is 215°C and nylon 66 is 250°C. Nylon 6 has a greater affinity for certain dyestuffs than nylon 66. Nylon 6 is claimed to have better elastic recovery and fatigue resistance than nylon 66.

Nylon 6 filaments blend more readily than those of nylon 66.

8.4.3   Manufacture of nylon 66

8.4.3.1   Starting from cyclohexanol

Most of the synthetic fibres starting point are petroleum products. In case of nylon, it can be started from benzene which is distilled from coal tar or petroleum. Benzene is converted to phenol. (1) The phenol is reduced to cyclohexanol by hydrogenation in the presence of a catalyst (A). (2) The cyclohexanol can be made starting from benzene, by first reducing to cyclohexane (B) and then the latter is then oxidized by air in the presence of catalyst (C), forming a mixture of cyclohexanol and cyclohexanone:

1. Phenol $\xrightarrow{A}$ Cyclohexanol

2. Benzene $\xrightarrow{B}$ Cyclohexane $\xrightarrow{C}$ Cyclohexanol + Cyclohexanone

3.

$$\text{Cyclohexanone} \xrightarrow{D} \underset{\text{Adipic Acid}}{HOOC\,(CH_2)_4\,COOH} \xrightarrow{E} \underset{\text{Adipamide}}{NH_2\,CO\,(CH_2)_4\,CONH_2}$$

4.

$$\underset{\text{Adipamide}}{NH_2\,CO\,(CH_2)_4\,CONH_2} \xrightarrow{F} \underset{\text{Adiponitrile}}{CN\,(CH_2)_4\,CN} \xrightarrow{G} \underset{\text{Hexamethylene Diamine}}{H_2N\,CH_2(CH_2)_4CH_2NH_2}$$

(3) Cyclohexanol or the mixture of cyclohexanol and cyclohexanone produced by the second reaction, on oxidation in presence of a catalyst (D) converted to adipic acid and adipamide. (4) Hexamethylene diamine, the second starting material, is made from adipic acid. First adipic acid is reacted with ammonia (E) to form adipamide. Adipamide is dehydrated to adiponitrile (F). Further adiponitrile is reduced to hexamethylene diamine with hydrogen in the presence of a catalyst (G).

8.4.3.2   Starting from butadiene
Butadiene is chlorinated with chlorine gas, to form dichlorobutene which is further treated with hydrocyanic acid, forming 1,4-dicyanobutene. Dicyanobutene is hydrogenated in the presence of catalyst, to form adiponitrile. Adiponitrile is then converted into adipic acid or hexamethylene diamine by hydrolysis or reduction, respectively.

$$\underset{\text{Butadiene}}{CH_2 = CH - CH = CH_2} + Cl \longrightarrow \underset{\text{Dichlorobutene}}{Cl\,CH = CH - CH_2\,CH_2\,Cl}$$

$$\underset{\text{Dichlorobutene}}{Cl\,CH = CH - CH_2\,CH_2\,Cl} + \underset{\substack{\text{Hydrocyanic}\\\text{acid}}}{HCN} \longrightarrow \underset{\text{Dicyanobutene}}{NC\,CH_2 = CH - CH_2\,CH_2\,CN} \xrightarrow{A} \underset{\text{Adiponitrile}}{NC\,(CH_2)_4CN}$$

8.4.3.3   Starting from furfural
Nylon can be manufactured from furfural which can be extracted from corn cobs and oat hulls. It is converted to adiponitrile by the following reactions:

Furfural is converted to furan (A) and this is reduced to tetrahydrofuran (1). Treatment with hydrochloric acid converts tetrahydrofuran to 1:4-dichlorobutane which is converted to adiponitrile by treatment with sodium cyanide (2). Hexamethylene diamine or adipic acid may then be made from the adiponitrile as described above.

1. Furfural $\xrightarrow{\quad A \quad}$ Furan $\xrightarrow{\text{Reduction}}$ Tetrahydrofuran

2. Tetrahydrofuran + HCl ----> Cl (CH2)4 CL + NaCN ------> NC (CH2)4 CN

                                                  1:4 Dichloro Butane   Adiponitrile

8.4.4   Manufacturing process

The first part of the process consists of mixing the two components in exact proportions in methanolic solution (condtions: 220–230°C for up to 2 h, at a pressure of about 17.5 kg/cm^2) from which the nylon salt settles out. A concentrated solution of this salt is first heated under pressure in an inert atmosphere to about 270°C; steam is then bled off, the pressure being at about 17.5 kg/cm^2 and the residue is heated further under vacuum to complete the polymerization. This is to make sure the molecular weight more than 12,000. The molecular weights less than 5000 cannot form into fibres and polymers of molecular weight between 5000 and 10,000 will make only weak fibres. Above 20,000 molecular weight create other problems like it will be difficult to melt and spin.

8.4.5   Stabilization

Hence, it is important to control the polymerization to the correct point. One of the important points is to start the process mixing exact proportions (mentioned above). This is practically done by stoichiometric proportions of the two components, and adding a small proportion of a monofunctional ingredient like acetic acid, which serves as a chain growth stopper in the same way as the extra proportion of a component. The amount of acetic acid is calculated to block the ends of the polymer chains after the desired average molecular weight has been reached. This technique is called 'stabilization'.

Once the polymerization is blocked at the desired point the melt nylon 66 is extruded through a slit and made into ribbons which is further cut into pieces and stored, which can be spun into filaments as required (Fig. 8.17).

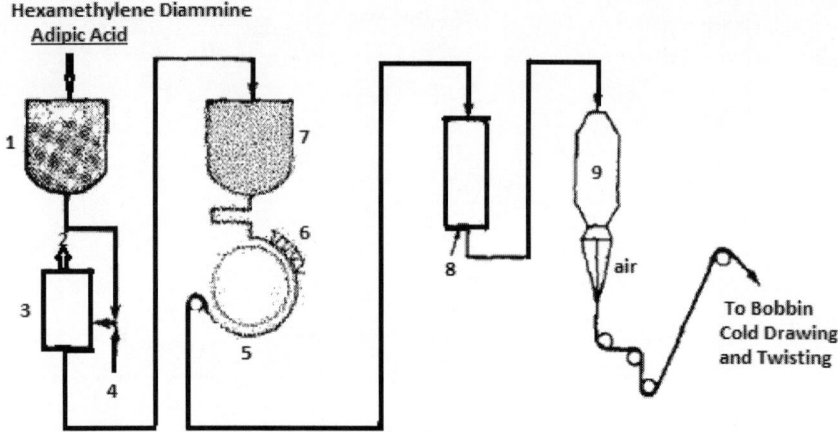

1- Polymeriser Reactor, 2- Water, 3- Evaporator, 4- Water, 5-Washer
6- Water, 7- Autoclave, 8 - Chipper, 9- Pressure Chamber

Figure 8.17. Nylon manufacturing (continuous) process

8.4.6   Spinning

As explained earlier nylons are thermoplastic, i.e., they progressively soften on heating and eventually melt nylon 6 at 210°C and nylon 66 at 250°C. This enables them to be extruded in their molten state through small holes in a plate called a spinneret to form very fine jets of molten polymer that quickly solidify to continuous filaments as they are transported down a cooling chimney to the wind-up positions.

The thermal stability and light fastness of polyamide fibres may be considerably increased by introducing during the manufacturing process small amounts of certain substances probably acting as inhibitors of the radical oxidizing processes that take place in thermal and photochemical oxidation of polyamides. Such substances include copper salts, naphthol, a number of organic compounds of the type of aromatic amines and phenols, which are added for increasing the fibre thermal stability, and manganese and chromium salts for improving its light fastness. Polyamide fibres are resistant to the attack of microorganisms and moths which is a very valuable quality. Apart from the terminal groups, polyamides do not contain active functional groups which would make them reactive.

The hydrogen atoms of amido groups, however, exhibit certain mobility, so that some polyamide derivatives may be obtained and the fibres may be subjected to chemical modification for imparting new properties to them. For instance, hydrogen in the –NH group may be substituted by the –CH$_3$ group; this reduces the intensity of intermolecular interaction and as a result the melting point is lowered and the elasticity of polyamide fibres is increased. Interaction with formaldehyde is also possible, with the formation of a methylol derivative:

$$
\begin{array}{c}
\text{—C}=\text{O} \\
| \\
\text{NH} \\
|
\end{array}
\;+\,CH_2O
\;\longrightarrow\;
\begin{array}{c}
\text{—C}=\text{O} \\
| \\
\text{N—CH}_2\text{OH} \\
|
\end{array}
$$

$$
\begin{array}{c}
\text{—C}=\text{O} \\
| \\
\text{NH} \\
|
\end{array}
+\,CH_2O+
\begin{array}{c}
\text{C}=\text{O—} \\
| \\
\text{NH} \\
|
\end{array}
\;\longrightarrow\;
\begin{array}{cc}
\text{—C}=\text{O} & \text{C}=\text{O—} \\
| & | \\
\text{N—CH}_2\text{—N} \\
| & |
\end{array}
+\,H_2O
$$

or cross linking of macromolecules and cyanurchloride, with the formation of dichloro derivatives and spatially cross-linked monochloro derivatives of cyanurichloride:

$$
\begin{array}{c}
\text{—C}=\text{O} \\
| \\
\text{NH} \\
|
\end{array}
+
\text{[triazine ring: Cl—C(N)(N)—C—Cl, with Cl]}
\;\longrightarrow\;
\text{[O=C—N(triazine ring)—C—Cl]}
+\,HCl
$$

$$
\begin{array}{c}
\text{—C}=\text{O} \\
| \\
\text{NH} \\
|
\end{array}
+
\text{[triazine ring Cl—C···C—Cl, with Cl]}
+
\begin{array}{c}
\text{C}=\text{O—} \\
| \\
\text{NH} \\
|
\end{array}
\;\longrightarrow\;
\text{[O=C—N(triazine ring)C—N—C=O]}
+\,2HCl
$$

The formation of cross linkages between macromolecules leads to an increase in the melting point, density, modulus of elasticity, heat resistance and thermal stability; to a decrease in the solubility and elongation of the polyamide fibres. Modification of polyamide fibres may also be effected by graft copolymerization.

8.4.6.1   Melt spinning

Once the polyamide is made by the polymerization process, it can be made in to fibres from the reaction vessel where it is available in the molten form. Otherwise the polymer can be converted into chip form and later on can be melted and extruded into fibres. Once the chip is melted in melt tank, it is delivered to an accurate metering device called a spin pump. The spinning temperature is usually 30°C higher than the polymer melting point. For example, for nylon 6 with a melting point of 220°C, the spinning temperature is generally targeted at 260–270°C, and around 290°C for nylon 6, with a melting point of 260°C. It is very important to dry the polyamide chip to a consistent moisture level (about 0.12%) for successful spinning. Otherwise, poor spinning yield and poor fibre quality may result from polyamide degradation.

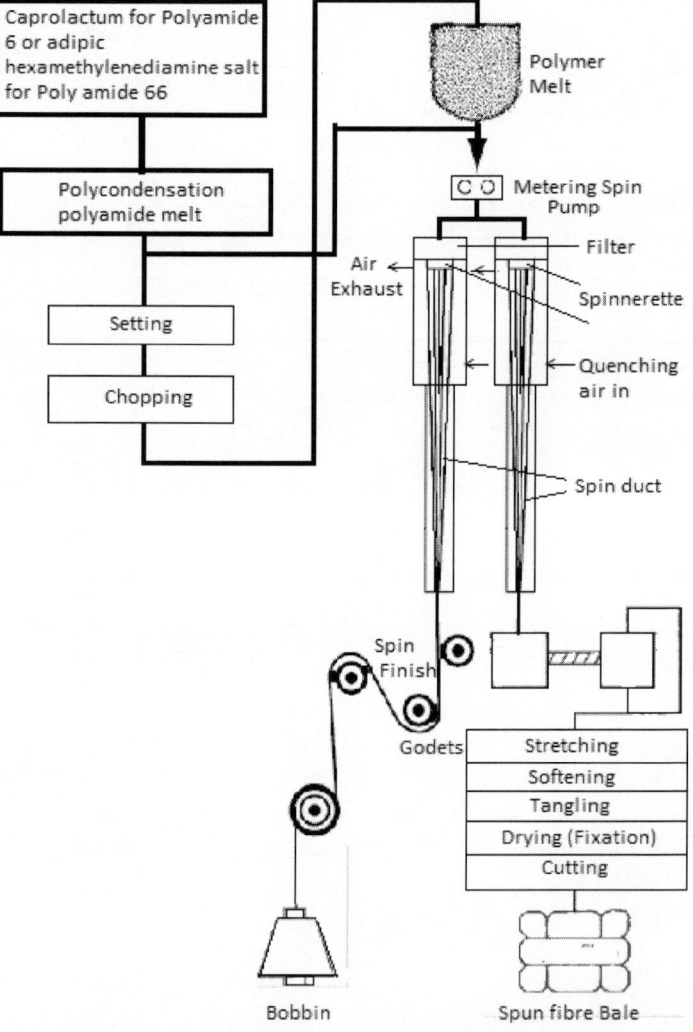

Figure 8.18. Schematic diagram of melt spinning

A schematic diagram of melt spinning is shown Fig. 8.18. The classic melt spinning process usually encompasses several process steps: polymer melting, transporting, spinning through a spinneret to form multiple filaments, quenching, finish application and take up.

The molten polymer (either from reaction vessel or melted chip) is forced into the spinnerets assembly through a metering pump (spin pump). The spinnerets assembly consists of a top cap, a breaker plate, a filtration media, and the actual spinnerets, which is usually in the shape of a showerhead and they together distribute polymer evenly to the spinnerets. The filtration media contains either different layers of special sand, layers of different size of stainless steel screens, or sintered metals. In addition to removing the foreign particles, gel particles and undesirable conglomerate additives, filtration may also improve the polymer melt homogeneity due to its torturous path and high shear of the filtration media.

The cross section shape of the fibre generally will depend on the shape of the hole on the spinnerets. Even though most nylon fibre products like high strength in tire cord, rope and cordage, and sling airbag requires round cross sections, there are other products calls for different cross section. However, for textile and carpet applications, fibres with modified cross sections have been developed to achieve different aesthetics such as lustre, opacity, insulation, walk resistance, etc. A fibre with a complex cross section can be used as a filtration medium for air or liquid.

Once the fibre is extruded through the spinneret, it has to be cooled. Cooling can be done either by air or water, but in most cases air is used – air at 18–20 °C and relative humidity of 55–65%. The temperature and humidity has to be strictly controlled since it has effect on the orientation and crystallization (Fig. 8.19).

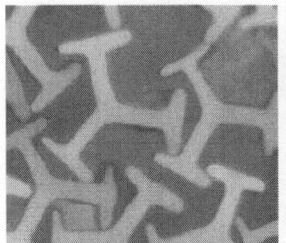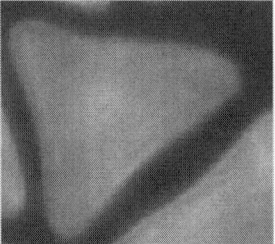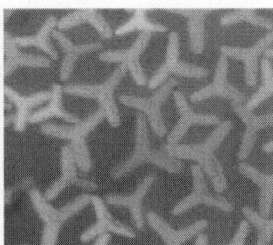

Figure 8.19. Examples of cross sections of fibres other than round

8.4.7   Drawing

This is followed by fibre drawing, comingling (interlacing) and package formation. In many cases, the drawing process is integrated with spinning to form a one-step process. By changing the operating conditions of various

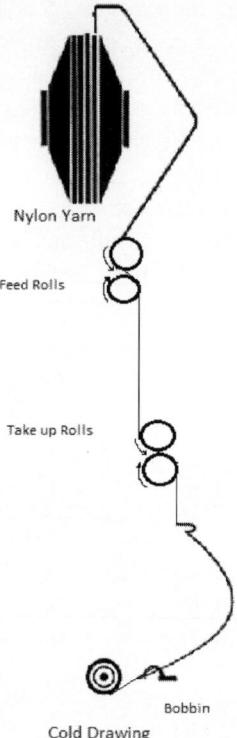

Nylon Yarn

Feed Rolls

Take up Rolls

Bobbin

Cold Drawing

Figure 8.20. Cold drawing

process steps in a manufacturing process, different fibre structures and properties can be obtained for different applications. During this manufacturing
process at high speed and the friction through air generates static electricity.
To help in the following process like drawing, a spin finish is applied on the
fibre bundle using a kiss roll or meter finish device. Most of nylon 6 and
nylon 66 spin finishes are water based. The amount of finish (water) applied
on the yarn may act as a plasticizer to lower the glass transition temperature
and change the rate of crystallization once the yarn is wound up on the package. In general, the 'wet pick up' of the finish is about 10% by weight of the
fibre to achieve the equilibrium moisture level for the downstream process,
and the finish on yarn is in the range of 0.2–1% of the fibre weight (Fig. 8.20).

Melt spinning and drawing can be combined as one operation. Thus,
there can be three different spinning arrangements:

1. Separate spinning and drawing: a two-step process

2. Stack-draw process: a one-step process

3. Spin-draw process: a one-step process

In the first process, melt spinning and drawing/texturizing is done as two entirely different operations. In this process after melt spinning at around 1000 m/min take-up speed to make 'undrawn' yarn of low orientation. This yarn is 'lagged' in storage for 4–12 h and then drawn at around 3:1 draw ratio.

In the second stack-draw one-step process, the take-up speed of 3000 m/min or more. The product is known as the partially oriented yarn (POY). The combination of drawing and textile finishing seems suited particularly for the production of POY at high spinning speeds.

In the third spin-draw one-step process, melt spinning is followed immediately by drawing and lagging. The take-up speed of this process ranges from 600 to 3000 m/min, coupled with immediate drawing on the panel. The resulting product is a fully drawn yarn.

8.4.8   Other polyamides

8.4.8.1   Nylon 6

Nylon 6 is also made almost same way as nylon 66. The nylon developed first is made by thermal polymerization of caprolactam in an inert atmosphere at temperatures up to 270°C, followed by the extraction with water of about 10% residual monomer before the polymer can be used. As the repeating unit in the polymer contains six carbon atoms, it is called nylon 6.

Starting from cyclohexanone

See above (nylon 66) for the production of cyclohexanone from basic chemicals. Cyclohexanone is reacted with hydroxyl amine (in the form of its sulphate) to form cyclohexanone oxime which is further treated with sulphuric acid gives caprolactum (Beckmann reaction):

Starting from benzene (or cyclohexane)

Benzene may be hydrogenated to cyclohexane, which is then nitrated. The nitro compound is reduced, forming cyclohexanone oxime, which is converted to caprolactam by Beckmann transformation:

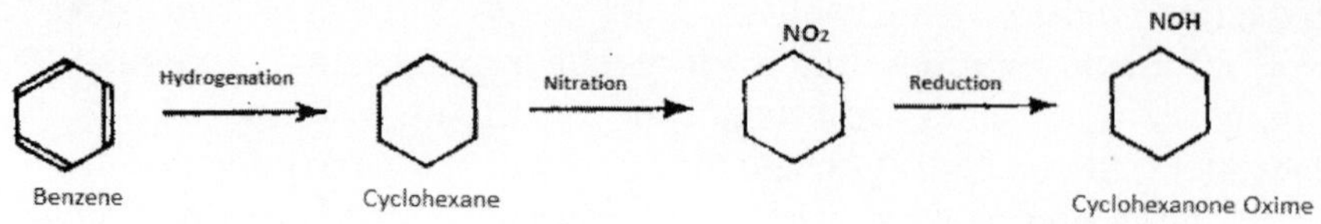

Starting from aniline

Nylon 6 can be manufactured starting from aniline. Aniline is hydrogenated to convert it to cyclohexyl amine. Cyclohexyl amine is reacted with hydrogen peroxide to get an addition compound which is converted to cyclhexanone oxime by reaction with ammonium tungstate, which can be converted to caprolactum, by Beckmann reaction:

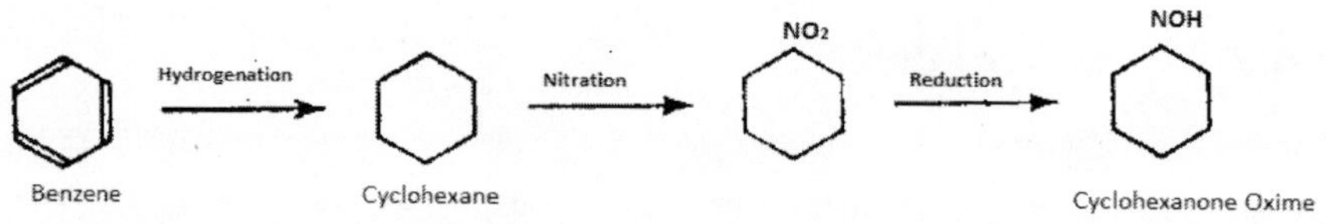

Starting from toluene

Nylon 6 can be made from toluene. Toluene is first oxidized to Benzoic acid and on hydrogenation of this produces hexahydrobenzoic acid. This compound on reaction with nytrosyl sulphuric acid in the presence of oleum gives caprolactum:

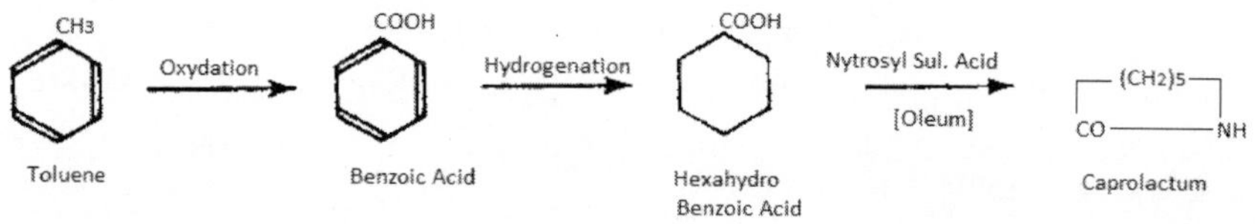

Polymerization

Next caprolactum has to be polymerized. The polymerization can be carried out by a nonhydrolytic process of an aqueous process.

Nonhydrolytic process

Carprolactam on heating in the presence of catalysts like alkali metals and their salts, to a temperature of 280°C it undergoes polymerization. Caprolactam rings open and the linking of the opened rings into polymer molecules. The reaction is rapid, and high molecular weight polycaproamide may be produced with a DP in the region of 200 or so. This process is generally not used for manufacturing fibres but may be used for manufacture of polyamide plastics.

Aqueous process

This polymerization process is more used for producing fibres. Here water takes part in the reaction. In this process, caprolactum mixed with about 10% water along with acid catalyst and an acid chain stopper (see above in nylon 66 manufacture). The process is continuous and the above mixture is fed into the top of a SS column of about 6 m height and 45 cm diametre. And the column is heated to 250–270 °C. As the mixture flows down, the polymerization takes place and polycapromide is formed. Polymerization takes place either by polycondensation or by polyaddition. Both reaction ends up in the creation of polymer molecules. Polyaddition predominates over the polycondensation. By the time it reaches the bottom, an equilibrium of the reaction takes place and a mass containing 89–92% polcapromide and 11–8% caprolactum.

Spinning

The molten polycaproamide may be spun directly at this stage, without any intermediate isolation of solid polymer.

Otherwise, the melted polycapromide is extruded and can be made into chips (around 6 mm diametre) and separately spun which is more practiced method as, after making these chips, they are washed with water to remove the caprolactum present in the mixture and hence the final nylon produced will be more uniform even though still it may contain around 1% caprolactum.

The melt is made in both cases at a temperature of around 250–260°C. The melted polycapromide is fed to the spinnerets through a metering pump and through filters while the atmosphere in the closed system is maintained with nitrogen gas to avoid any oxidation. The spun nylon is applied with antistatic agents and lubricating agent and wound on bobbins. The filaments are further drawn, twisted as per the final usage or cut into staples.

8.4.9   Identification of yarn as nylon 6 or nylon 66

The following test is used to distinguish between nylon 6 and nylon 66.

8.4.9.1   *Preparation of solution*

A 50% formic acid solution is prepared by dilution of the 90% formic acid solution commonly available, e.g., 1 l of acid is diluted with enough cold water to bring the total volume to 1800 cc.

8.4.9.2   *Procedure*

The 50% formic acid solution is heated carefully to 80°C. Several pieces of yarn or individual filaments are dropped into the solution. Nylon 6 will shrivel or ball up and dissolve almost immediately, very little agitation being necessary. Nylon 6.6 will float in the solution and appear not to be affected.

The temperature control is very important. At temperatures several degrees lower than 80°C, neither nylon 6 nor nylon 66 will appear to be destroyed. If the temperature is at about 90°C, both nylon 6 and nylon 66 will disintegrate. When making this test, it is advisable to carry out preliminary tests first on known samples of nylon 6 and nylon 66.

8.4.10   Nylon 11

Nylon 11 is made by the self condensation of 11-aminoundecanoic acid:

$$NH_2 (CH_2)_{10} COOH \quad ----> \quad ---NH (CH_2)_{10} CONH (CH_2)_{10} \quad CONH (CH_2)_{10}-$$

Omega Undecanoic Acid   Nylon 11

Undecanoic acid can be manufactured by one of the three methods explained below.

8.4.10.1   From castor oil

Castor oil which is extracted from castor beans contains 85% triglyceryl ricinoleate. Triglyceryl ricinoleate is reacted with methyl alcohol to form methyl ricinoleate which is pyrolyzed at high temperature to give heptaldehyde, methyl undecylenate and a small amount of fatty acids. Heptaldehyde and methyl undecylenate which are required are isolated by distillation and are hydrolyzed to undecylenic acid. Undecylenic acid is aminated by reaction with ammonia, to form 11-aminoundecanoic acid.

$$C_{17}H_{32} (OH) COOCH_3 \xrightarrow{\text{Pyrolysis}} C_6H_{13} CHO + CH_2= CH (CH_2)_8 COOCH_3$$

Methyl ricinoleate   Heptaldehyde   Methyl Undecylenate

$$C_6H_{13} CHO + CH_2 = CH (CH_2) COOCH_3 \xrightarrow{\text{Hydrolysis}} CH_2 = CH (CH_2)_8 COOH \xrightarrow{\text{Ammonia}} NH_2(CH_2)_{10} COOH$$

Heptaldehyde   Methyl Undecylenate   Udecylenic Acid   11-amino Undecanoic Acid

8.4.10.2   From carbon tetrachloride and ethylene

When carbon tetrachloride and ethylene are heated together at high temperature in the presence of benzyl peroxide (catalyst) ethylene undergoes polymerization until such time as the ends of the polymer chain are blocked by Cl and CCl_3 radicals formed from the carbon tetrachloride. The reaction conditions are controlled in such a way as this occurs after only 1, 2, 3, 4 or 5 ethylene molecules, e.g., have linked together (telomerization). Thus we get a mixture of compounds of the structure $Cl(C_2H_4)_nCCl_3$, where n is 1–5. One of the products, which may be separated by distillation, is 1-chloro-11-trichloroundecane, i.e., $Cl(CH_2)_1CCl_3$. Hydrolysis of this by aqueous sulphuric acid yields 11-chloroundecanoic acid, which is reacted with ammonia to produce 11-aminoundecanoic acid.

$$n\ C_2H_4 + CCl_4 \xrightarrow{H2SO4} Cl\ (C_2H_4)n\ CCl_4 + H_2O \longrightarrow Cl(CH_2)_{10}\ COOH \xrightarrow{NH3} NH_2\ (CH_2)_{10}\ COOH$$

Ethylene	1-chloro-trichloroundecane	11-Chloro und-	11-Amino- Unde-
n= 1,2,3,4,5		ecanoic acid	canoic acid

8.4.10.3   Polymerization

Aqueous suspension of 11-aminodecanoic acid is fed into the reaction vessel, and the water is removed and the vessel heated to 215°C. At this stage, polycondensation starts and allowed to continue till the desired DP is reached. Once the molecular weight distribution is even, the molten polymer is transferred to storage tanks from where it can be pumped to the spinnerets.

$$nH_2N\ (CH_2)_{10}\ COOH \rightarrow H[HN(CH_2)_{10}\ CO]_n OH + (n-1)\ H_2O$$

8.4.10.4   Spinning

The molten nylon 11 polymer can be stored for a considerable time in an inert atmosphere like nitrogen for avoiding any oxidation. Hence the spinning is done from this stage and not necessary to make into chips. The spinning is done in the same way as explained under nylon 6 and drawn/twisted as per final fibre requirements.

A comparison of properties of different nylons are given in Table 8.7.

Table 8.7. Comparison of properties of different nylons

Property	Nylon 6	Nylon 6.6	Nylon 11
Tensile strength	5110–5880 kg/cm²	4550–5950 kg/cm²	6860 kg/cm²
Tenacity, cN/tex	39.7–51.2 cN/tex (4.5–5.8 g/den)	6.2 cN/tex (0.7 g/den)	44.15–66.23 cN/tex (5.0–7.5g/den)
Elongation, %	23–42.5	26–32	25
Elastic recovery	100% at 8% stretch	100% at 8% stretch	100% at 6% stretch
Specific gravity	1.14	1.14	1.04
Moisture regain	4–4.5	4–4.5	1.18
Melting point, °C	215	250	189

Property	Nylon 6	Nylon 6.6	Nylon 11
Ironing	Iron below 150°C	Adequate resistance heat	Safe ironing between 80 and 100°C
Dry cleaning	Resistant dry cleaning agents	Resistant dry cleaning agents	Resistant dry cleaning agents
Heat setting, °C	Dry 190–193, wet 105–115	Dry 205, wet 120	Softening temp. 170°C, but fibre yellows at 150°C
Flammability	Good resistance to burning	Good resistance to burr.ing	Good resistance to burning
Therm. conductivity		1.7 BTU/h/ft²/for 1 inch thickness	
Effect of sunlight	Yellowing and strength loss on prolonged exposure	No discolouration, strength loss	Same as nylon 66
Effect of acid	Diluted acid no much effect, conc. mineral acid decompose	Diluted acid no much effect, conc. mineral acid decompose	Diluted acid no much effect, conc. mineral acid decompose
Effect of alkali	Excellent resistance, can be boiled in strong caustic soda	Excellent resistance, can be boiled in strong caustic soda	Nylon 11 has a high resistance to alkalis
Solvents	Conc. formic acid, 50% acid at 80°C, conc. hydrochloric, sulphuric and nitric acid, 25% $ZnCl_2$ soln. in methanol at 50°C, phenol	Conc. formic acid at RT (27 °C), phenolic compounds, saturated calcium chloride in methanol, hot zinc chloride in methanol	Phenols, 100% formic and acetic acids
Cross section enlarged			
Fibre enlarged			

8.4.11   Nylon 6,10

Produced by condensation of hexamethylene diamine and sebacic acid:

$H_2N\ (CH_2)_6NH_2\ +\ HOOC\ (CH_2)_6\ COOH\ \text{-->}\ \text{--}HNCO\ (CH_2)_6\ NHCO(CH_2)_6\ CONH\ (CH_2)_6\ CONH\text{---}$

Hexamethylene   Sebacic Acid   Nylon 6.10
   Diamine

8.4.11.1   Manufacture

Sebacic acid is produced from castor oil. It is condensed with hexamethylene diamine, the process being similar to that used in the production of nylon 66. The product, polyhexamethylene sebacate, may be melt spun without difficulty.

8.4.12   Modified nylon

The modified nylons can take several forms: changing cross section of shape and changing the physical and chemical properties to improve dyeability, handle, tenacity, heat stability, etc.

8.4.12.1   Cross-sectional modification – multilobal nylons

As in the case of other synthetic fibres, the nylons are also made with multilobal cross section thereby increasing the surface area and achieve advantages like:

1. Increased cover
2. Crisp, silk like, firm handle
3. Reduced pilling in spun yarn fabrics
4. Increased bulk
5. A sparkle or high light effect
6. Resistance to soiling, etc.

The main disadvantage of these multilobal fibres are, due to the increase in the surface area, they consume more dyestuff in dyeing than round cross section fibres.

8.4.12.2   Bicomponent and biconstituent type modifications

Bicomponent technology has been used to introduce functional and novelty effects other than stretch to nylon fibres. The bicomponent type modifications involve two polymers of the same generic class, e.g., nylon 6 and nylon 6,10. A biconstituent modification consist of two dissimilar generic polymers, e.g.,

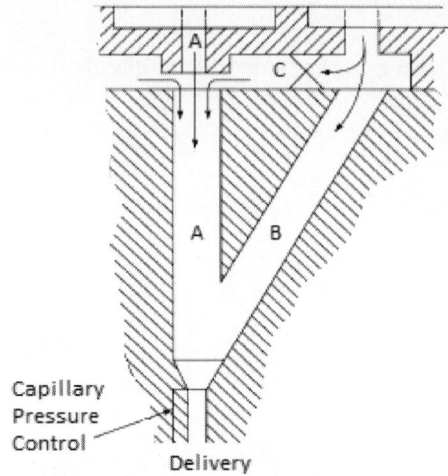

Bicomponemt Spinnerette- A = Copolymer; B=
Homopolymer; If side by side fibre is required the
interconnecting duct C is closed at X

Figure 8.21. Separation of two different polymers

nylon 66 and polyester. These modification is done to make up some disadvantage of one fibre with another fibre or to improve some of its qualities.
Both fibre types are made separately by melt spinning the two different polymers through a common, specially designed spinneret. The spinneret hole
and block channels can be designed so that the two polymers emerge side
by side, as sheath–core or conjugate fibres. The fibres are processed through
conventional drawing or spin–draw operations. The original intent of the
side-by-side bicomponent yarn was to impart stretch in tricot knit and hosiery
applications. Nylon 6 and nylon 66 homopolymers were paired to a copolyamide on the basis of shrinkage difference as measured on the individually
spun polymers (Fig. 8.21).

Depending on the differential shrinkage, the bicomponent yarn would
assume a level of helical crimp in dyeing or steaming. Examples of earlier
bicomponent stretch yarns were spun with a nylon 66 sheath and core compositions of nylon 66/6,10 (50/50) and nylon 66/6 (80/20). Stretch was also
achieved by spinning side-by-side two nylon 66 polymers having a relative viscosity difference of at least 15 units, and by spinning nylon 6 with a
melt-spinnable polyurethane.

Most of the stretch hosiery today are made from textured POY and with
elastomeric spandex fibres. For instance, antistatic yarns are made by spinning a conductive carbon-black polymer dispersion as a core with a sheath
of nylon and as a side-by-side configuration. At 0.1–1.0% implants, these

conductive filaments give durable static resistance to nylon carpets without interfering with dye colouration.

Dye selectivity can be altered by spinning a high amine-end nylon core with a sulphonate-containing nylon sheath. In standard dyeing conditions, the core accepts levelling acid dyes, but not critical or basic dyes; the sheath accepts basic dyes. The effect creates a third colour when dyed in combination with acid- and base-dyeable yarns.

8.4.13   Copolymers

The copolymers are of two forms – block and random. A nylon block copolymer can be made by combining two or more homopolymers in the melt. The composition of the melt is a function of temperature and more so of time. Two homopolyamides in a moisture-equilibrated molten state undergo amide interchange where amine ends react with the amide groups. As time progresses, the two homopolyamides in the melt form a block and eventually a random copolymer as a result of amide interchange. Block copolymerization is a way of introducing a new variant into a base polymer without grossly affecting the spinning performance and physical properties of the yarn. The process requires careful control, however, to maintain the desired composition to ensure uniform product and spinning continuity. Examples of this technology are the block copolymers of poly(4,7-dioxadecamethyleneadipamide) and poly(dioxa-arylamide) with nylon 6 for hydrophilic nylons.

Random copolymers are made by combining two or more monomers in the polymerization process. The melting temperature of random copolymers are lowered as the regularity with which the monomer groups are spaced along the backbone is reduced. Hydrogen bonding between the amides is also reduced. Crystallinity and orientation are also reduced with increasing randomization. However, in the case of nylon 66/6T, the crystallinity is not reduced because the copolyamide segment is similar in size to that which is replaced. Such monomers, which can be exchanged or replaced each other in the crystal lattice, are often termed isomorphic monomers.

Random copolyamides that melt below 200°C are difficult to process through melt spinning because they are difficult to crystallize and consequently remain tacky through the windup. Copolyamides that melt above 300–310°C are also difficult to process because of their higher melt viscosities and greater susceptibility to thermal oxidation. The general properties of random copolyamides are high dyeability (especially with nonlevelling large dye molecules), lower melting and softening points, reduced dry and wet strength properties, high creep failure and high shrinkage. The last property led to the use of the copolyamides in bicomponent selfcrimping yarns. Copolyamides are also suitable for thermal bonding of fibre and fabrics because of their low softening temperature. Random copolyamides are also used to modify the dyeability of nylon 66. The addition of 0.5–2% comonomer increases the dyeing rate to almost that of nylon 6. Comonomers are also used as assists in the high-speed godetless spinning of draw-texturing feed yarns. Nylon 66 copolyamides with ethyltetramethyleneadipamide, pentamethyleneadipamide or 2-methylpentamethyleneadipamide units lower the gelation rate of nylon 66 at the spinning temperature of nylon 66.

8.4.14   Graft polymers

It was found that the chemical or physical properties of nylon could be drastically changed by grafting polymers on to it. Vinyl monomers can be grafted to almost any nylon by ionizing radiation usually from a Co-60 source, high-energy UV radiation, or chemical means. Radiation grafting of acrylonitrile on nylon 6 gives basic dyeability, and that of styrene, hydrophobicity. Nylon 66 can also be made basic dyeable by chemically grafting acrylic or methacrylic acid with a water-soluble formaldehyde sulphoxylate salt. The friction of nylon fibres can be changed by grafting poly(dimethylsiloxane) macromolecules to the surface.

Nylon carpet fibres can be made to resist stain from certain acidic artificial and natural colourants found in soft drinks, juices, coffee and red wines, by grafting a phenyl–vinyl and ether–maleic anhydride copolymer to the fibre surface using UV light and a photoactivator in the solution. However, there are other techniques for blocking the stain-sensitive amine ends that are more commercially applicable and efficient. They include (1) the post-treatment of nylon carpets with an alkali metal silicate in a phenol–formaldehyde product or with a sulphonated naphthol or sulphonated phenol–formaldehyde product at a specific pH, (2) manufacturing of carpets with only cation-dyeable nylon yarn and dyeing or printing them to shade and acceptable light fastness with premetallized or acid dyes and (3) melt-spinning of fibres – pigmented or producer-coloured fibres – from nylon polymer containing a high level of a sulphonated derivative.

8.4.14.1   Chemical modifications

Chemical changes in the fibre for improving the quality is another technique. The long chain molecules of nylon may be linked together by reacting them with chemicals carrying an active group at each end of the molecule. The isocyanate group, e.g., will react readily with amine or carboxylic groups, such as may be present at the ends of polyamide molecules. Reaction of polyamides with a di-isocyanate, therefore, would be expected to link up adjacent polyamide molecules. This technique has been used successfully in attempts to modify nylon fibres with a view to improving their flat-spotting characteristics when used in tyres.

Grafting is another technique used to improve the quality or to bring in a character which is required for a final useage. The acrylic acid is can graft on to nylon which after converting to their sodium salt will give better moisture regain than the parent fibre and better wet crease recovery. Graft polymerization of ethylene oxide on to the fibre also increases the moisture absorption.

8.4.15   Tactile

Tactile is a polyamide microfibre. It includes a wide range of nylon 66 yarns, which can be altered during the finishing processes to create many effects. DuPont is the leading producer of polyamide microfibre with brand name Tactel.

Unique properties of Tactel

1. Anti-wrinkle

2. Texture like silk

3. Softness and drape

4. Good stability and shape retention

5. Have a good wicking property compared to normal nylon fibre

6. Shrink resistance

7. Presents insulation and breathability

8. Less sweaty than other synthetic fibres in warm weather

Its anti wrinkle property makes it suitable for shirting, coats, dresses, trousers, T-shirts, etc. It is also used in lingerie, sweatshirts, pullovers, stockings, socks, tracksuits underwear, vests, etc.

8.4.16   Quina fibre

It is a type of polyamide fibre made from bis-para-aminocyclohexyl and made in fine filament form. It is a polycondensate of diaminodiphenylmethane and decane dicarboxylic acid. The melt spinning is done such a way as to get a tri-lobal shape for the fibre cross section. It is distinguished by its silk-like handle and regarded as synthetic substitute for pure silk. The trilobal structure gives a good lustre and bulky feel. It has moisture regain 2–2.5%. The crease resistance and pleat resistance is good in this fibre/fabrics. Resistance to UV light makes the fabrics more attractive. It is used for ladies dress wear, blouses and suitings.

8.4.17   Stiff and strong fibres

Researches were undergoing based on structure–property relations in polymers to produce strong, stiff fibres, which were required for special enduses. Since the covalent carbon–carbon bond is a very strong bond, one would expect that linear chain polymers such as polyethylene (PE) would be potentially very strong and stiff. Conventional, isotropic polymers show a Young's modulus, E, of about 10 GPa. To obtain high stiffness and strength polymers, one must extend these polymer chains and pack them in a parallel array. The orientation of these polymer chains with respect to the fibre axis and the manner in which they fit together (i.e., order or crystallinity) are controlled by their chemical nature and the processing route. Molecular chain orientation coupled with molecular chain extension is required for high stiffness and strength. It is possible to make fibre of Young's modulus greater than 70 GPa, which is the modulus of aluminium and glass. This needs high draw ratios, i.e., a very high degree of elongation must be carried out under such conditions that macroscopic elongation results in a corresponding elongation at a molecular level. It turns out that the Young's modulus, E, of a polymeric fibre increases linearly with the deformation ratio (draw ratio in tensile drawing or die drawing and extrusion ratio in hydrostatic extrusion).

Two such fibres are oriented PE and aramid fibres. Two very different approaches have been taken to make high modulus organic fibres. They are as follows:

1. Processing of the conventional flexible-chain polymers in such a way that the internal structure takes a highly oriented and extended-chain arrangement. Structural modification of 'conventional' polymers such as high modulus PE was developed by choosing appropriate molecular weight distributions, followed by drawing at suitable temperatures to convert the original folded-chain structure into an oriented, extended chain structure.

2. The second approach, radically different, involves synthesis, followed by extrusion of a new class of polymers, called liquid crystal polymers. These have a rigid-rod molecular chain structure. The liquid crystalline state, as we shall presently see, has played a very significant role in providing highly ordered, extended chain fibres.

8.4.18   Aromatic polyamides – Aramid

The term aramid in aramid fibre is a short form for aromatic polyamide. As described above, conventional polyamides, e.g., nylon, contain mostly aliphatic and cycloaliphatic units in the macromolecular chain structure.

Aramid is a generic term that represents an important class of fibres. Aramid fibres such as Nomex or Kevlar are ring compounds based on the structure of benzene, as opposed to linear compounds used to make nylon. The basic chemical structure of Kevlar aramid fibres consists of oriented para-substituted aromatic units, which limits conformational freedom and makes them rigid rod-like polymers.

Aromatic diamines condensed with terephthalic acid provide polyamides with exceptional resistance to high temperatures. The intermolecular bonding and chain stiffness are such as to confer high thermal stability on the polymer molecules.

Aromatic Polyamides

Reaction completion is when all the phenylene units in the polyamide are para-substituted and in such cases, the melting and decomposition temperature will be around 555°C. If all the phenylene units are meta-substituted, the melting pint will be around 410°C. Fully aromatic polyamides are prepared with both substituted polymers, in any proportions mixed in a manner to get the desired end properties. Polyamides where 85% of the amides has been substituted are called 'aramids'.

8.4.18.1   Manufacture

Manufacturing of aramid fibres involves solution polycondensation of di-amines and diacid halides at low temperatures, e.g., low temperature poly-condensation of p-phenylene diamine (PPD) and terephthaloyl chloride (TCI) in a dialkyl amide solvent. The amide solvents used are N-methyl pyrrolidone and dimethyl acetamide, separately or mixed, and generally in the presence

Chemical synthesis formula for Aramid fibre

of inorganic salts such as LiCl or $CaCl_3$. The polymer is precipitated with water, neutralized, washed and dried. PPTA polymer is insoluble in ordinary solvents but dissolves in strong acids such as concentrated sulphuric acid. The route followed for manufacturing aramid fibre is mostly the second approach mentioned above to making high stiffness and high strength polymeric fibre is the liquid crystal route. This involves synthesis and extrusion of rigid-rod molecular chain polymers. This rigid rod-like structure results in a high glass transition temperature and poor solubility, which makes fabrication of these polymers, by conventional drawing techniques difficult. Instead, they are spun from liquid crystalline polymer solutions as described below. Many aromatic polymers can be converted to high modulus fibres via the liquid-crystal route. In theory, it was expected that such a rigid-rod molecular structure should result in a highly oriented fibre structure in the as-spun state, i.e., without any need to resort to drawing after spinning. There were, however, two practical problems:

1. These aromatic polyamides have relatively high melting points or they degrade thermally at high temperatures. Thus, they could not be melt spun.

2. The viscosity of their isotropic solutions is very high which renders them unsuitable for spinning.

A liquid crystal has a structure intermediate between a three-dimensionally ordered crystal and a disordered isotropic liquid. There are two main classes of liquid crystals: lyotropic and thermotropic. Lyotropic liquid crystals are obtained from low viscosity polymer solutions in a critical concentration range while thermotropic liquid crystals are obtained from polymer melts where a low viscosity phase forms over a certain temperature range. Aromatic polyamides and aramid type fibres are lyotropic liquid crystal polymers. These polymers have a melting point that is high and close to their decomposition temperature. One must therefore spin these from a solution in an appropriate solvent such as sulphuric acid. Aromatic polyesters, on the other hand, are thermotropic liquid crystal polymers. These can be injection moulded, extruded or melt spun. The parallel arrays of polymer chains in liquid crystalline state become even more ordered when these solutions are subjected to shear as, e.g., in extruding through a spinneret hole. It is this inherent property of liquid crystal solutions which is exploited in the manufacture of aramid fibres. Para-oriented aromatic polyamides form liquid crystal solutions under certain conditions of concentration, temperature, solvent and molecular weight. Such a liquid crystal shows the anomalous relationship between viscosity and polymer concentration described above. Initially, there occurs an increase in viscosity as the concentration of polymer in solution increases, as would be expected in any ordinary polymer solution. At a critical point where the solution starts assuming an anisotropic liquid crystalline shape, there occurs a sharp drop in the viscosity. Starting from liquid crystalline spinning solutions containing highly ordered arrays of extended polymer chains, fibres can be spun directly into an extremely oriented, chain extended form.

Para-oriented rigid diamines and dibasic acids give polyamides which yield, under appropriate conditions of solvent, concentration and polymer

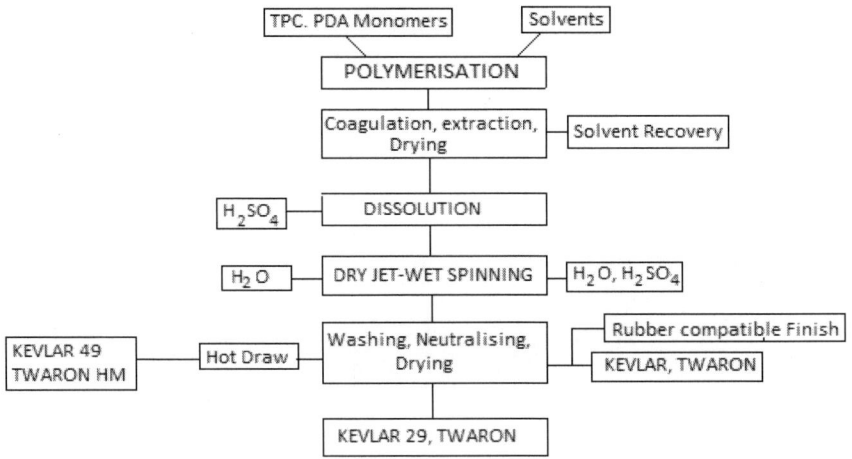

Figure 8.22. Flow diagram of manufacture of different types of aramid fibres

molecular weight, the desired nematic liquid crystal structure. One would like to have, for any solution spinning process, a high molecular weight in order to have improved mechanical properties, a low viscosity to ease processing conditions, and a high polymer concentration for high yield. For para-aramid, PPTA, trade name Kevlar, the nematic liquid crystalline state is obtained in 100% sulphuric acid at a polymer concentration of about 20%. The polymer solution, often referred to as the dope, has concentrated sulphuric acid as a solvent for PPTA. Five moles of sulphuric acid are needed per PPTA amide bond, which translates into about 4 kg of sulphuric acid per kilogram of polymer. The spent acid after spinning is converted to calcium sulphate (gypsum). For every kilogram of fibre, 7 kg of gypsum is produced. Fig. 8.22 shows the flow diagram for making different types of aramid fibres (Kevlar, Twaron, etc.).

The polymer is pulverized, washed and dried, mixed with concentrated H_2SO_4 and extruded through a spinneret at about 100°C. The jets from the orifices pass through about 1 cm of air layer before entering a cold water (0°C) bath. The fibre precipitates in the air gap, and the acid is removed in the coagulation bath. The spinneret capillary and air gap cause alignment of the domains resulting in highly ordered, crystalline and oriented as spun fibres. The air gap also allows the dope to be at a higher temperature than would be possible without the air gap. The higher temperature allows a more concentrated spinning solution to be used and higher spinning rates are possible. Spinning rates of several hundred metres per minute are not unusual. Fig. 8.23. compares the dry jet-wet spinning method used with nematic liquid crystals and

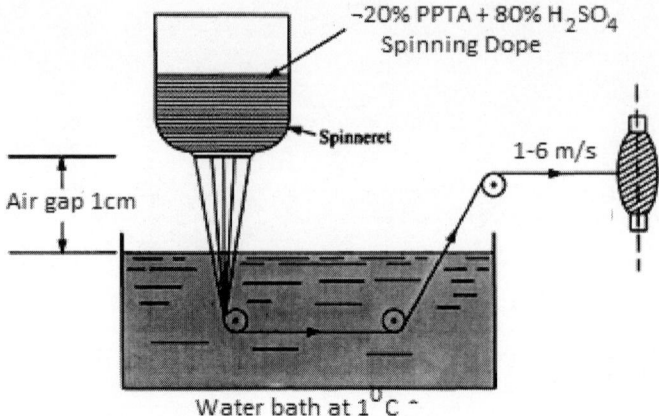

Figure 8.23. Dry jet-wet spinning method

the spinning of a conventional polymer. The oriented chain structure together with molecular extension is achieved with dry jet-wet spinning. The conventional wet or dry spinning gives precursors that need further processing for a marked improvement in properties. The as-spun fibres are washed in water, wound on a bobbin and dried. fibre properties are modified by the use of appropriate solvent additives, by changing the spinning conditions, and by means of some post-spinning heat treatments.

8.4.18.2   Structure and properties

Structure of aramid fibre can be seen from the equation given under manufacture section. The aromatic rings impart the rigid rod-like characteristics

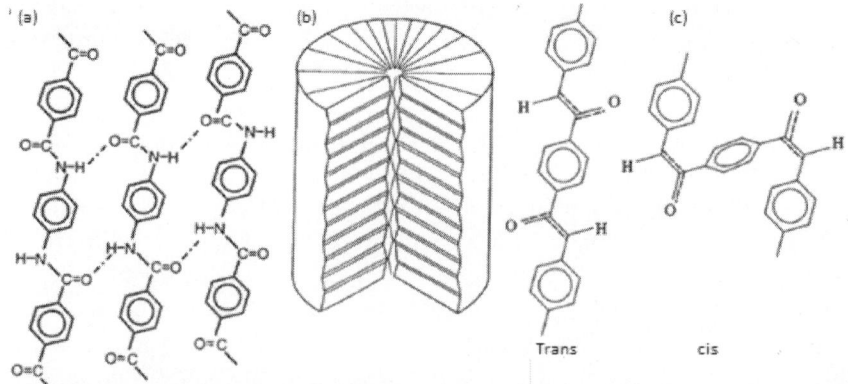

Figure 8.24. (a) Bonding in aramid fibre – strong covalent bond along fibre axis and weak hydrogen bonding across (b) Schematic diagram of Supramolecular structure of aramid fibre (c) Molecular rotation of the amide carbon-nitrogen bond

for aramid. These chains are highly oriented and extended along the fibre axis with the resultant high axial modulus. Aramid has a highly crystalline structure, and the linearity of the polymer chains results in a high packing efficiency. Fig. 8.24 shows how the chains are arranged along the axis of the fibre. They are in radially arranged, axially pleated crystalline supramolecular sheets. The molecules form a planar array with interchanging hydrogen bonding. The stacking sheets form a crystalline array, but between the sheets the bonding is weak. Each pleat is about 500 nm long and the pleats are separated by transitional bands. The adjacent components of a pleat make an angle of 170°. Table 8.8 gives the main properties of different aramid fibres.

Table 8.8. Properties of different aramid fibres

Property	Different types of aramid fibre					
	Kevlar 29	Kevlar 49	Kevlar 68	Kevlar 119	Kevlar 129	Kevlar 149
Density (g/cc)	1.44	1.45	1.44	1.44	1.45	1.47
Diameter (mm)	12	12	12	12	12	12
Tensile strength (Gpa)	2.8	2.8	2.8	3	3.4	2.4
Tensile strain to fracture (%)	3.5–4.0	2.8	3	4.4	3.3	1.5–1.9
Tensile modulus (GPa)	65	125	101	55	100	147
Moisture regain at 25°C, 65RH (%)	6	4.3	4.3			1.5
Coefficient of expansion (10^{-6} K^{-1})	-4	-4.9				

Like other polymers, aramid fibres when exposed to UV light for an extended period, they discolour from yellow to brown and lose mechanical properties. Radiation of a particular wavelength can cause degradation because of absorption by the polymer and breakage of chemical bonds.

8.4.18.3   Oriented PE fibres

The PE can be processed to achieve ultra-high molecular weight and highly crystalline fibre with very high stiffness and strength. Drawing of melt crystallized PE (molecular mass 104–105) to very high draw ratios can result in moduli of up to 70 GPa. In all the normal drawing processes, the polymer chains become merely oriented without undergoing molecular extension. In general, molecular orientation is achieved together with chain extension, increase the modulus of the fibres. This can be achieved by gel spinning which has been exploited commercially also. Such a spinning of molecular

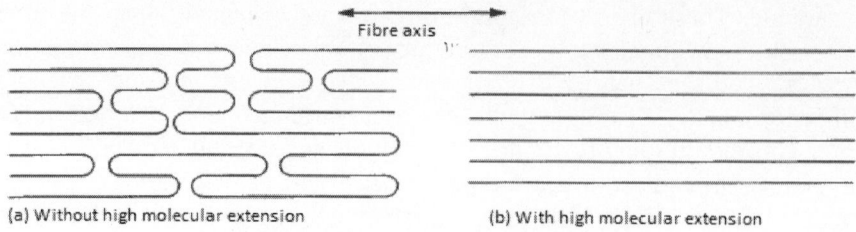

(a) Without high molecular extension   (b) With high molecular extension

Figure 8.25. Molecular orientation

orientation without high molecular extension – See Fig. 8.25 (a) to a molecular orientation with high molecular extension, Fig. 8.25 (b) is obtained.

The PE is a particularly simple, linear macromolecule, with the chemical formula $-(-CH_2-CH_2-)-$. This chemical formula is easier to obtain an extended and oriented chain structure in PE. High-density polyethylene (HDPE) is preferred to other types of PE because HDPE has fewer branch points along its backbone and a high degree of crystallinity. These characteristics

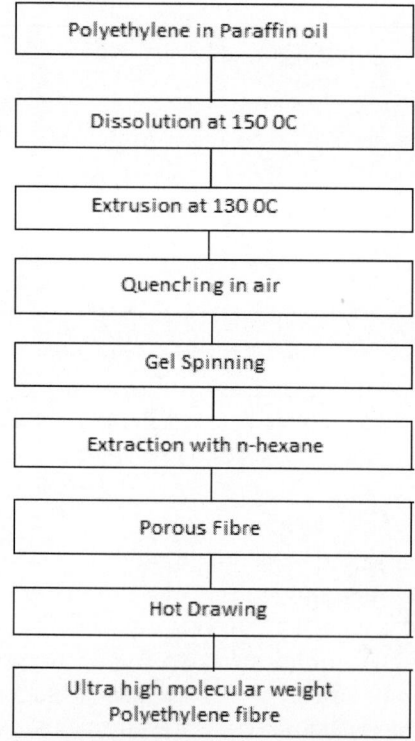

Figure 8.26. Flow diagram of the manufacture of ultra high molecular polyethylene fibre

of linearity and crystallinity are important to obtain a high degree of orientational order and an extended chain structure in the final fibre. In the manufacturing process, three solvents are used – decalin, paraffin oil and paraffin wax. Using one of these solvents, a 2–10% solution was obtained at about 150°C. A dilute solution ensures less chain entanglement, which makes it easier for the final fibre to be highly oriented. A PE gel is produced when the solution coming out of the spinneret is quenched by air. The as-spun gelled fibre enters a cooling bath. At this stage the fibre is thought to have a structure consisting of folded chain lamellae with solvent between them and a swollen network of entanglements. These entanglements allow the as-spun fibre to be drawn to very high draw ratios; draw ratios can be as high as 200. The maximum draw ratio is related to the average distance between the entanglements, i.e., the solution concentration. The gelled fibres are drawn at 120°C. One problem with this gel route is the rather low spinning rates of l.5 m/min (Fig. 8.26).

8.4.18.4   Structure and the properties of PE fibres

The following figure shows a three-dimensional figure of unit cell of a single crystal of PE. The PE fibre (UHMWPE – ultra high molecular weight polyethylene) has a density of 0.97 g/cc. Its strength and modulus are slightly lower than those of aramid fibres on a per-unit-weight basis, i.e., specific property values are about 30–40% higher than those of aramid. It should be pointed out that both PE and aramid fibres, as is true of most organic fibres, must be limited to low temperature (less than 150°C) applications (Fig. 8.27).

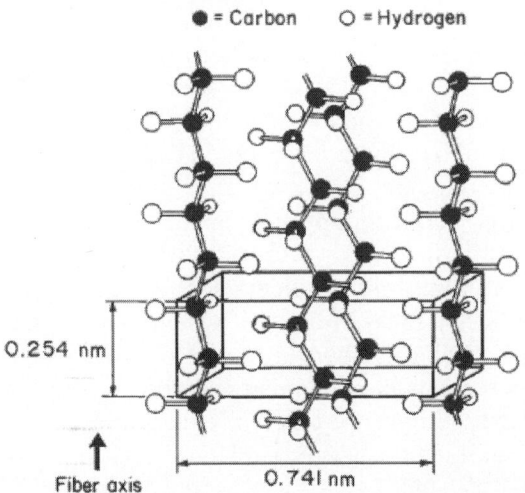

Figure 8.27. Unit cell of a single crystal of polyethylene. The sides of the unit cell have dimensions 0.741, 0.494 and 0.254 nm

8.4.19   Other polyamides

8.4.19.1   Nylon 6T

It is product of the condensation of hexamethylene diamine with terephthalic acid.

Polyhexamethylene Perephthalamide (Nylon 6T)

Nylon 6T melts at 370°C. This temperature is too high to permit an effective melt spinning. Hence, usually it is spun from solutions in sulphuric acid. One can see that the structure of nylon 6T appears to be a combination of the structures of nylon 66 and a PET fibre (polyester). Thus nylon 6T has many common properties of nylon 66 and polyester. They have the low density, moderate moisture regain, high abrasion resistance, easy dyeing, high alkali resistance and excellent elastic properties associated with nylon, and the high initial modulus, especially at elevated temperatures, outstanding resistance to stretch and a very high recovery at high temperatures, of the polyester fibres. Nylon 6T retains its full strength after 5 h at 185°C. It discolours after 1 h at 220°C.

8.4.19.2   Nomex, Kevlar, etc.

Nomex is practically flameproof. Kevlar, poly (*p*-phynyleneterephthalamide), represents a breakthrough in high modulus aromatic polyamide fibre and mainly used as reinforcement tyres, conveyers belt, etc., PBI is obtained by reacting diaminobenzidine and diphenylisophthalate and is used as an alternate to asbestos in high temperature filtration and thermal protection clothing. Its moisture regain is high at 14.4%, which permits high degree of comfort.

Kermel is a polyamide-imide fibre made from either trimillitic anhydride chloride and a diamine or a diisocyanate. These polymers have reasonable thermal stability, very resilient and have excellent flame resistance. The Monsanto's X-500 class of fibre, of which that prepared from polyamide-hydrazide is typical. Developments in melt spinning have led to filaments with novel cross sections or containing cavities to obtain improved properties, such as increased cover, a crisp, silk-like, firm hand, reduced pilling, increased bulk, sparkle effects and heightened resistance to soil. Generally, normal nylon has a rounded cross section and Quina has trilobal cross section. Cadon (Monsanto) has a rounded square cross section with four cavities (Fig. 8.28).

Repeating Unit	Fibre
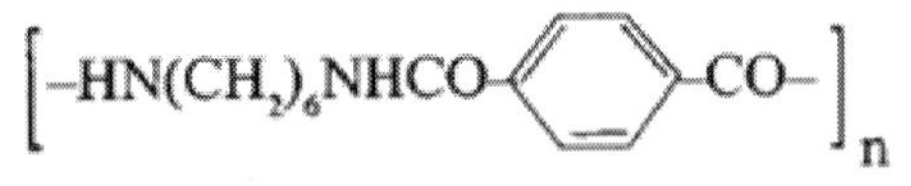 $\left[-HN(CH_2)_6NHCO-\bigcirc-CO-\right]_n$	Nylon 6T
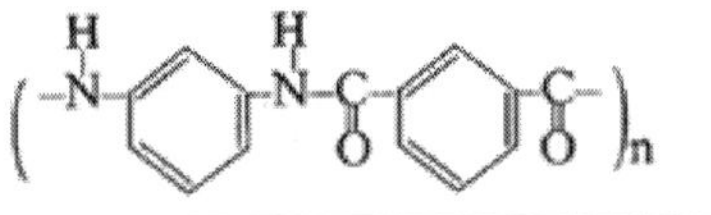 $\left(-\overset{H}{N}-\bigcirc-\overset{H}{N}-\overset{O}{\underset{}{C}}-\bigcirc-\overset{}{\underset{O}{C}}-\right)_n$	Nomex, Kevlar (du Pont)
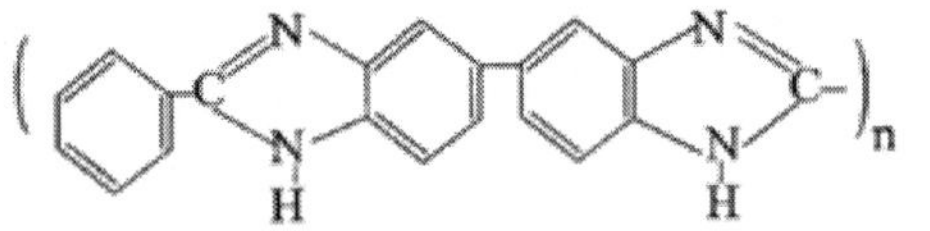 PBI structure	PBI (Celanese Corpn)
 Quina structure	Quina (du Pont)
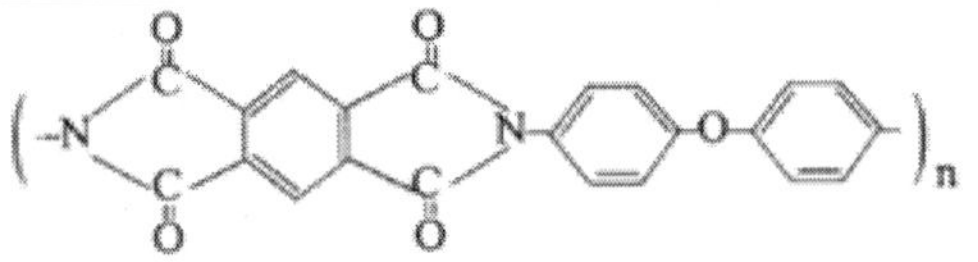 Polyimide structure	Polyimide
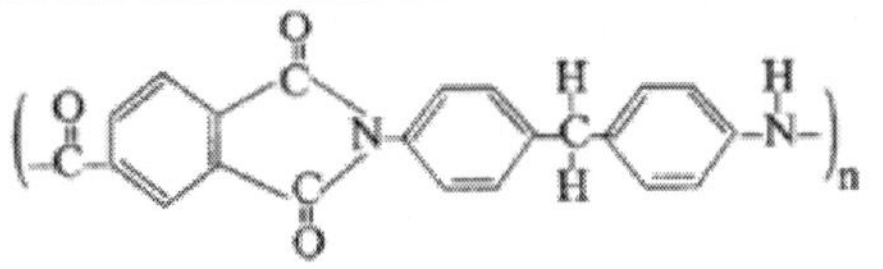 Kermel structure	Kermel (Rone Poulene)
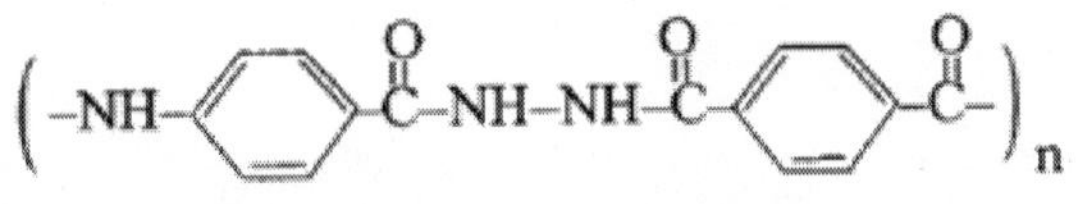 $\left(-NH-\bigcirc-\overset{O}{\underset{}{C}}-NH-NH-\overset{O}{\underset{}{C}}-\bigcirc-\overset{O}{\underset{}{C}}-\right)_n$	X-500 (Monsanto)

Figure 8.28. Repeating units of different fibres

8.4.19.3   Microfibres

The development of technology to manufacture by extruding through spin-neret holes of extremely small size, producing microfine filaments of less than 1.0 denier were a breakthrough in fibre technology. Such filaments were earlier made by making bicomponent filaments and on further processing split into ultrafine fibres. These ultrafine fibres are generally called microfibres. The microfibres have been manufactured maintaining the same the strength, uniformity and processing characteristics expected by textile manufacturers and consumers. The definition of microfibre, as accepted in the trade, is a fibre finer than 1.2 dtex for polyester and finer than 1.0 dtex for polyamide. These fibres are finer than luxury natural fibres such as silk. The most popular technology to manufacture such fine fibres entails the bicomponent spinning technique called the island in the sea as explained earlier in this book. A bi-component fibre containing the ultrafine filament in the core (island) and a sheath material (sea) is first spun and processed as in conventional fibre spinning. Later, the 'sea' is either dissolved or peeled away by drawing, leaving a very fine denier filament (the island). The ultrafine denier microfibre has unusual properties. Since the small filaments pack closely together and trap air pockets, they provide insulation and barrier to loss of body heat and assure comfort on chilly days. Thus, a polyester microfibre raincoat or jacket is much lighter and more comfortable than one made from conventional denier fibre. The close packing of fibres also gives the fabric the ability to repel rain, but at the same time allows the fabric to 'breathe'.

Other Synthetic Fibres

9.1 Polyvinyl alcohol fibres

Polyvinyl alcohol fibre cannot be produced from vinyl alcohol, as the latter does not exist in the free state and at the moment of its formation is isomerized into acetaldehyde:

$$- CH_2 - CHOH \longrightarrow CH_2 - C\!\!\begin{smallmatrix}\nearrow O\\ \searrow H\end{smallmatrix}$$

Vinyl Alcohol   Acetaldehyde

Usually the initial substance for producing polyvinyl alcohol fibre is vinyl acetate, from which polyvinyl acetate is obtained:

$$nH_2C = CH \longrightarrow (- H_2C\ CH -)n$$
$$\qquad OCOCH_3 \qquad\qquad\qquad OCOCH_3$$

Then, by polyvinyl acetate saponification, polyvinyl alcohol is obtained:

$$(- H_2C\ CH -)n \xrightarrow{+ nH_2O} (- H_2C\ CH -)n\ nCH_3COOH$$
$$\quad OCOCH_3 \qquad\qquad\qquad OH$$

9.1.1 Manufacture

The vinyl acetate is dissolved in methanol and is polymerized with the help of a catalyst (e.g., peroxide or azo-compound), forming polyvinyl acetate. Caustic soda is added to the methanol solution, bringing about saponification of the polyvinyl acetate to polyvinyl alcohol. This is precipitated from the methanol solution, pressed and dried.

9.1.2   Spinning

9.1.2.1   Wet spinning

Polyvinyl alcohol is soluble in water and its aqueous solution is used for fibre formation. The polymer is dissolved in water to form a 14–16% solution, which is filtered and pumped through spinnerets. The jets emerge into an aqueous coagulating bath containing sodium sulphate solution.

9.1.2.2   Dry/melt spinning

Polyvinyl alcohol fibres may also be spun by a process which combines features of dry and melt spinning. The polymer is dissolved in water under pressure and made into a highly concentrated solution (30–50%). The hot molten mass is forced through spinnerets, and the jets emerge into a hot air stream which evaporates the solvent to leave solid filaments of polyvinyl alcohol. These are hot drawn to increase the tenacity (Fig. 9.1).

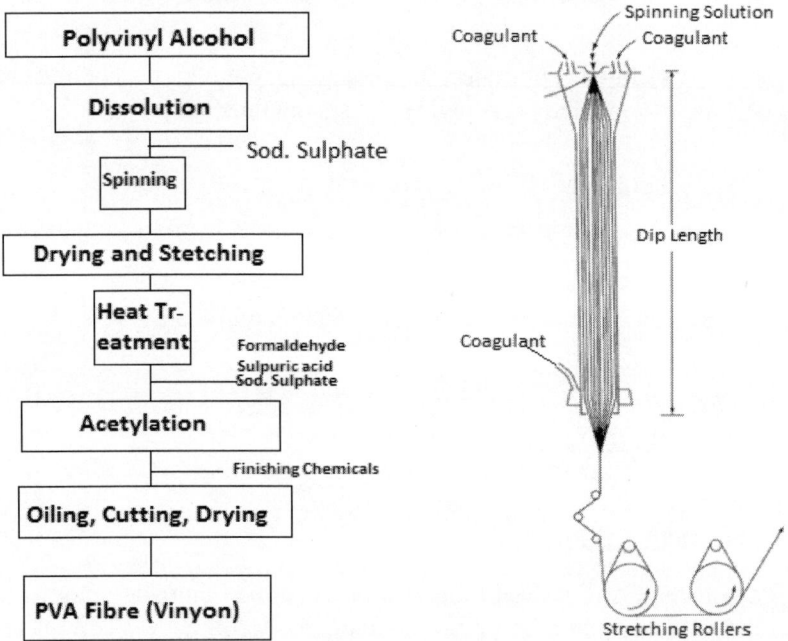

Figure 9.1. Flow diagram of PVA fibre production spinning process

9.1.3   Insolubilization

For eliminating this undesirable property of the fibres, it is subjected to additional treatments to form chemical cross links between the macromolecules. Most often, macromolecule cross linking is achieved by acetalization, using aldehydes and particularly formaldehyde for this purpose. In this case, both intramolecular and intermolecular acetal linkages are formed. The resulting product has the following structure:

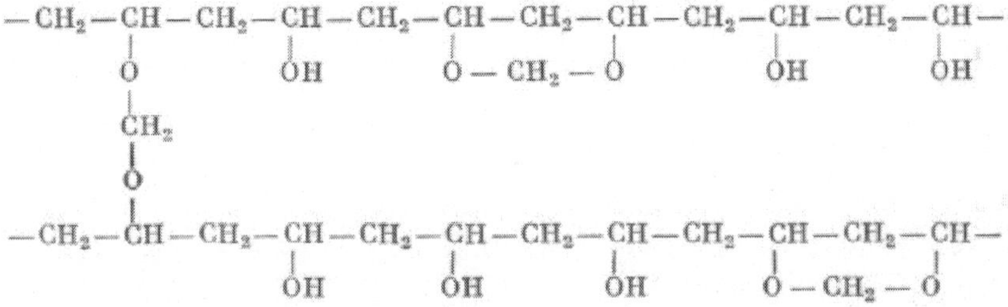

$$-CH_2-CH-CH_2-CH-CH_2-CH-CH_2-CH-CH_2-CH-CH_2-CH-$$

$$\begin{array}{cccccc} & | & | & | & | & | \\ & O & OH & O-CH_2-O & OH & OH \\ & | \\ & CH_2 \\ & | \\ & O \\ & | \end{array}$$

$$-CH_2-CH-CH_2-CH-CH_2-CH-CH_2-CH-CH_2-CH-CH_2-CH-$$

$$\begin{array}{cccccc} | & | & | & | \\ OH & OH & OH & O-CH_2-O \end{array}$$

The treatment with formaldehyde is carried out in such conditions that about 30–40% of the total amount of hydroxyl groups contained in polyvinyl alcohol fibre interact with formaldehyde. Among other aldehydes used for acetylization are benzaldehyde, $C_6H_5 - C \overset{\displaystyle O}{\underset{\displaystyle H}{\diagdown}}$, chloracetaldehyde,

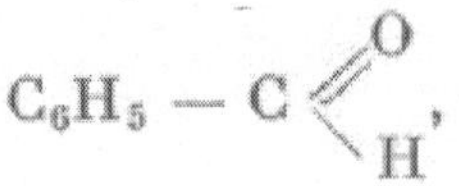

$ClCH_2 - C \overset{\displaystyle O}{\underset{\displaystyle H}{\diagdown}}$ and some dialdehydes.

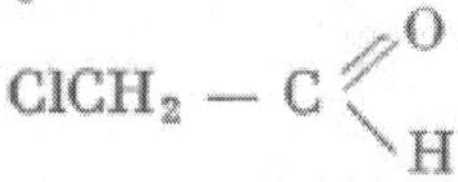

It should also be noted that polyvinyl alcohol fibre can be insolubilized not only by cross linking of macromolecules but also by increasing their orientation, with the corresponding intensification of intermolecular interaction. Thus, for instance, when polyvinyl alcohol with a stereospecific structure is used for manufacturing the fibre, the resulting fibre will be insoluble in water and will not require acetalization.

These fibres have a breaking length of 25–50 km. The breaking length of fibres produced from the stereospecific polymer is up to 60–80 km. The elastic properties of polyvinyl alcohol fibre are inferior to those of most of the other synthetic fibres, but better than those of natural and artificial fibres. At a relative humidity of 65%, polyvinyl alcohol fibre absorbs about 5% of moisture, i.e., its hygroscopicity is close to that of cotton fibre. This fibre is somewhat hydrophilic owing to the considerable number of –OH groups contained in it, which distinguishes it favourably from all other carbon–carbon chain fibres. But because the fibre is hydrophilic, there is a certain loss of strength in the wet state (10–15%). Polyvinyl alcohol fibre is characterized by a high abrasion resistance; it softens at a temperature of 200°C and is resistant to the action of acids and alkalies as for instance, to a 20% sulphuric acid solution at a temperature of 20°C and 5% sulphuric acid solution at a temperature of 65°C and to boiling diluted caustic soda solutions. However, concentrated solutions of mineral acids at an elevated temperature may cause the hydrolysis of intermolecular acetal linkages which leads to the dissolution of the fibre. The fibre is soluble in formic acid at a temperature of 55°C as well as in phenol and cresol.

In addition to hydroxyl groups found in the usual 1,3-position of the

$$- \underset{\underset{OH}{|}}{CH} - CH_2 - \underset{\underset{OH}{|}}{CH} -$$

macromolecules of the PVA fibre, it may also contain a small amount (up to 2% of the total amount of –OH groups) in glycol groupings.

$$- CH_2 - \underset{\underset{OH}{|}}{CH} - \underset{\underset{OH}{|}}{CH} - CH_2-$$

This has the disadvantage of reducing the degree of crystallinity of the fibre and decreases the resistance to the action of oxidizing agents (Table 9.1). Polyvinyl alcohol fibres are not attacked by microorganisms. Fibres containing reactive alcohol hydroxyl groups can form different derivatives, owing to which there are great possibilities for the development of further chemical modifications of these fibres.

Table 9.1. Properties of polyvinyl alcohol

Properties	Polyvinyl alcohol	
	Staple	Filament
Tenacity, cN/tex (g/den) dry, wet	33.5–54.7 (3.8–6.2) 28.3–44.1 (3.2–5.0)	53.0–75.1 (6.0–8.5) 44.1–67.1 (5.0–7.6)
Tensile strength	3.8	
Elongation, % dry (wet)	13–26 (14–27)	9–22 (10–26)
Elastic recovery	65–85 at 3% stretch	70–90 at 3% stretch
Specific gravity	1.26–1.30	1.26–1.30
Effect of moisture	Moisture regain 4.5–5.0%	Moisture regain 3.0–5.0%
Thermal properties	Starts shrinking at 220–230°C. At 220°C yellow and softens after 230–250°C	Starts shrinking at 220–230°C. At 220°C yellow and softens after 230–250°C
Flammability	PVA fibres do not burn readily	PVA fibres do not burn readily

| Properties | Polyvinyl alcohol | |
	Staple	Filament
Effect of sunlight	Do not get affected easily, 100 h or more exposure weakens the fibre, no colour change	Do not get affected easily, 100 h or more exposure weakens the fibre, no colour change
Effect of acids	It has a good resistance, normal conditions	
Effect of alkalis	Excellent resistance to most alkalis	Resistance is generally good
General	They have excellent resistance to chemicals, bleaches, urine, perspiration, reducing and oxidizing agents, etc.	They have excellent resistance to chemicals, bleaches, urine, perspiration, reducing and oxidizing agents, etc.
Effect of organic solvents	Swells or dissolves by phenol, cresol, formic. Not affected by animal, vegetable and mineral oils, and to most common organic solvents	Not affected by most of the solvents, oils
Cross section (enlarged)		
Enlarged view of fibres (longitudinal)		

9.2   Polyvinyl chloride fibres (vinyon)

(a) The polyvinyl chloride (PVC) fibres are produced from the polymers of vinyl chloride.

$$n\ CH_2 = \underset{Cl}{CH} \longrightarrow \ - CH_2 - \underset{Cl}{CH} - CH_2 - \underset{Cl}{CH} -$$

(b) It can also be produced from copolymers obtained from the following chemicals:

- From vinyl chloride (10–60%) and vinylidene chloride (saran):

$$- H_2C - CHCl - CH_2 - CCl_2 -$$

- From vinyl chloride and vinyl acetate (15%) (vinyon):

$$- H_2C - CHCl - CH_2 - \underset{OCOCH_3}{CH} -$$

- From vinyl chloride (40–60%) and acrylonitrile (vinyon H, dynel):

$$- H_2C - CHCl - CH_2 - \underset{CN}{CH} -$$

- From vinylidene chloride (60%) and acrylonitrile:

$$- H_2C - CCl_2 - CH_2 - \underset{CN}{CH} -$$

(c) The PVC subjected to additional chlorination:

$$- CHCl - CHCl - CH_2 - CHCL - CHCl - CHCl - CH_2 -$$

9.2.1   Manufacture

The PVC is produced by the polymerization of vinyl chloride.

1. *Starting from acetylene and hydrogen chloride*: Acetylene can be reacted with hydrochloric acid in the presence of a catalyst like mercuric chloride. The reaction gives vinyl chloride

$$CH \equiv CH + HCl \xrightarrow[Chloride]{Mercuric} H_2C - CHCl$$
Aceylene   Vinyl Chloride

2. *Starting from ethylene and chlorine*: In this method, ethylene and chlorine are reacted to form dichloroethane, and vinyl chloride monomer is prepared by thermal decomposition of dichloroethane in the presence of an appropriate catalyst; hydrogen chloride is liberated as a by-product in this reaction.

$$H_2C = CH_2 + Cl_2 \longrightarrow ClH_2C - CH_2Cl \xrightarrow[Catalyst]{Pressure/Heat} H_2C = CHCl + HCl$$

Ethylene   Ehtylene Dichloride   Vinyl Chloride

Hydrogen chloride is reacted with ethylene and oxygen to form dichloroethane and dichloroethane thus formed is also decomposed to vinyl chloride. Water is liberated as a by-product.

$$H_2C = CH_2 + 1/2 \; O_2 + 2HCl \xrightarrow[(-H_2O)]{} CH_2Cl - CH_2Cl \longrightarrow CH_2 = CHCl + HCl$$

9.2.2   Polymerization

The aqueous solution of vinyl chloride id heated to 65°C in an autoclave and pressurized to 45–50°C when the polymerization takes place. The polymerization takes place and will be suspended in water. The polymer can be recovered by spray drying.

9.2.3   Spinning

Vinylidene chloride is copolymerized with small quantity of vinyl chloride to produce powdered resin.

9.2.3.1   *Dry spinning*

Dry spinning is usually employed for manufacturing filaments for textile use. The polymer separated as above is dissolved in acetone/carbon disulphide mixture at 70–100°C and pumped through the spinneret. The filaments come out in a hot air atmosphere whereby the solvents are evaporates and dry filament emerges. The evaporated solvents are recovered.

9.2.3.2   *Melt spinning*

Melt spinning has certain problem when used for spinning fine filaments. The poor thermal stability where by the polymer starts softening at 90–100°C. At 180–190°C, it starts decomposing. Hence the spinning temperatures cannot be raised high enough to permit of the production of fine deniers needed for staple and tow. Below the decomposition temperature, the viscosity of the molten PVC restricts the fineness of the extruded filaments to a diameter of about 0.2 mm. However, the resin is melted and extruded through a spinneret into water-cooling bath. While still being cooled, it is drawn out 400% to improve crystallinity and molecular orientation, increasing strength in a manner similar to nylon.

These fibres are characterized by good physical and mechanical properties, high resistance to acids, alkalies and oxidizing agents. The chemical stability, of vinyon is somewhat lower due to the presence of units containing residues of acetyl groups. The number of solvents which can be used for each fibre is limited. This particularly concerns PVC fibres and fibres produced from copolymers of vinyl chloride and vinylidene chloride which are not readily soluble in easily available solvents. Fibres produced from copolymers of vinyl chloride with vinyl acetate and acrylonitrile as well as chlorinated PVC fibre are soluble in acetone.

All the fibres belonging to this group do not contain hydrophilic functional groups, are hydrophobic and do not swell in an aqueous medium. A disadvantage of these fibres is their low thermal stability; most of them begin to deform, shrink and soften at temperatures from 60 to 100°C. Fibres containing acrylonitrile in the macromolecules (vinyon , dynel) exhibit somewhat greater thermal stability. Considering the heat instability and difficulties encountered in dyeing, the possibilities of applying PVC fibres for manufacturing consumer goods are rather limited. These fibre are principally used for manufacturing fabrics for industrial uses.

The PVC (85–86.5%) and polyvinyl acetate (15–13.5%) are copolymerized. Resultant vinyl resin copolymer fibre is dissolved in acetone, filtered, deaerated and stored in heated tanks as a viscous solution. Later extruded through spinneret down a hollow tube in the presence of a current of warm air to evaporate the acetone, producing filaments in tow form. Tow is passed through lubricating bath and cut into staple. They are relatively weak, highly extensibile (stretchable), water repellent, highly resistant to acids and alkalies. They do not take dyes well, are nontoxic, do not mildew or support bacterial growth, do not attacked by moth/larvae/beetles and do not support combustion. They are used for bonding fibre for nonwovens, heat-sealable papers, filled and needled felts and bonded fabrics used for moulding, embossing, bulking, not suitable for spinning into yarn, however, can be blended into fabrics for industrial applications.

Its natural colour pale gold or straw. It has excellent abrasion resistance and stretchable, rather stiff but essentially nonabsorbent. It has good resistance to bleaches, alkalis, acids and resistant to mildew, bacteria and insects. It is nonflammable (Fig. 9.2).

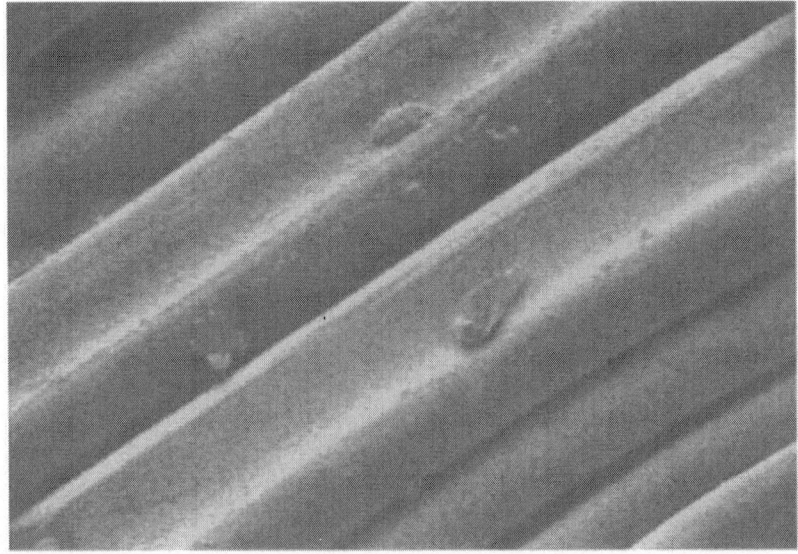

Figure 9.2. SEM view of vinyon (×1400)

Some of the properties of PVC are summarized in the Table 9.2.

Table 9.2. Properties of PVC filament

Properties	PVC filament
Tenacity	24–27 cN/tex (2.7–3.0 g/den), dry/wet
Tensile strength	32–36 kg/mm^2
Elongation	12–20%, wet or dry
Specific gravity	1.38–1.4
Effect of moisture	Water absorption is virtually nil, and the fibre does not swell in water
Thermal properties	Shrinks on heating above 70°C. Decomposition begins at about 180°C
Flammability	PVC fibres are inherently nonflammable
Thermal conductivity	74×10^{-2} (Wm^{-1}K^{-1})
Effect of sunlight	Excellent resistance
Effect of acids	No effect
Effect of alkalis	No effect

Properties	PVC filament
General	They have excellent resistance to chemicals, bleaches, urine, perspiration, reducing and oxidizing agents, etc.
Effect of organic solvents	Fibres are swelled by toluene, trichlorethylene, benzene, carbon disulphide, ethyl acetate, acetone, chloroform, methylene chloride and nitrobenzene
Cross section (enlarged)	

9.3   Saran

The generic term 'saran' is used for a manufactured fibre in which the fibre forming substance is any long chain synthetic polymer composed at least 80% by weight of vinylidene chloride units. It is manufactured from polymers or copolymers of vinylidene chloride:

$$CH_2 = CCl_2 \quad ----> \quad -CH_2 - CCl_2 - CH_2 - CCL_2 - CH_2 - CCl_2 -$$
Vinylidene Chloride   Polyvinylidene Chloride

They may contain a small (<15%) proportion of other monomers, commonly vinyl chloride. Other monomers may be included in minor proportions, such as acrylonitrile.

9.3.1   Manufacture

It is mainly manufactured by polymerization of vinyl chloride and vinylidene chloride. The production of vinyl chloride has been explained earlier.

Vinylidene chloride may be prepared the following way: chlorination of 1,2-dichloroethane yields trichloroethane. This can also be produced by the chlorination of vinyl chloride or ethylene.

$$CH_2Cl - CH_2Cl + Cl_2 \longrightarrow CH_2Cl - CHCl_2$$

1, 2 Dichloroethane   Trichloroethane

$$CH_2 = CHCl + Cl_2 \longrightarrow CH_2Cl - CHCl_2$$

Vinyl Chloride   Trichloroethane

$$H_2C = CH_2 + Cl_2 \longrightarrow CH_2Cl - CHCl_2$$

Ethylene   Trichloroethane

Trichloroethane is converted to vinylidene chloride, either by pyrolysis at 400°C, or by treatment with lime with the removal of hydrochloric acid.

$$CH_2Cl - CHCl_2 \xrightarrow[\text{(-HCl)}]{\text{Pyrolysis, 400 C}} CH_2 = CCl_2$$

Trichloroethane   Vinylidene chloride

$$\xrightarrow{\text{Lime (- HCl)}} CH_2 = CCl_2$$

9.3.2   Polymerization

Polymerization of vinyl chloride and vinylidene chloride in aqueous emulsion in the presence of a catalyst yields the polymer of saran. Reaction is allowed continue to get a molecular weight of 20,000–22,000 and separated.

9.3.3   Spinning

Saran is usually melt spun through spinnerets at 180°C. The filaments, as soon as they come out of the spinnerets, are quenched and cold drawn to achieve the required tenacity.

Main properties of saran are summarized in Table 9.3.

Table 9.3. Properties of saran

Properties	Saran
Tenacity	20.3 cN/tex (2.3 g/den), Dry/wet
Tensile strength, kg/cm^2	1050–3150
Elongation	15–30%, wet/dry
Elastic recovery	98.5% at 3% elongation

Properties	Saran
Specific gravity	1.1–1.7
Effect of moisture	Regain: 0.1–1.0%
Thermal properties	Softening point: 115–160°C. Sticking point: 99–104°C. Melting point: 171°C.
Flammability	Saran is almost nonflammable
Effect of sunlight	Good resistance, but discolours slightly on prolonged exposure
Effect of acids	Excellent resistance to most acids
Effect of alkalis	Excellent resistance to most alkalis. Ammonium hydroxide causes discolouration
General	They have excellent resistance to chemicals, bleaches, urine, perspiration, reducing and oxidizing agents, etc.
Effect of organic solvents	Soluble in cyclohexanone, dioxan and tetrahydrofuran. Not affected by alcohols or aliphatic hydrocarbons
Cross section (enlarged)	
Enlarged view of fibres (longitudinal)	

Used as monofilament yarn in Hosiery, car seat covers, filter fabrics, outdoor furniture tape, insert screening, grille fabrics and as multifilament yarn: filter cloth, upholstery, drapery and rope. The chemical resistance and nonflammability of PVC fibres have enabled them to find important uses in the industrial field. Typical applications include: waddings, filter cloths, braiding, piping and other uses in the chemical industry; battery fabrics; protective clothing; tarpaulins, awnings, curtains, fishing nets, etc. Fairings and canvas

Repeat Unit	Fibre
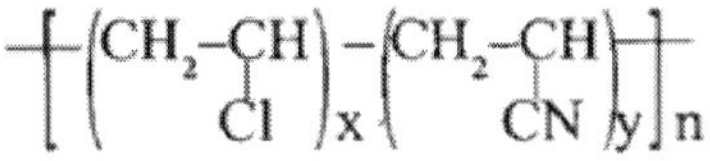 $-\!\!\left[\left(CH_2\text{--}\underset{Cl}{CH}\right)_x\!\!\left(CH_2\text{--}\underset{CN}{CH}\right)_y\right]_n-$ x = 60% y = 40%	Dynel (Union Carbide)
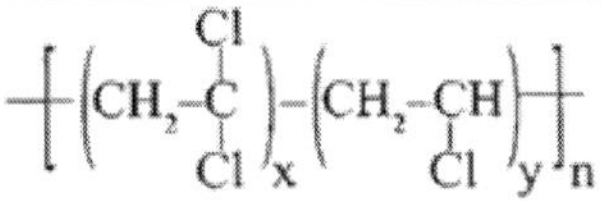 $-\!\!\left[\left(CH_2\text{--}\underset{Cl}{\overset{Cl}{C}}\right)_x\!\!\left(CH_2\text{--}\underset{Cl}{CH}\right)_y\right]_n-$ x > 80%, y < 20%	Saran (Dow Chem. Corpn)
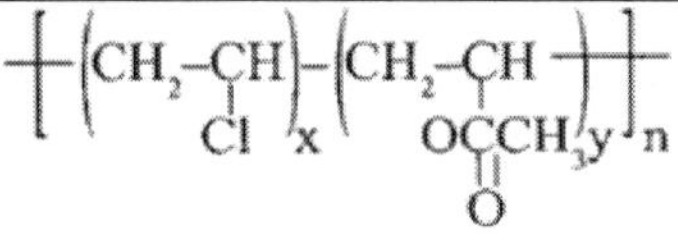 $-\!\!\left[\left(CH_2\text{--}\underset{Cl}{CH}\right)_x\!\!\left(CH_2\text{--}\underset{\underset{O}{O\overset{\shortparallel}{C}CH_3}}{CH}\right)_y\right]_n-$ x = 85% y > 15%	Vinyon (Avtex Fibres Inc.)

Figure 9.3. Repeat units of some of the chlorofibres

awnings for aircraft, gliders, boats, buoys, etc.; orthopaedic materials, artificial limbs, saddlery, etc.; accessories for textile machinery, billiard cloths (Fig. 9.3).

9.4 Novoloid (KYNOL)

A generic term for manufactured fibre containing at least 55% by weight of cross-linked Novolac manufacturing process. A cross-linking agent is blended with Novolac resin at temperatures below 40°C, rapidly melting the blend. Melted blend is fiberized before it can cure. Resulting fibres are cured by exposing to acidic gas. Its principal characteristics are it is flame resistant and natural colour gold. Novoloid has fair abrasion resistance and resilience and excellent resistance to mildew, insects and ageing. Used for flame retardant garments and protective clothing, home furnishings. It is blended with other fibres to enhance properties, while retaining much of the flame retardant characteristics.

9.5 Polycarbonate fibres

In a chemical sense, these are polyesters derived from carbonic acid. They are liner polymers containing the characteristic grouping –O–CO–O– as part of the repeating unit.

9.5.1   Manufacturing process

An aromatic dihydroxyl compound is condensed with suitable carbonic acid derivative such as carbonyl chloride or esters of carbonic acid. They are commercially produced from carbonyl chloride (Phosgene) and carbonic acid esters.

Fibre is extruded through a spinneret into a monofilament. Chemical composition can vary so significantly that specific properties cannot be provided. Presently, made only for very limited and specific purposes like basting for tailoring men's suits.

9.6   Polybenzimidazole fibre (PBI)

A manufactured fibre in which the fibre forming substance is a long-chain aromatic polymer having recurring imidazole groups as an integral part of the polymer chain.

9.6.1   Manufacturing process

The PBI is prepared from tetra-aminobiphenyl and diphenyl isophthalate spun via a dry spinning process using dimethyl acetamide as the solvent. It has no melting point and have good strength (stretch about 30% of its length), abrasion resistant. It is more absorbent than other man-made fibres and resistant to inorganic acid and organic solvents. It is used thermal protective clothing and used in application in aircraft, aerospace, industrial and medical uses.

9.7   Polyolefin fibres

Polyolefin fibres produced from polymers of unsaturated hydrocarbons of the type of olefins, such as ethylene, propylene, polyethylene and polypropylene, have not received wide application due to their heat instability. Recently, with the development of the synthesis of stereo-specific polymers which are characterized by high crystallinity, a closely packed structure, a relatively high energy of intermolecular interaction and, consequently, better physico-mechanical properties (softening and melting points), new ways have been found for producing polyolefin fibres with improved properties. Polyethylene and polypropylene fibres produced from stereo-specific polymers exhibit a great breaking length (35–80 km for propylene and 60–70 km for polyethylene) and high elasticity, which is inferior only to that of polyamide fibres.

The two important polyolefin fibres are polyethylene and polypropylene fibres.

9.7.1   Polyethylene fibres

Polyethylene has a simple, linear chain structure consisting of a carbon backbone and small hydrogen side groups. Such a structure makes it easy to crystallize. There are three common grades of polyethylene: low-density polyethylene (LDPE), high-density polyethylene (HDPE) and ultra-high molecular weight polyethylene (UHMWPE). It is the UHMWPE variety that is used to make high modulus (HM) fibres.

They are polymers or copolymers of ethylene. The polyethylene fibre is produced by either high pressure polymerization of ethylene with a peroxide-catalyzed process or low pressure polymerization of ethylene using new catalysts systems. The molecular structure of polyethylene is a linear polymer of ethylene units with repeat unit of:

$$\left[\begin{array}{c} \overset{\displaystyle H}{\underset{\displaystyle H}{|}} \;\; \overset{\displaystyle H}{\underset{\displaystyle H}{|}} \\ -C - C - \\ \end{array} \right]_n \qquad n = c.\,550$$

9.7.1.1   Manufacture

Generally, there are four processes:

1. High pressure processes
2. Ziegler processes
3. The Phillips process
4. Standard oil (Indiana) process

High pressure process: The conditions required for the polymerization is usually very severe like pressure of 1000–3000 atmospheres and temperature of about 80–300°C. Free-radical initiators, such as benzoyl peroxide or oxygen, are generally used, and conditions need to be carefully controlled to prevent a runaway reaction, which would generate hydrogen, methane and graphite rather than polymer. The polymerization may not take place as straight linear molecules but the molecules are branched, and polymer produced by this method may have as many as 30 branches for every 1000 carbon atoms in the molecular chain. Unfortunately, this method of polymerization results in branching and restricts the ability of polymer molecules to pack together, and prevents them aligning themselves into the orderly patterns that make for regions of crystallinity. In general, high-pressure processes tend to yield lower density poly(ethylenes), typically in the range 0.915–0.945 g/cc, which also have relatively low molar masses. This is reflected in its properties, especially with respect to the characteristics of filaments spun from it. The melting point of polyethylene made by this process, for example, is comparatively low – about 110–120°C.

This disadvantage and process difficulties made researchers to find processes less tedious and which can make less branching in the polymer chain. Thus methods were developed which is called low temperature which needs only temperature below 100°C and lower pressures in the presence of organometallic catalysts (e.g., of lithium, sodium and aluminium, in conjunction with a small amount of transition metal compound, e.g., titanium tetrachloride).

Ziegler processes: Ziegler processes are based on co-ordination reactions catalyzed by metal alkyl systems. Such reactions were discovered by Karl Ziegler in Germany and developed by G. Natta at Milan in the early 1950s. A typical Ziegler–Natta catalyst is the complex prepared from titanium tetrachloride and triethylaluminium. It is fed into the reaction vessel first, after which ethylene is added. Reaction is carried out at low pressures and low temperatures, typically no more than 70°C, with rigorous exclusion of air and moisture, which would destroy the catalyst. The poly(ethylenes) produced by such processes are of intermediate density, giving values of about 0.945 g/cc. A range of relative molar masses may be obtained for such polymers by varying the ratio of the catalyst components or by introducing a small amount of hydrogen into the reaction vessel.

Phillips and the standard oil (Indiana) process: These both yield high-density poly(ethylenes), using relatively low pressures and temperatures. This process can produce polymers with some four to five branches per 1000 carbon atoms in the molecular chain or even less. They are able to pack together more effectively and hence crystalline and due to this, their physical properties also differ from the high-temperature-pressure polymer. The density of low-temperature polymer is higher because the long unbranched molecules of the can pack closer together, so that the weight per unit volume is increased with a density of 0.95–0.96. The density of the earlier type of polyethylene is 0.92. The polymers made by the high-pressure, high-temperature method are usually called low-density polymers and the low-temperature process is called high-density polymers/fibres.

The details of these two processes are summarized in Table 9.4.

Table 9.4. Phillips and the standard oil (Indiana) process

Process	Catalyst	Pressure /Atm	Temperature, °C	Density of Product g/cc
Phillips	5% CrO_3 in finely divided silica/alumina	15–35	130–160	9.6
Standard oil (Indiana)	Supported MoO_3 with Na, Ca metal or hydride promoters	40–80	230–270	9.6

9.7.1.2   Spinning

Spinning methods differ slightly in case of high- and low-density fibres.

Low-density polymers are spun by melt spinning at a temperature of 205°C and spun as monofilaments as in the case of polyamide fibres. The molten polymers are extruded through spinnerets of round holes or other required shapes and the emerging monofilaments are immediately cooled to 15–60°C and cold drawn suitably to 4–40 times its length and wound on spools or tubes.

High-density polyester also is melt spun at 210°C and the extruded as monofilament. After spinning, the filament is drawn to 10 times its length at higher temperature, of the order 115–120°C, the filaments being heated by hot water, air or steam.

9.7.1.3   Properties

Table 9.5. Properties of polyethylene

Properties	Polyethylene	
	Low density	High density
Tenacity, cN/Tex, g/den	8.8–13.2, 1.0–1.5	6.5–8.0 g/den
Tensile strength, kg/cm²	15,000	90,000
Elongation % at break	45–50	10–20
Specific gravity	0.92	0.96
Effect of moisture	Moisture absorption of polyethylene is virtually nil	Moisture absorption of polyethylene is virtually nil
Softening point, °C	85–96	126–132
Thermal properties	Polyethylene does not degrade readily on heating. It can be heated to 315°C, in the presence of stabilizers	Effect is same as low-density fibre

Flammability	Polyethylene burns slowly in air, but fine filaments tend to melt and drop away	Same as low-density polyethylene
Effect of sunlight	The polyethylene molecule is attacked by oxygen, the reaction being stimulated by UV light	The effect is almost same as low-density polyethylene with more resistance to oxygen
Effect of acids	Polyethylene fibres have a high resistance to acids at all concentrations, even at comparatively high temperatures	More resistant than low-density polyethylene
Effect of alkalis	Polyethylene fibres have a high resistance to alkalis at all concentrations, even at comparatively high temperatures	More resistant than low-density polyethylene
General	Except oxidizing agents the polyethylene fibres is resistant to wide range of chemicals at normal temperatures	They have excellent resistance to chemicals, bleaches, urine, perspiration, reducing and oxidizing agents, etc.
Effect of organic solvents	They dissolve in some chlorinated hydrocarbons and aromatic solvents, e.g., benzene, toluene, xylene, etc at 70–80°C. They are insoluble in the common solvents	Not affected by most of the solvents. More resistant to solvents than low-density polymer
Enlarged cross section		
Fibre enlarged longitudinally		

The LDPE fibres are used for many industrial applications, including ropes and cordage, filtration fabrics and protective clothing. The HDPE fibres are used in aerial tow targets, in high-altitude balloons and in equipment to be used in Arctic conditions, where the retention of strength and flexibility at low temperatures are invaluable characteristics, twines and nets, etc. The various grades of poly(ethylene) have variety of other uses like pipes, packaging, components for chemical plant, crates and items for electrical insulation.

9.7.2   Polypropylene fibres

Polyethylene has many applications in the plastic industry, but its low melting point renders it unsuitable for fibre formation. Its homologue polypropylene exists in various forms according to the disposition of the substituent methyl groups; the isotactic polymers, in which these groups are all attached on the same side of the main carbon chain, can be spun and drawn into fibres. Researchers discovered catalysts promoting formation of isolactic polymers.

Propylene is catalytically polymerized to form linear using special catalysts, flexible-chain polymer according to the following equation:

$$nCH_3\text{-}CH = CH_2 \xrightarrow{\text{Catalyst}} \text{-}[CH_2\text{-}CH\text{-}]n\text{-}$$
$$CH_3$$

Propylene   Polypropylene

The polymerization process involved in addition, as the double bond is broken in the propylene molecule, the monomer or single molecules join or add together. The filaments are produced by the melt-spinning process in a similar manner to polyester. The polymer contains a repeat unit of $-CH_2-$ $HC(CH_3)-$ from the opening of its double bond, and the formation of a pendant methyl group. The reaction is exothermic. The polymerization of propylene is catalyzed by either Ziegler–Natta catalyst or metallocene catalyst. Polypropylene fibres and hydrophobic and resist chemical attack, but they are not readily dyed. Certain disperse dyes can be applied, but only pale and medium shades are obtainable. Many methods have been described in the patent literature whereby the fibre may be modified to confer affinity for acid or basic dyes. Side chains carrying polar groups may be grafted to the main polymer chain, or basic substances may be included in the melt from which the fibre is spun. Much attention has been paid to a method whereby a compound of a polyvalent metal such as nickel, zinc or aluminium is incorporated in the fibre, which can then be dyed with metalizable dyes.

Polypropylene is an important fibre, although it does not possess a very HM. The degree of crystallinity achievable in polypropylene is generally less than that in polyethylene. This is because polypropylene has side groups while polyethylene is a highly linear polymer. In general, bulky side groups make it difficult to obtain an orderly arrangement of molecular chains, i.e., crystallization. The spatial arrangement of groups (tacticity) is also important in this regard. Polyolefin fibres are spun from polymers or copolymers of olefins such as ethylene and propylene. Polypropylene fibres are made by melt spinning, involving extrusion through an extruder followed by thermal and mechanical treatment.

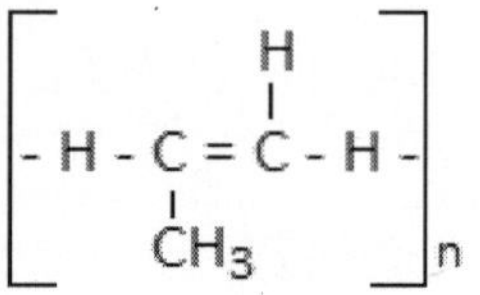

$$\left[- H - C = \overset{\overset{\displaystyle H}{|}}{C} - H - \right]_n \qquad n = c.1800$$
$$\qquad\quad \underset{\displaystyle CH_3}{|}$$

When the propylene (monomer) polymerizes, the pendent $-CH_3$ can lie in either direction. The randomly disposed CH_3 group (atactic form) does not form good fibre. In the isotactic form, the methyl groups are on the same side of the polymer backbone, but in syndiotactic arrangement the methyl groups lay alternately on either side of the polymer chain (see Figs. 9.4–9.6). To produce this form of polypropylene, special catalysts are used.

In isotactic polypropylene molecular structure bulky methyl groups take up angles of 120° with respect to neighbouring methyl groups. This makes possible a regular helical or spiral structure as shown Fig. 9.7.

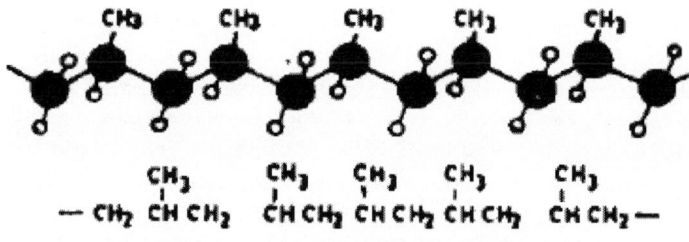

Figure 9.4. Isotactic polypropylene

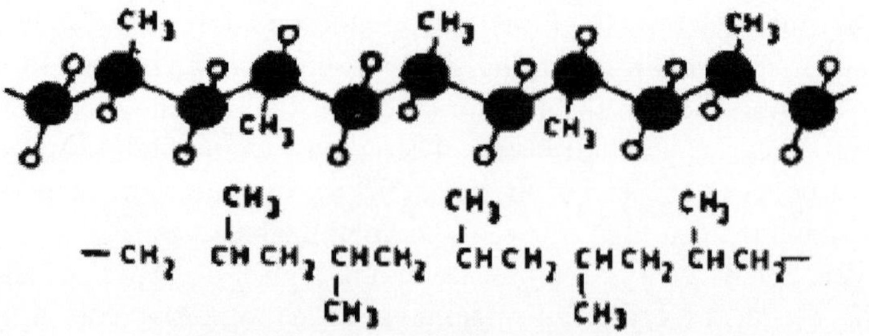

Figure 9.5. Syndiotactic polypropylene

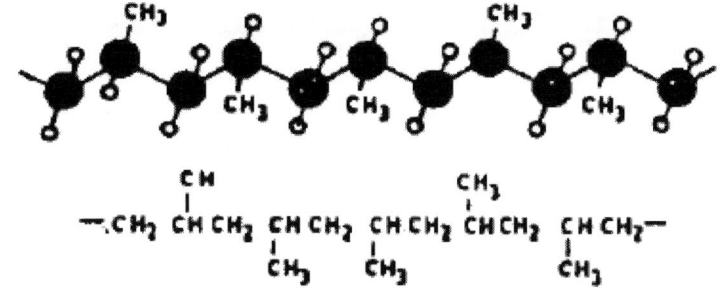

Figure 9.6. Atactic polypropylene

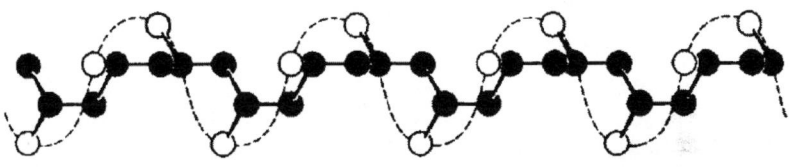

Figure 9.7. ◯ Methyl Group   ● Carbon Atom

9.7.2.1   Manufacture

The starting material polypropylene is available in plenty and at low cost from thermal and catalytic cracking processes in the petroleum industry.

9.7.2.2   Polymerization

The polymerization process is not well revealed in public. It is generally known that the polypropylene is polymerized under high pressure like 10 atmospheres and catalysts used is organometallic catalysts of aluminium in the presence of titanium chloride at temperatures less than 80°C. The catalyst in this reaction controls the way in which the polymer is built up, feeding each monomer molecule to the end of the polymer chain in such a way that it adds on in the desired position.

9.7.2.3   Spinning

Melt spinning process is followed where by molten high viscosity polymer is pumped to the spinneret and the emerging filaments are cooled by air and collected. This is further hot drawn to a filament size of 2.2–16.5 dtex (2–20 den) and used as multifilament. The extruder has a spinneret at the one end. As in other cases, the spinneret has orifices generally arranged in a circular fashion. The melt temperature is about 250°C. The filaments go through a water filled quench tank to pull rolls and then on to a draw oven, followed by draw rolls, an annealing oven and finally to a windup drum. As with any

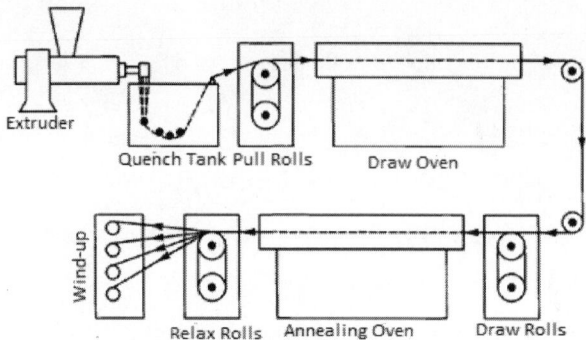

Figure 9.8. Schematic diagram of melt-spinning of polypropylene

other polymer, the degree of chain orientation in the polymer is a function of the draw ratio. In this case, the draw ratio is simply the ratio of the speed of the draw roll to that of the pull rolls. Commonly, this draw ratio is 9, which gives a high strength fibre. The annealing treatment serves to relieve any processing induced residual stresses (Fig. 9.8).

For staple form hundreds of multifilaments are joined into tow and drawn, crimped and cut to required size. Care has to be taken during melt spinning due to two main reasons. Compared to other polymers the polypropylene is crystallizes so rapidly that the undrawn filament itself is highly crystalline. Hence the cooling of the fibre, drawing conditions , drawing have high impact on the characteristics of the fibre. Secondly, the degradation of the fibre during these processes happens mainly due to oxidation in the presence of heat and sunlight. Hence, in the spinning stage itself stabilizers, which are mainly compounds capable of deactivating free radicals which promote the oxidative chain reaction, have to be added. (This stabilizers can be leached by solvents, especially by repeated dry cleaning, laundering, etc. and hence special stabilizers has to be used.)

Due to their chemical constitution, polyolefin fibres are hydrophobic and do not swell in an aqueous medium; their hygroscopicity is close to zero which renders dyeing difficult. One of the characteristic features of these fibres is their density, which is the lowest as compared with the density of other known fibres, being equal to 0.90–0.91 for polypropylene.

These fibres are capable of swelling and dissolving in a limited number of nonpolar organic solvents, such as tetralin (tetrahydronaphthalene), decalin and white spirit. Polyolefin fibres are highly resistant to the action of acids and alkalies in a wide range of concentrations and temperatures. Their resistance to the action of oxidizing agents is not so good; this particularly concerns polypropylene fibres containing the tertiary carbon atom, which is more easily oxidized. This is the reason why these fibres, especially polypropylene fibres, are insufficiently resistant to thermal and photochemical oxidation; considering the aforesaid it is advisable to use antioxidants when manufacturing these fibres. The thermal stability of polyolefin fibre is not as yet sufficiently high: polypropylene fibre softens at a temperature of 140°C and melts at a temperature of 160–165°C; but at 120°C, even when very small loads are applied, the fibre is liable to plastic deformation. The softening point of polyethylene fibre is still lower (127–132°C). Polyolefin fibres may find application in manufacturing consumer goods and, especially, industrial fabrics (Fig. 9.9).

Properties of polypropylene is summarized in Table 9.6.

Table 9.6. Properties of polypropylene

Properties	Polypropylene
Tenacity, cN/Tex, g/den	26.5–44.1 cN/tex (3–5 g/den)
Tensile strength, kg/cm²	2450–4200
Elongation % at break	15–25 (Monofil), 20–30 (Multifil)
Elastic recovery	90–98 at 5% stretch
Specific gravity	0.90–0.94
Effect of moisture	Moisture absorption of polyethylene is virtually nil
Softening point, °C	150
Melting point, °C	160–170
Thermal properties	Mechanical properties of the fibre deteriorate with increasing temperature even below softening point
Thermal conductivity	6 (compared to air = 1)
Flammability	Polypropylene is attacked by atmospheric oxygen and the reaction is stimulated by sunlight
Effect of sunlight	oxygen, and the reaction is stimulated by sunlight
Effect of acids	Excellent resistance, similar to polyethylene
Effect of alkalis	Excellent resistance, similar to polyethylene
General	Polypropylene is inert to a wide range of chemicals
Effect of organic solvents	Excellent resistance, generally similar to polyethylene
Enlarged cross section	

Figure 9.9. SEM view of polypropylene (×1000)

Properties of olefin fibres:

1. Able to give good bulk and cover

2. Abrasion resistant

3. Low moisture regain, 0.01%

4. Resistant towards chemicals, mildew, rot, weather, perspiration, etc.

5. Stain and oil resistant

6. Very light weight. It has the lowest specific gravity of all fibres

Used mainly for sportswear. Other uses include socks thermal underwear.

9.8   Polytetrafluoroethylene fibres (PTFE)

This fibre was first introduced by DuPont Company, in the name of Teflon, which is supposed to be an extremely inert fibre with extraordinary resistance to chemicals. It is produced by the polymerization of tetrafluoroethylene.

9.8.1   Manufacture

Calcium fluoride, which is available as a mineral, is reacted with sulphuric acid to produce hydrogen fluoride, which on reaction with chloroform yields chlorodifluoromethane. Tetrafluoroethylene can be produced by the pyrolysis of chlorodifluoromethane.

$$CaF_2 + H_2SO_4 ----> 2HF + CaSO_4$$

$$2HF + CHCl_3 ----> CHClF_2 + 2HCl$$

Hydrogen   Chloroform   Chlorodifluo-
Fluoride   ro methane

$$2CHClF_2 \xrightarrow[\text{600-800 C}]{\text{Pyrolysis}} CF_2 - CF_2 + 2HCl$$

Chlorodifluo-   Tetrafluoroethane
ro methane

9.8.2   Polymerization

The tetrafluoroethane formed as above is separated, purified and heated under pressure in autocalave to undergo polymerization, as given below in the presence of a peroxide catalyst. The polytetrafluoroethylene formed as a powder which can be separated and dried. But for spinning purposes, the polymerization is done in such a way as to form a dispersion of PTFE of about 15%.

$$n \, CF_2 - CF_2 \quad ------> \quad \begin{array}{ccccc} F & F & F & F & F \\ | & | & | & | & | \\ -C & -C & -C & -C & -C- \\ | & | & | & | & | \\ F & F & F & F & F \end{array}$$

Tetrafluoroethane   Polytetrafluoroethylene

9.8.3   Spinning

The polymerized PTFE cannot be melt spun because it decomposes before melting or wet spun since it is insoluble. Hence, the polymer is manufactured as dispersion and pumped through the spinnerets and the filaments emerge into a coagulating bath of dilute hydrochloric acid. At this coagulated state, the filaments are weak since the polymer stay separate and not coherent. To make the filament strong, they are rapidly heater to 385°C and held at this temperature for few seconds to allow it to get sintered and fused into coherent fibres and strong. Further, it is cooled immediately and drawn three to four times its length.

9.8.4   Properties

The PTFE has an extraordinary resistance to chemicals and is attacked only by molten alkali metals, chlorine trifluoride and fluorine gas, hot and at high pressure. There is no known solvent for the polymer at normal temperatures, and it withstands elevated temperatures much better than any other organic plastic. It begins to decompose at about 300°C without melting. It is practically undyeable. They are tan to light brown as it is manufactured but can be bleached to white using oxidizing strong mineral acids and are very strong and having high density due to its structure – close packing of fluorine atoms around carbon atoms. The major properties of the fibre is summarized in (Table 9.7).

Table 9.7. Properties of PTFE -- multifilament

Properties	PTFE – multifilament
Tenacity, cN/tex (g/den) dry, wet	10.6–12.4 (1.2–1.4), 10.6–12.4 (1.2–1.4)
Tensile strength, kg/cm^2	2205–2625
Elongation, %, dry (wet)	15–32 (15–32)
Specific gravity	2.1
Effect of moisture	PTFE does not absorb moisture
Thermal properties	They have best thermal stability, flexible, tough and chemical-resistant. PTFE fibre loses its fibre properties 327°C. It retains a useful strength up to 205°C, and perform at temperatures as high as 288°C.
Flammability	Nonflammable; melts with decomposition
Effect of sunlight	Negligible
Effect of acids	PTFE fibre is completely inert, even to boiling sulphuric acid, to fuming nitric acid or to aqua regia
Effect of alkalis	They completely inert even to all strong, hot alkalis
General	In PTFE, fluorine atoms are packed around the carbon atoms and protecting them, it can be broken by only by fluorine gas at high temp. and pressure or chlorine fluoride. The carbon-to-carbon bonds are extremely strong, the only reagents that will break them being molten alkali metals
Effect of organic solvents	The only known solvents for PTFE are certain perfluorinated organic liquids at temperatures above 299°C.

Enlarged cross section

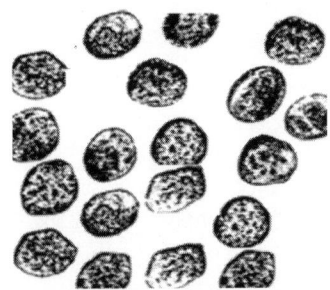

Fibre enlarged longitudinally

9.9   Elastomeric fibres/polyurethene fibres

Several elastomeric fibres developed in the United States are based on polymeric structures containing urethane (–NHCOO–) linkages. Full details of the processes used are not available but complex cross-linked polymers with rubber-like properties are obtained from polyesters or polyethers, containing terminal hydroxyl groups by means of a series of reactions involving diisocyanantes and diamines. Typical examples of such fibres are Lycra (Du Pont) and Vyrene (U.S. Rubber Co.), which are extensively used for foundation garments and swimsuits. These fibres have advantages over rubber in strength, resistance to oxidation, perspiration and cosmetic oils, also in whiteness and affinity for dyes. They are readily dyed by acid, basic and disperse dyes, but fastness properties are in general rather low. In short, elastomers are materials that show a very high degree of elastic, nonlinear strain. Most elastomers have a very low Young's modulus, E. Elastomeric materials, although they have rather low stiffness, possess very good damping properties. Their main characteristic is that they have a long range of nonlinear, reversible elastic strain. For example, an elastomer can be elastically stressed to a strain of 700% while steel can be elastically stressed only to about 0.1% strain. This is because in metals and ceramics, elasticity corresponds to small elastic displacements of atoms from their equilibrium position. Microstructurally, elastomers are long chain polymers with a glass-transition temperature, T, less than room temperature (293 K). Their unstrained state corresponds to a random coil structure of the macromolecular chains while in the strained state, these macromolecules become extended chains. Thus, straining results in chain orientation, which results in a decrease in entropy, S. Entropy is a measure of disorder in the chain configuration. We can write the free energy change on straining as (Gibbs equation):

$$\Delta G = \Delta H - T\Delta S$$

where ΔH is the change in enthalpy, T is the temperature and ΔS is the change in entropy. The enthalpy does not change when an elastomer is strained. Therefore, the free energy change resulting from straining an elastomeric band can be written as:

$$\Delta G = -T\Delta S$$

Consider a polymer chain extending between two points. The flexible chain can have many configurations between these points. For the configurational entropy, we can write an expression from statistical mechanics as:

$$S_{config} = k \ln p$$

where k is Boltzmann's constant and p is the number of configurations. When we stretch a rubber band, we move the two fixed points further apart and the number of chain configurations possible will be reduced compared to the number of configurations possible in the unstrained state. This means that the entropy, S, is reduced on straining. Since T is always positive and ΔS is negative on straining, we have a positive ΔG. That is to say that the strained state of a rubber band is thermodynamically not favoured! That is why when we release a strained rubber band (a high entropy state), the unstrained state is regained. The essential characteristic of an elastomeric fibre is that the material can be stretched several times its original length and when unloaded, it returns to its original dimensions. This characteristic can be imparted to a polymer in two ways:

1. Chemically cross-link flexible polymeric chains. For example, this is done by vulcanization of rubber by sulphur.
2. Make a two-phase, multiblock polymer, more commonly known as a segmented polymer. An example of such a polymer is the segmented polyurethanes.

Polyurethene polymers are made by a reaction taking place between small molecules, in which the linkage of the molecules occurs through the formation of urethane groups ($-NHCOO-$). Reaction of a glycol with a diisocyanate can give a linear polyurethane in which, in the reaction between isocyanate group ($-N=C-O$), and the hydroxyl group ($-OH$), the hydrogen atom of the hydroxyl group migrates to the nitrogen atom of the isocyanate and the residue of the alcohol is transferred to the carbon atom of the isocyanate group resulting in the formation of a urethane group, –NHCOO–. For example, toluene diisocyanate and butanediol:

The isocyanate-terminated prepolymer is then coupled to form the segmented polyurethane. Soluble polyurethanes that are essentially linear can be dissolved in any number of solvents and subjected to wet or dry spinning. As described above, in the wet spinning process, the fibres from the spinneret pass through a coagulating bath while in the dry spinning process they pass through an atmosphere that removes the solvent. Dry spinning is by far the most common method of making elastomeric fibres. The spinning dope has a variety of additives such as TiO_2 for delustering, stabilizers, antioxidants, lubricants, antitacking agents, dye stuff receptive agents, etc. The dope is extruded through a spinneret and the solvent is removed by flowing hot gases. Single filaments are coalesced to form a fused multifilament assembly. The spinning rates are from 200–800 m/min. Generally, a post-spinning stretching treatment is given at 150°C to obtain enhanced orientation of the hard segments parallel to the fibre axis and thus enhanced strength.

If the polyurethane molecule is allowed to form a three-dimensional structure, it becomes insoluble and neither wet nor dry spinning processes can be used. A chemical spinning process must be used in this case. The trick is to spin the isocyanate-terminated prepolymer at a stage where it forms a viscous melt, and with the jets produced in an environment that contains a chain extender. This chain extender diffuses into the fibres and reacts to couple the prepolymer molecules.

The importance of polyurethane today is that they are the basis of a novel type of elastomeric fibre which is known generically as *spandex*. During the beginning of the development of polyurethane researchers were able to produce fibres of the characteristics of nylon (e.g., Perlon U) with elastomeric character to the to the extent expected. What was probably needed was a polymer molecule in which longer segments of amorphous type structure which were separated by segments of molecule capable of developing powerful hydrogen bonds. The solution to these problems was found in the development of polyurethane polymers in which segments of the molecule are deliberately tailored to perform the desired functions. The lack of flexibility of the pliable constituent (with low melting point) was overcome by using preformed segments of molecule of considerable size (the 'soft' segments); the cohesion of the interchain bonding constituents ('hard segments') was assured through the use of urethane or urea groups. These engineered segmented fibre has provided the textile trade with an entirely new type of elastomeric fibre – elastane fibres or spandex fibres.

9.10   Elastane fibres

Elastane fibres are synthetic fibres, consisting of at least 85% polyurethane, built up of linear macromolecules, arranged in segments. In the United States, these fibres are generally referred to as 'Spandex'. Elastane fibres, together with rubber fibres ('elastoides'), are classified as stretch fibres. The most important characteristic of this group of fibres is their elasticity (they can stretch to up to eight times their length) and their ability to return almost to their original length when the deforming force is removed.

In addition, they are responsible for the behaviour at higher temperatures, the physical shape and the heat setting characteristics. The soft segments – long, unstructured and flexible – define the rubber-like, elastic behaviour of the fibre.

Elastic hard fibres may be prepared from polypropylene, poly(3-methylbut-1-ene), poly(oxymethylene) and poly(isobutene oxide) and largely based on 4,4′-methylenebis(4-phenylisocyanate), utilizing diamine or glycol extenders.

Spandex fibres have dog bone to rounder cross section and tend to be larger in diameter. Spandex have a density of 1.0–1.2 g/cm^3, moisture absorption is 0.3–3%, while elongation at break amounts 450–900% and 95% elastic recovery from 200% extension. This fibre has glass-transition temperature below 0°C so that soft components have a considerable freedom of movement at room temperature and above.

NO [CH2 CH2 CH2 CH2-O] ON O = C = N ◯ CH2 ◯ = N = C = O –>   O = C = N〰〰〰N = C = O

Polymer reaction

O = C = N〰〰〰N = C = O + Diamine   –>   〰HS〰〰〰HS〰

Chain extension reaction

9.10.1   Properties of polyurethane fibres

Polyurethanes that form the basis of these elastomeric fibres are made by reacting glycols and diisocyanates. Such elastomeric fibres are block copolymers. Their main characteristic is, as we said above, that they can be stretched repeatedly. Structurally, these block copolymers consist of a soft and flexible chain segment (random-coiled aliphatic polyethers or copolyesters) and a hard segment (aromatic-aliphatic polyureas). Urethane linkages connect the hard segment blocks to the soft segments. Fig. 9.10 shows schematically how an elastomeric fibre consisting of hard and soft segments can show high stretchability. The soft segment domains, comprising 65–90% by weight, are unoriented in the relaxed state of the fibre but straighten out when stretched. The soft segments allow the polyurethane to stretch while the hard segments act as anchors that allow the fibres to recover to their original shape when the load is removed. One can control the amount of stretching by controlling the ratio of hard to soft segments. Spandex fibres have linear density between 1 and 500 tex and a strain to failure in the range 400–800 (Table 9.8).

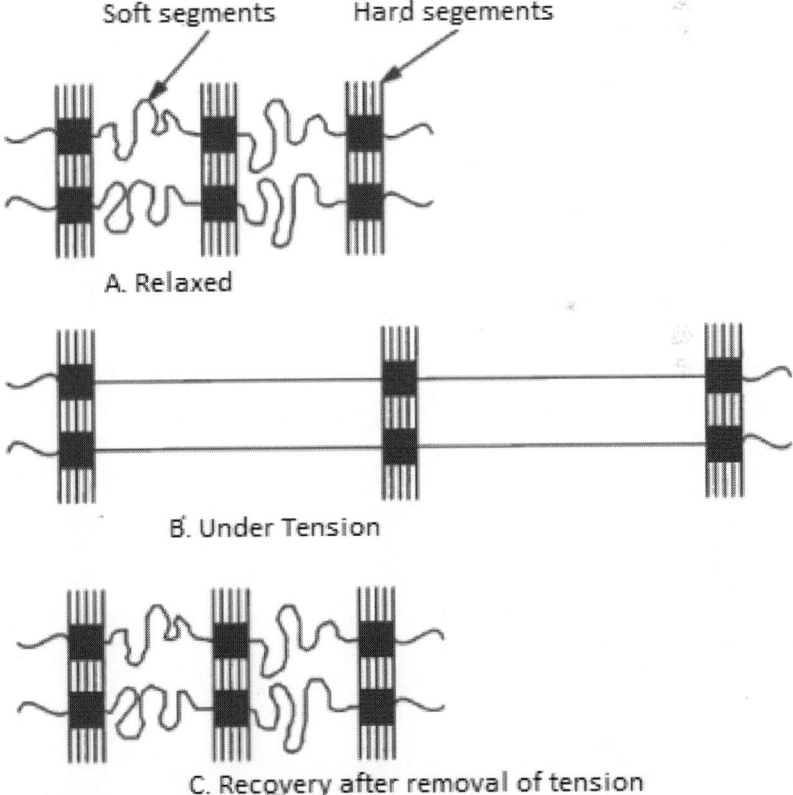

Figure 9.10. General structure of elastane fibres

Table 9.8. Physical properties of elastane

Properties	Elastane
Thread fineness (tex)	1–5000
Max. extension (%)	420–570 (normal modulus types)
Tensile strength (g/den)	0.7
Fusion temperature (°C)	From 170
Long-term stability at 130°C	Acceptable
Heat setting (thermofixation)	Possible
Dyeability	Possible
Specific gravity	1.2
Moisture regain	<1%
Decomposition temperature	Around 232°C (450°F)
Chlorine resistance	No tensile degradation (AATCC 162-1991)
Resistance to	
Hydrolysis	Moderate
Bleaching	Good
UV	No effect on physical property
Cosmetic oils	Good
Perspiration	Good

On igniting, elastane fibres melt and burn with a bright flame, giving off an acrid smell. They are largely unaffected by organic solvents. They tend to dissolve in highly polar solvents, such as dimethylformamide or dimethylacetamide. Elastanes are resistant to acids, alkalis, oxidizing and reducing agents under 'mild' conditions. Concentrated acids or alkalis cause a greater or lesser loss of elasticity, depending on time and temperature. Elastane fibres are unaffected by washing or the usual dry cleaning agents. Chlorine bleaching agents cause loss of strength and discolouration. LYCRA fibres, in particular, are resistant to mercerizing and to a slight carbonization. The resistance to chlorinated water, containing activated chlorine in the concentrations usual in swimming pools, is good. The other commercially available bleaching agents do not cause any damage.

Long exposure to light, particularly light with a high proportion of the UV fraction causes discolouration and photochemical degradation of the fibre. In this respect, polyester urethanes show greater resistance when compared to polyether urethanes. In general, elastane fibres are much more resistant to ageing and abrasion than rubber fibres. At temperatures above 170°C, an appreciable thermal degradation sets in, which makes itself evident as a yellowing and a reduction in the elastic properties of the fibres.

There are two types of polyurethene fibres. The two types of polyurethene fibres and their characteristics are given in following sections.

9.11   Polyester polyurethene

Polyester polyurethanes have the following special characteristics. They are good resistance to

(a) ageing

(b) chlorinated bath water

(c) light

(d) skin grease

(e) dry cleaning

(f) lower tendency to yellowing

(g) good fastness to NO_x fumes

9.11.1   Manufacture

There are three stages in the manufacture of elastane fibres. First stage is the production process is to manufacture the low molecular weight prepolymer. This forms the soft segment (amorphous) of the molecule. They are macroglycols with reactive hydroxyl groups at each end of the molecule and of molecular weight of 500–4000.

The second stage is the reaction of the prepolymer with diisocyanate.

The third stage is the coupling of the isocyanate terminated (second stage) prepolymer to form segmented polyurethane. The prepolymer is of two different types – polyester and polyether. Thus, there are two types of polyurethane fibres.

Polyester prepolymers are made by condensation of dicarboxylic acids with a slight excess of glycol; the condensation takes place until there are glycol units at each end of the polymer molecule, which thus has hydroxyl end groups

HOOC - R - COOH + HO - R' - OH -----> HO [R'OOC - R - COO]n R' OH

Dicarboxylic Acid   Glycol   Pre-Polymer (Polyester-PUE)

Next, macroglycol is reacted with an excess of diisocyanate, the hydroxyl groups on the ends of the macroglycol molecules reacting with isocyanate groups to form urethane groups. If two moles of diisocyanate are used per mole of macroglycol, for example, a prepolymer is formed in which each molecule now has isocyanate end groups. The viscosity of this isocyanate-terminated prepolymer may be adjusted by adding small amounts of an inert solvent, and then extruded into a coagulating bath that contains a diamine so that filament and polymer formation occurs simultaneously.

HO [R'OOC - R - COO]n R' OH + 2OCN - R - NCO ---> OCN - R - NHCOO - [R'OOC - R - COO]n R' OCONH - R - NCO
Pre-Polymer (Polyester-PUE)

OCN - R - NHCOO - [R'OOC - R - COO]n R' OCONH - R - NCO + 2OCN - R - NCO + H$_2$N - EX - NH + (TT)$_2$NH -->
(Excess)

(T)$_2$N(OCN - R - NHCOO - [R'OOC - R - COO]nCNO - R - NCON - Ex - NCON - R - NCON (T)$_2$
S(Soft segment ~85%)   Hard seg ~15%

The third step in making the segmented polyurethane consists in the creation of the hard segment by 'chain extension', i.e., coupling of the isocyanate-terminated prepolymer by reaction with low molecular weight bifunctional compounds, such as glycol, diamine or water. The reaction product is a polymer having hydrogen bonding sites in the form of urethane or urea groups, at least two of which will occur in the resulting 'hard segment'. In case of water, the reaction is little different. The exact quantity water is added so as to react with a proportion of the terminal isocyanate groups, forming prepolymer molecules with an isocyanate group on one end and an amine group on the other end. When this polymer is heated, the amine and the isocyanate groups react to bring about further polymerization and cross linking of the molecules.

·OCN - POLY- NCO + HO - R - OH ----> --OCONH - POLY - NHCOOROCONH - POLY- NHCOO---
Glycol

·OCN - POLY- NCO + 2HN-R'-NH$_2$ ----> - NHCONH - POLY - NHCONH - R'NHCONH - POLY - NHCONH--
Diamine

OCN - POLY - NCO + H$_2$O ---> OCN - POLY - NH$_2$ - CO$_2$

OCN - POLY - NH$_2$ + OCN - POLY - NH$_2$ ---> OCN - POLY - NH CO NH - POLY - NH$_2$

If branching or cross linking occurs during the polymerization process, it will render it insoluble or incapable of melting.

9.11.2   Spinning

Four different processes are currently used to produce spandex fibres commercially: melt extrusion, reaction spinning, solution dry spinning and solution wet spinning. These processes involve different practical applications of basically similar chemistry. If the diol or diamine(s) reaction with the prepolymer is carried out in a solvent, the resulting block copolymer solution may be wet- or dry-spun into fibre. Alternatively, the prepolymer may be reaction-spun by extrusion into a bath containing diamine to form a fibre, or the prepolymer may be permitted to react in bulk with a diol and the resulting polymer melt-extruded in fibre form (Fig. 9.11).

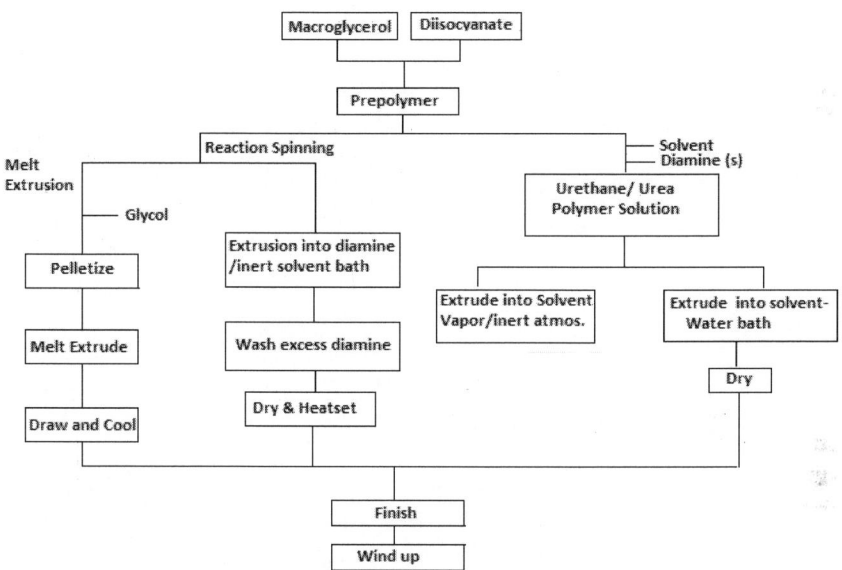

Figure 9.11. Flow chart of spandex manufacture by different processes

9.11.2.1   *Wet spinning*

The spinning process has to be selected based on whether the polymer is soluble or insoluble. Soluble polyurethane is dissolved in a suitable solvent and spun as per usual hard fibre spinning. Spinning solution is pumped by precision gear pumps through spinnerets into a solvent–water coagulation bath. The individual filament size is maintained at about 0.6–1.7 tex (5–15 den) to optimize solvent removal rates. At the exit of the coagulation bath, filaments are collected in bundles of the desired tex. A false twist may be imposed at the bath exit to give the multifilament bundles a more rounded cross section.

After the coagulation bath, the multifilament bundles are countercurrently washed in successive extraction baths to remove residual solvent, then dried and heat relaxed, generally on heated cans. Finally, a finish is applied and the multifilaments wound on individual tubes. A typical spinning line may produce 100–300 multifilaments at side-by-side filament spacings of less than 5 mm. Water is continuously added to the last extraction bath and flows counter currently to filament from bath to bath. Maximum solvent concentration of 15–30% is reached in the coagulation bath and maintained constant by continuously removing the solvent–water mixture for solvent recovery. Spinning solvent is generally recovered by a two-stage process in which the excess water is initially removed by distillation followed by transfer of crude solvent to a second column where it is distilled and transferred for reuse in polymer manufacture (Fig. 9.12).

In wet-spinning processes, spinning speeds are limited to about 100–150 m/min by hydrodynamic drag of the bath medium. It is this limitation that has apparently caused most spandex fibre producers to choose dry-spinning techniques. However, this limitation has been minimized by subjecting the spandex filament to drawing as much as three to four times after the spinning bath. Temperatures and residence times are selected so that the filaments are brought to temperatures above their second-order transition points, i.e., the hard segment melting points. This allows the molecular

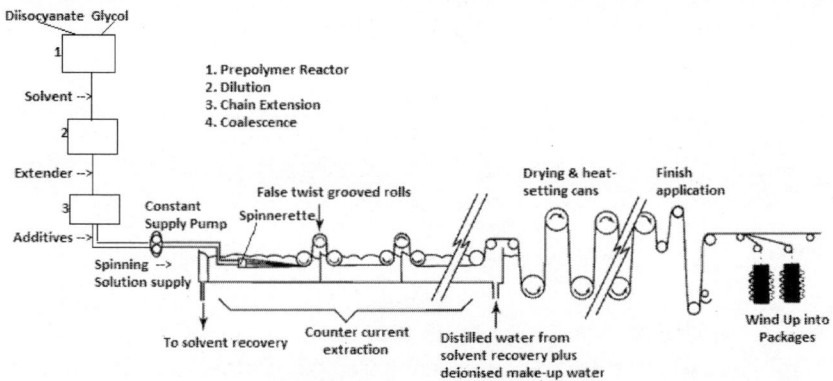

Figure 9.12. Wet spinning process

chains to move freely to relieve stresses and results in filaments of fine tex but with similar mechanical properties as the heavier tex feed. Thus it is possible to windup fibres from a wet-spinning process at speeds in excess of 300 m/min by continuously drawing and heat relaxing the filaments after drying.

9.11.2.2   Dry spinning

The insoluble type three-dimensional polyurethanes are spun in different way. The isocyanate-terminated prepolymer is spun at a stage when it forms a viscous dope, the jets emerging into a gaseous or liquid environment containing a chain extender which diffuses into the fibre and reacts. The prepolymer molecules are linked into their final form, producing the branched or cross-linked polyurethane in fibrous form (Fig. 9.13).

Any urethane–urea polymer that can be wet spun may also be dry spun; however, the productivity constraints of wet-spun processes have limited their utility. On a worldwide basis, greater than 90% of all spandex fibres are produced by various adaptations of dry spinning. The polymer spinning solution is metered at a constant temperature by a precision gear pump through a spinnerets into a cylindrical spinning cell 3–8 m in length. Heated cell gas, made up of solvent vapour and an inert gas, normally nitrogen, is

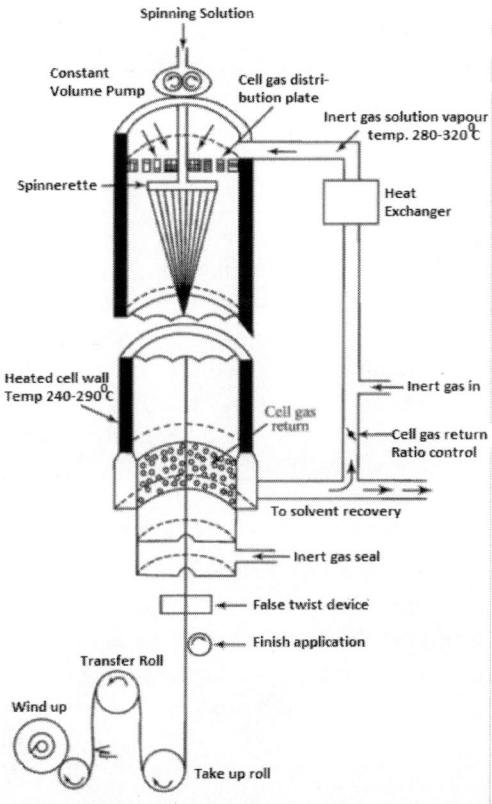

Figure 9.13. Dry spinning process

introduced at the top of the cell and passed through a distribution plate behind the spinnerets pack. Because both cell gas and cell walls are maintained at high temperatures, solvent evaporates rapidly from the filaments as they travel down the spinning cell. The spinning solvent is then condensed from the cell gases, purified by distillation and returned for reuse. Individual filament size is normally maintained in the range of 0.6–1.7 tex (5–15 den) to maximize, within operable limits, surface area-to-mass ratio and solvent removal rate. Individual filaments are grouped into bundles of the desired final tex at the exit of the spinning cell by a coalescence guide. A commonly used guide employs compressed air to create a minivortex which imparts a false twist and rounded cross section to the filament bundle. Solution dry-spun spandex fibres are normally referred to as continuous multifilaments or coalesced multifilaments. However, the individual filaments do not coalesce into larger structures but remain discrete; they adhere to one another because of natural elastomer tack at their surface.

After coalescence, a finish is applied to the multifilament bundle before it is wound onto a tube. Commonly used finishing agents include polydimethylsiloxane and magnesium stearate which provide lubrication for textile processing and prevent fibres from sticking together on the package. Windup speeds are in the range of 300–500 m/min, depending on tex and producer.

9.11.2.3   Reaction spinning

To produce spandex by reaction spinning, a 1000–3500 molecular weight polyester or polyether glycol reacts with a diisocyanate at a molar ratio of about 1:2. The viscosity of this isocyanate-terminated prepolymer may be adjusted by adding small amounts of an inert solvent, and then extruded into a coagulating bath that contains a diamine so that filament and polymer formation occurs simultaneously. Reactions are completed as the filaments are cured and solvent evaporated on a belt dryer. After application of a finish, the fibres are wound on tubes or bobbins and rewound if necessary to reduce interfibre cohesion.

Trifunctional hydroxy compounds, e.g., glycerol or 2-ethyl-2-(hydroxymethyl)-1,3-propanediol, may be added with the macroglycol to produce covalent cross links in the reaction-spun spandex fibre. Also, covalent cross links may result from allophanate and/or biuret formation during curing by reaction of free isocyanate end groups with urethane or urea NH groups along the polymer chain. A multiplicity of filaments is normally extruded from each spinneret of about 1.1–3.3 tex (10–30 den), and then collected in bundles of the desired tex at the exit of the reaction bath. This approach makes the surface area-to-mass ratio and diamine diffusion into the prepolymer cross section substantially constant irrespective of the final tex produced, thus minimizing condition changes required in changing tex. Because the individual filaments have reacted incompletely and are in a semiplastic state at the exit of the diamine bath, they interbond quite tightly into a fused multifilament. Production speeds in reaction spinning are limited by filament weakness in the bath along with hydrodynamic drag. Take-up speeds are limited to about 100 m/min. Stabilizers and pigments are normally slurried with macroglycol and added to the polymeric glycol charge, prior to diisocyanate addition. Therefore, care must be taken to avoid additives that react significantly with diisocyanates or diamines under processing conditions. Also, stabilizers should be chosen that have no adverse catalytic effect on the prepolymer or chain-extension reactions.

Reaction-spinning equipment is quite similar to that of solution wet spinning. It differs principally in the use of fewer wash baths and in the use of belt-type dryers instead of heated cans.

9.11.3   Fields of application

Elastane fibres are used in flat textile goods in those cases where high stretch properties are required with good recovery of the original dimension when the deforming force is removed. Compared to rubber fibres, elastane fibres posses greater resistance to abrasion and greater resistance to ageing. Elastanes are usually used in combination with other fibres, either naked or with a spun casing (core yarn), presented as elastic twisted yarns, or elastic yarns with other fibres twisted core spun around them (elastic combination yarns).

Typical items are summarized in Table 9.9.

Table 9.9. Usage of elastane in different industries

Material	Elastane content, %
Woven goods	2–8
Lingerie/under wear	2–5
Swimwear/sports goods	12–20
Foundation garments	10–45
Fine hosiery	2–12
Surgical stockings	35–50

Elastane fibre blends are, as already mentioned, textiles with particular elastic properties. To retain these properties to the desired extent in the finished article, the processing parameters in the finishing and the desired characteristics in the finished article must be carefully balanced one against the other. Parameters such as temperature, concentrations of chemicals and auxiliary, duration of treatment and cloth tension must be controlled and maintained at as Iowa level as possible. Since elastane fibres, especially in the heat-stretch processes, suffer irreversibly somewhat in their powers of recovery from stretch, it is important that the stretching during high temperature processes is kept to a minimum (i.e., during relaxing, thermofixing, heat post treatment and drying).

Elastane in the stretch fabric is knitted or woven under tension. When relaxed, it tends the fabric to compact in the length and width to a 'jam point'. The hard fibre buckles when the fabric is 'jammed' and it limits the extension of the fabric when it is stretched. Elastic fabrics are very often required wider and lighter than their width and weight at jam point. To ensure complete control of the desired physical changes, the processor must determine the parameter of the fully jammed (fully relaxed) fabric namely:

(a) Narrowest width highest weight

(b) Corresponding wale and course or end and pick counts

(c) Greatest shrinkage and stretch in length and width

For this purpose, a marked grey sample is boiled for 10–15 min. then it is dried and relaxed and measured. This check is especially important for the development of new fabric styles.

Elastane yarn requires careful control of processing conditions to preserve the intrinsic elastic properties of the fibre, while obtaining the required fabric characteristics. The dyeing and finishing conditions should be chosen because the performance of elastane can be changed due to prolonged hot/wet treatments certain chemicals, excessive tensions and high temperatures. Any process settings have to be done after proper study and preferably after a trial run.

Tension, temperature, concentration of chemicals and duration of treatment has to be kept to a minimum, because they can affect the elastic properties and the appearance of the finished goods. In particular, the tension must be kept to a minimum during those steps which are carried out at elevated temperatures, i.e., relaxation, hot wet processing, drying and curing. Although the use of high temperatures and tensions does not degrade elastane but it can loose the stretchability and stretch recovery.

There are different types of elastane yarn available in the market as per the end uses. There may be slight difference in the heat setting temperature in each case.

9.12   Polyether polyurethane elastomer (PUE)

Polyether spandex is designed to provide improved protection from damage caused by chlorine, UV rays, suntan oil and perspiration. It is mainly used in warp knitted elastic fabrics and circular knitted elastic fabrics. It is ideal for swimwear and active wear to minimizing the damage caused by chlorine water (swimming pool). It helps the garment to maintain their whiteness and vivid colours longer. While this product is primarily intended for swimwear, it is also suitable for fabrics, which needs chlorine resistance.

Polyether PUE fibres have the following properties:

1. Excellent strength, elongation and modulus
2. Flexlife
3. Compression power
4. Abrasion resistance
5. Good acid dyeability for union dyeing to prevent and reduce grin through
6. Good resistance to discolouration from atmospheric contamination (gas fading)
7. Excellent resistance to oxidation and UV light exposure, suntan oil and perspiration
8. Excellent resistance to discolouration in fabric moulding process
9. Excellent resistance to chlorine bleaching.
10. Dyeing at high temperature
11. Alkaline processes
12. Good light fastness

Polyether PUE is produced by the ring-opening polymerization of epoxides or cyclic ethers.

$$R - CH_2O + H_2O \longrightarrow HO (R - CH_2O)n - CH_2OH$$

Polyether Prepolymer

After polyether prepolymer formation, further reactions and processes are the same way as in the case of polyester PUE (Table 9.10).

Table 9.10. Average physical properties of polyether PUE

Properties	Value
Breaking elongation (%)	550
Modulus (g/den)	0.045–0.70
Moulding temperature	195°C, 60 s
Specific gravity	1.04–1.06
Softening temperature	220°C
Moisture regain	<1

Moulding operation can be done for garments with polyether PUE fibres wherever sewn seam is undesirable, such as bra. Spandex has enough fabric stretch in wale and course (warp and weft) direction to prevent the nonelastic fibres (like nylon polyester) from cutting and rupturing the spandex fibres as the fabric is stretched during moulding.

Resistance for moulding chemicals: Polyether PUE is resistant to moulding chemicals as soft acrylic or polyvinyl and acetate thermoplastic resins. However, resins based on melamine or triazine formaldehyde condensate discolour and are therefore not recommended. Spandex is resistant to silicone softeners and hand modifiers along with resins. For moulding, acid-dyed spandex is recommended (including acid-dye-based OBA – is Optical Brightening Agent), and disperse-dyed spandex is not suitable as the latter may decolourize during moulding.

Resistance to chemicals: Polyether PUE is resistant to chlorine bleaches and wide range of chemicals usually used in the wet processing. However, there are certain chemicals, finishing chemicals and machine oils, which are not suitable on these spandex. Peracetic acid, formic acid and oxalic acids are not recommended. Lubricants or oils containing unsaturated fatty acids or esters can degrade spandex through oxidative degradation. Hence, any such lubricants and oils should be avoided. Dyeing characteristics are mostly same as polyester PUE explained earlier.

9.13   Carbon fibres

Carbon fibres are manufactured from rayon (viscose fibres with/without drawing) and synthetic fibres (polycarbonate, polyvinyl alcohol, polyacrylonitrile [PAN]). The acrylic fibres at 200–300°C under oxidizing conditions (stabilization). This is followed by a second stage when the oxidized fibre is heated in an inert atmosphere (nitrogen) to temperatures around 1000°C. Hydrogen and nitrogen atoms are expelled, leaving the carbon atoms in the form of hexagonal rings which are arranged in oriented fibrils. In the last, the carbonized filaments are heated to temperatures of up to 3000°C, again in an inert atmosphere whereby the orderly arrangement of the carbon atoms takes place which are organized into a crystalline structure similar to that of graphite ('graphitization'). The atoms are in layers or planes which lie virtually parallel to each other. The planes are well oriented in the direction of the fibre axis, this being an important factor in producing HM fibres. The carbon fibres so formed consist of 80–90% carbon, C atoms forming ribbon-like C hexagonal planes.

Carbon fibres can be heated up to 1500°C and contains up to 95% of elemental carbon. Graphite fibres can be heated above 2500°C with 99% carbon. The formation of carbon fibres from PAN is outlined in the Fig. 9.14.

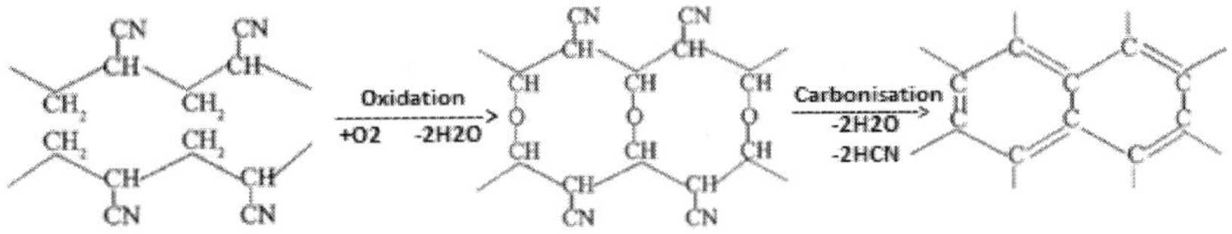

Figure 9.14. Reaction route of PAN-based carbon fibres

Carbon fibres are used in the aerospace industry, in compressor blade to jet engines, helicopter rotor blades, aircraft fuselage structures, golf-club shafts, cross bows for archery and in high speed reciprocating parts in loom.

Carbon fibres are classified into two broad categories: high performance (HP) and general performance (GP). HP fibres are further classified into high tensile (HT) and HM.

9.13.1   Manufacturing

The majority of commercial carbon fibres are produced from PAN fibres. PAN-based fibres are the strongest commercially available carbon fibres and dominate structural applications. Mesophase pitch-based carbon fibres represent a smaller but significant market niche. These fibres develop exceptional moduli and excel in lattice-based properties, including stiffness and thermal conductivity. Rayon-based fibres are used in heat shielding and in missile nosecones. Carbon fibres made from HP polymers or from chemical vapour deposition of hydrocarbons, such as benzene or methane, display unique properties that make them potentially attractive future alternatives.

9.13.2   PAN-Based carbon fibres

The PAN fibres for making carbon fibres are mostly made by air-gap spinning (see under PAN fibres in this book). Before the acrylic fibres (qv) produced, it can be subjected to the elevated temperatures of carbonization, they must be converted to a thermally stable form that will not melt through oxidative stabilization. Stabilization is usually performed in air at temperatures between 200 and 300°C and under tension to prevent fibre shrinkage.

Production of carbon fibres has essentially the following processes:

1. *Stabilization* (180–300°C): oxidative and/or chemical treatment with or without fibre drawing as a precondition for solid phase pyrolysis. A medium temperature 'stabilization' stage in air renders the fibre infusible. The polymer composition generally has about 98% acrylonitrile and 2% of a weak acid such as itaconic or acrylic acid. The function of the acid is to provide a site to initiate the 'ladder'

formation in the stabilization step, and thus lower the temperature at which the reaction occurs and reduce the exotherm. Studies of stabilization at elevated temperatures (350–400°C) show the development of additional intermolecular cross linking, resulting in improved strengths.

2. *Carbonization* (300–1600°C): Thermal decomposition of the stabilized, unmeltable fibre intermediate product. A high-temperature 'carbonization' treatment in an inert atmosphere, where the fibre is converted into nearly pure carbon.

3. *Graphitization* (1600–3000°C): High-temperature treatment with or without fibre drawing. So-called HS carbon fibres and isotropic carbon fibres are obtained as products of the second stage. The third stage produces HM carbon fibres.

Heating rate plays a major role in controlling the release of gases. A decrease in tensile strength is generally observed at temperatures above 1500°C, which corresponds to the final major release of nitrogen. As a result, many of the highest strength PAN-based carbon fibres contain residual nitrogen. Although heat treatments of PAN-based carbon fibres above 1700°C are referred to as graphitization, these fibres are not graphitizable and do not develop highly ordered graphene planes.

The PAN-based carbon fibres are well suited to developing HT strengths, but are not likely to develop the high level of three-dimensional order associated with HM precursor materials such as mesophase pitch. As such, PAN-based carbon fibres are considered nongraphitizing and maintain a highly turbostratic organization of the graphene layer planes, even when exposed to very high treatment temperatures.

Low temperature carbonized fibres have found many biological uses today like, in ligament and tendon prostheses and in surgical sutures. These fibres are heat treated to temperatures below 1300°C and are far less crystalline than traditional carbonized PAN fibres.

9.13.3   Mesophase pitch-based carbon fibres

The other method is based on mesophase pitch which consists of a heavy fraction of predominantly aromatic hydrocarbons that is the residue of petroleum or coal tar distillation following the removal of creosote oil and anthracene. The pitch consists of hundreds of thousands of different species with an average molecular weight of several hundreds. Many of the species are heterocyclic and are formed by a complex variety of chemical reactions, including thermal decompositions, hydrogen transfers and oligomerization reactions.

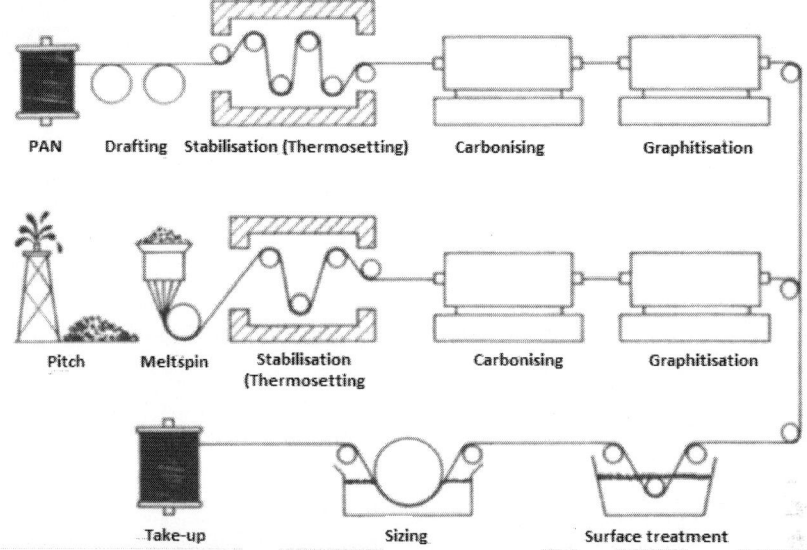

Figure 9.15. Processes involved in the manufacturing of carbon fibres

Pitch is a glassy solid at room temperature, but softens upon heating to form a viscous liquid.

The manufacturing steps consists of thermal treatment, melt spinning and oxidation, carbonization and graphitization annealing. The carbonization and graphitization annealing serve as surface treatments. The graphitization treatment in the last step is done at temperatures of up to 3000°C. The main objective of both production methods is to arrange graphite layers in the fibre direction. For carbon fibres, the carbon content is supposed to be at least 90%. The typical processing steps are depicted below. Thermosetting stands for stabilization and oxidation; the step 'graphitization' is optional and only necessary to achieve a high E-modulus (Fig. 9.15).

Once the mesophase pitch has been prepared, the next step in processing is spinning it into fibres. This process is particularly important because when mesophase pitch is carbonized, the morphology of the pitch is the dominant factor in determining the microstructure of the resulting graphitic fibre. This results from the stacking behaviour of the mesophase molecules. Mesophase pitch fibres are produced through melt spinning that is essentially the same as that used to spin commercial polymers. The spun fibres next has to undergo stabilization prior to carbonization. During this process, oxygen reacts with side groups in the pitch fibre, creating cross linkages and adding weight. Typically, a 6% weight gain is needed to completely stabilize mesophase pitch fibres, while an 8% weight gain is more typical for PAN. Lower heating rates and lower temperatures during stabilization result in a more uniform stabilization, but also add to processing time.

The carbonization of mesophase pitch-based fibres closely resembles that which is previously described for PAN, but occurs in two stages. Typically, fibres are heated to approximately 1000°C and held there allowing for the evolution of off-gases. The rate-limiting step in this low temperature carbonization is the free-radical breaking of carbon–hydrogen bonds. Following the release of gases, the fibres are heated to their final treatment temperature, which may be as high as 3000°C. Unlike PAN-based carbon fibres, mesophase pitch-based fibres experience significant graphitization during which dislocations in the turbostratic carbon stacks are gradually annealed, resulting in the formation of a three-dimensional lattice.

9.13.4   Viscose-based carbon fibres

Carbon fibres can also be produced from viscose fibres. This process has little market relevance at present and is mainly used to manufacture textiles for medical applications. Stabilized rayon fibres are carbonized and then activated with air, steam or carbon dioxide, much as in granular carbon activation. The extent of pyrolysis governs the pore structure, carbon yield and surface area of the fibre, while activation impacts the presence of functional groups on the pore surface. Rayon carbonization consists of a blend of depolymerization and dehydration reactions. The depolymerization involves the formation of *l*-glucosan and volatile products, while the dehydration reaction inhibits *l*-glucosan and volatile production, resulting in a higher carbon yield. The inclusion of Lewis acids such as $AlCl_3$ and $ZnCl_2$, during low temperature carbonization enhances the carbon yield by favouring the dehydration reactions.

The PAN- and pitch-based carbon fibres have replaced rayon-based fibres in most HP applications; however, they continue to find use as ablative materials in missile nosecones and heat shielding. Additionally, the combination of low cost, ease of handling and high natural porosity makes rayon an attractive precursor for activated carbon fibres.

9.13.5   Properties and uses

Carbon fibres have greater thermal stability (fireproofness, infusibility). Among the exceptional properties of carbon fibres are their high tenacity, HM of elasticity, high brittleness, low creeping tendency, chemically inert behaviour, low heat expansion and good electrical conductivity. It has strength (see comparison with steel and metalo fibres), approximately three times the strength of glass fibres and steel (and tear strengths, starting from polycarbonates, up to 2500 N/mm², starting from viscose filaments even up to 3000 N/mm²), and also absolute chemical resistance and hydrophobic characteristics.

Table 9.11. Properties of carbon fibres compared to glass and steel fibres

Properties	Carbon fibres	Glass fibres	Steel fibres
Density, g/cm³	1.95	2.54	7.9–8.1
Strength, cN/tex	120–180	36–90	30–35
E-modulus, kN/mm²	220–500	45–75	140–200
Elongation, %	0.5–1.3	3–4	1–1.8

Carbon fibres made from viscose filament retain its microfibrillar structures of the cellulose structure, despite the extraordinarily high pyrolysis temperature. Carbon fibres are black, lustrous, slippery handle, greater stiffness than glass fibres. The most important textile technological characteristics result from the typical structures of the fibres (see Figs. 9.16 and 9.17).

They are used (in the textile sector) in the production of fireproof woven fabrics, filter cloths, electrically conductive woven fabrics for heating, thermally insulating felts, composites (CFK).

The inner structure of carbon fibres is shown Fig. 9.16. The carbon layers causing the extremely high values of tenacity and E-modulus are clearly

Figure 9.16. SEM view of carbon fibre

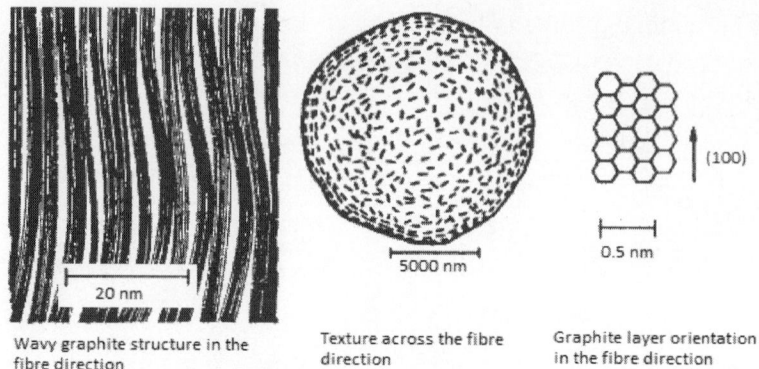

Wavy graphite structure in the fibre direction

Texture across the fibre direction

Graphite layer orientation in the fibre direction

Figure 9.17. Detailed structure of carbon fibre

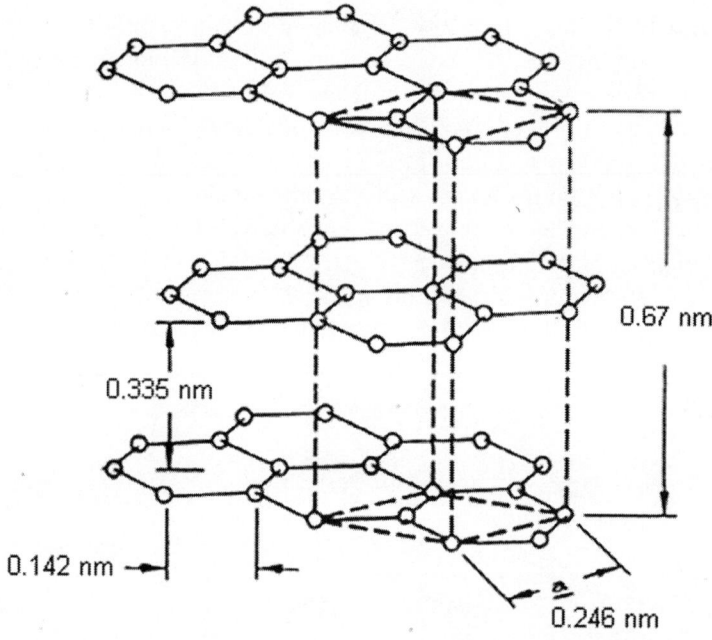

Figure 9.18. Inner structure of carbon fibre

visible. The free electrons between the layers are responsible for the excellent electric conductivity of carbon fibres.

Rayon-based activated carbon fabrics (ACFs) are used in the adsorption of many volatile organic compounds including formaldehyde, methyl ethyl ketones and benzene. ACFs are also finding uses in natural gas storage, electrodes for batteries, catalyst supports and NOx removal (Fig. 9.18).

9.14   Polyurea fibres

Polyurea is formed by condensation of diamines with urea and the polymers are having a typical urea group as a part of the repeating unit. The monomers used for the production are urea and nonamethylene diamine.

9.14.1   Manufacture

Nonamethylene diamine is manufactured by a series of reactions with oleic acid which is available in rice bran oil. Oleic acid on reaction with ozone forms pelargonic acid and azelaic acid, the latter is reacted with ammonia to form ammonium azelate. Ammonium azelate on hydrogenation gives nonamethylene diamine. Linoleic acid, also available in rice bran oil, gives azelaic acid on ozonation.

$$CH_3(CH_2)_7 CH = CH(CH_2)_7COOH + 2O_2 \rightarrow CH_3(CH_2)_7COOH + HOOC(CH_2)_7COOH$$

Oleic Acid   Pelargonic acid   Azelaic acid

$$CH_3(CH_2)_7 CH = CH- HC=CH(CH_2)_5 +2O_2 \rightarrow H \ C(CH_2)_5COOH+HOOC(CH_2)_7COOH$$

Oleic Acid   Caproic acid   Azelaic acid

$$HOOC(CH_2)_7COOH + 2NH_3 \rightarrow (CH_2)_7 \underset{COONH_4}{\overset{COONH_4}{<}} \xrightarrow{Dehydration} (CH_2)_7 \underset{CN}{\overset{CN}{<}} \xrightarrow[enation]{Hydrog-} (CH_2)_7 \underset{CH_2NH_2}{\overset{CH_2NH_2}{<}}$$

Azelaic acid   Ammonium azelate   Azelaic Dinitrile   Nonamethylene Diamine

9.14.2   Polymerization

For polymerization, urea and nonamethylene diamine (slightly molar excess) are dissolved in water and heated to 100°C. Again, it is heated to higher temperature in closed atmosphere to 250°C in 15–18 h, vacuum being applied in final stages, if necessary. Complete polymerization takes place with small amounts of the initial chemicals remaining.

9.14.3   Spinning

The melted milky white polymer as we get from the polymerization stage can be directly melt spun (MP 225–231°C). The spun filaments can be hot drawn, may be at 100°C at several stages and can be used as filaments or staple after cutting.

9.14.4   Properties

Table 9.12. Properties of polyurea fibres

Properties	Polypropylene
Tenacity, cN/Tex, g/den	39.7–48.6 cN/tex (4.5–5.5 g/den)
Tensile strength, kg/cm^2	
Elongation	15–20
Elastic recovery	70% recovery from 8% extension
Specific gravity	1.07
Effect of moisture	Hydrolysis occurs on exposure to steam at 130°C
Moisture regain	1.8%
Softening point, °C	205°C
Melting point, °C	240°C
Thermal properties	Some loss of strength on continued exposure to air at 150°C.
Thermal conductivity	6 (compared to air = 1)
Flammability	
Effect of sunlight	Similar to nylon.
Effect of acids	Good resistance.
Effect of alkalis	Good resistance
General	Good resistance to most common chemicals
Effect of organic solvents	Similar to nylon

Unconventional Fibres

10.1   Cocona fibres

It is known that the finest quality of activated carbon comes from coconut shells to provide superior odour adsorption and UV protection on a wide range of product applications. Cocona material is a natural fabric enhancer that is produced from activated carbon from coconut shells, which is infused into the textile fibre by Cocona Inc. patent method (Fig. 10.1). Activated carbon has a huge absorbing area – 1 g of it has a surface area the size of two tennis courts. Cocona fabric utilizes natural technology that outperforms other fabrics and yarns. Cocona fibres and yarns can be used in a wide range of knit and woven fabrics as well as non-woven fabrics that provide effective evaporative cooling, odour adsorption and UV protection. Fabrics made from Cocona yarns and fibres are lightweight, comfortable and retain all of the conventional product features, such as stretch and wash ability. In addition, because of our patented technology, Cocona fabrics will retain or improve their performance over the life of the garment. Cocona technology is contained inside the fibre and, unlike topical treatments, it will never wash off or wear out!

| Raw Coconut Shell | Activated Carbon | Master Batch | Cocona Raw Fiber | Cocona Performance Thread |

Figure 10.1. Coconut in textile – Cocona Inc.

This innovative technology of infusing activated carbon into natural fibres like cotton, wool and synthetic fibres like polyester, nylon, etc., which is then blended with other fibres to create performance fabrics is invented and patented by Cocona TrapTek, LLC. It is a natural, sustainable technology. The technology was developed in 2002 by Gregory W. Haggquist, the company's founder and made its consumer market debut in 2006 in knitted cycling apparel developed by United Knitting (Cleveland, TN, USA), and Cannondale Bicycle Corp. (Bethel, CT, USA).

10.1.1   Technology

As mentioned earlier, the activated carbon made from coconut shell has a porous structure. The activated carbon is used for water and air filtration, wastewater treatment and other such industries. The pores absorb odour molecules at the same time as enabling the moisture to escape from the skin's surface and absorb into the sock more quickly than normal due to the large surface area. Yarn made with Cocona has a surface area that is up to 10 times larger than conventional polyester yarn. The technology imbeds activated carbon from coconut shells into polyester or nylon polymers. The coconut shells are burned at 300°C followed by a 1000°C steam-activation process. The microscopic, ultrafine particles that are too small for those applications are just what TrapTek needs to incorporate into its fibre and yarn.

Due to the special features of the activated carbon, the fabrics made from the yarn using this technology offer superior comfort and enhanced performance in a variety of clothing applications. It provides protection from harmful UV rays, absorb odours, antistatic and moisture absorbent. It has been found that with this technology, a wet polyester fabric made with these special fibres can dry 50% faster than bare polyester fabrics. In addition, similar cotton and bamboo fabrics dry 92% and 96% faster, respectively, even after wicking treatment. By the special technology, the activated carbon is embedded inside the fibre, so that it cannot be washed off or worn out like ordinary surface treatments to fabrics.

10.1.2   Advantage of coconut shell fibre

It is environmental friendly as it utilizes recycled coconut shells that would have gone to landfills. A majority of coconut shells are converted into activated carbon, primarily for the air and water filtration industries. The particles that are too small to use in water and air filters, is converted to microscopic particles and combine with other fibre/fabrics. While this is not a completely 'green' process, it is much better than using chemicals to treat the materials.

They are inherently lightweight, comfortable and easy-care and create a fabric that provides highly effective evaporative cooling, odour control and superior UV protection, up to 50+ UPF (the highest possible rating).

10.1.3   Applications

Apparel made from the fibre helps spread and evaporate surface moisture rapidly, making it perfect for sportswear, undergarments, golf apparel and other active applications. They are used in garments ranging from shirts, pants, shorts, outerwear, underwear, footwear, travel wear and tank tops.

10.2   Coffee fibre

The technology is similar to Cocona fibre but uses the used coffee powder in place of activated carbon from coconut shells. The technology was developed by Singtex Industrial Co. (Taiwan, China). The fabric, called 'S. Cafe', is a result of 3 years dedication and efforts of the company to transform waste coffee grounds into eco-friendly fabric with the investment of about $1.7 million. The technology embeds the coffee powder in polyester and nylon primarily which can absorb the odour along with other qualities.

10.2.1   Technology

The technology uses coffee ground, which was generally treated as waste, through extracting, nano-grinding, micropolis and wicking material improvement, and all these patented methods recycle coffee grounds into tiny-sized particles. This is then processed to create a technical composite fibre. Later, this embedded coffee fibre converted into yarn and are then woven together to create the S. Café fabric, which can be used for both knitted and woven clothing. A single cup of coffee can make two T-shirts. The fabric is not made just from coffee grounds, but apparently a mixture of coffee grounds in a low percentage along with polyester or other more traditional material such as nylon. The benefits of this product are numerous. It is remarkable in odour control and offers some protection against harmful UV as well. Another benefit is that the coffee element also makes the fabric dry quickly. Of course this is attractive to many, including sports manufacturers. Most importantly, brewed coffee grounds which were treated as waste becomes a brand new material in fibre.

As in the case of activated carbon, when looked at under a microscope, coffee beans are found to have many pores. These pores increase greatly in number after the beans have been roasted, a characteristic that when blended into the fabric helps 'accelerate the moisture transportation process and provide better elimination properties' than any other PET materials. While roasting, the green bean expands, and this means the space inside the coffee bean become bigger. These spaces are crucial for the functions. After brewing, some materials had been washed off from these spaces by hot water, and through the patented process to maximize its capacity.

The advantages and applications are similar to that of Cocona fibres. The fabric made with coffee fibre does not use any chemicals and is free of harmful materials commonly generated during the production or transportation of other types of yarn. The manufacture of the new product does not require high temperature carbonization, so its manufacture is also energy efficient. In addition, more importantly process does not use any hazardous chemicals as well. They are eco-friendly fabrics and can be washed by using clean water without need for detergent.

Fabric made with coffee fibre is perfect for mid and base layers for adrenalin-powered sports like rock climbing, as well as walking, running and yoga. Coffee fibres can be used in active wear T-shirts and even sports bras. Currently, it is used primarily for clothing, but this material can be used as part of the structures used in interior design for coffee shops and home furnishing.

The super fine grounds added to S. Café yarn are only 1–2 microns in diameter and are added to be about 2% of the yarn. Any higher concentration, the yarn begins to lose strength.

10.3   Soya bean fibre

Soya bean fibre has the physical properties of the synthetic fibres but has the moisture absorption of cotton and softness of silk which makes it more desirable. Soya bean protein fibre is the only protein fibre derived from plants. It is made from soya bean cake.

These fibres have the unique properties like it is crease resistant, has inherent antibacterial capacity, lustre and handle of silk, absorbency of cotton and comfortable to wear. In its antibacterial property, it can resist *Staphylococcus aureus*, *Candida albicans* and *Colibacillosis*. The antibacterial property makes it suitable in the production of underwear, the softness and absorbency are advantages to be used as baby wear, the lustre and other qualities make it suitable for high fashion wear.

10.4   Bamboo fibre

It is a regenerated cellulose fibre. Bamboo is one plant with high cellulose content in their barks. With suitable technology, the cellulose from the bamboo barks are extracted and made into fibre. The specialty of this fibre is its natural antibacterial properties.

The advantages of these fibres are the UV protection capacity, gentle drape, luxurious softness and antibacterial property which are all inherent.

They are environment friendly because the garments using these fibres needs no special chemical treatments to achieve these properties, as it is done in case of many other fibres (Table 10.1).

The antibacterial properties make the fibres preferred for under garments but it is also used for socks, shirts and other summer clothings.

Characteristics of bamboo fibres are as follows:

1. Naturally antibacterial and hygienic effect
2. Biodegradable and sustainable material
3. Breathable and moisture absorbing character
4. Soft and cool fabric hand feel
5. Smooth fabric drape
6. Luxurious and shiny appearance
7. Low tensile strength (wet strength 60% lower compared with dry tensile strength)
8. Natural UV protection capacity

Physical parameters:

1. Dry tensile strength: 2.2–2.4 cN/dtex
2. Wet tensile strength 1.2–1.4 cN/dtex
3. Dry elongation at break 18–25%
4. Moisture regain approximately 8–9%

Test conditions:

1. Temperature: 20°C
2. Relative humidity: 65%

Table 10.1. Comparison of some properties of bamboo, viscose and cotton

Properties	Bamboo	Viscose	Cotton
Linear density (dtex)	1.67	1.67	1.5–1.7
Single dry tensile strength (cN/dtex)	2.2–2.5	2.5–3.1	2.5–3.1
Single wet tensile strength (cN/dtex)	1.3–1.7	1.4–2.0	1.5–2.1
Dry tensile elongation, %	14–18	18–22	8–10
Moisture regain, %	13	13	8.5
Absorbency rates, %	90–120	90–110	45–60
Specific density	1.32	1.32	1.5–1.6
Mass specific resistance	1.09×10^6	2.29×10^7	105

10.4.1   Manufacture

Two types of processing are done to obtain bamboo fibres: mechanical processing and chemical processing. Chemical processing is based on hydrolysis alkalization. The crushed bamboo is 'cooked' with the help of sodium hydroxide (NaOH) which is also known as caustic soda or lye into a form of regenerated cellulose fibre. Hydrolysis alkalization is then done through carbon disulphide combined with multi-phase bleaching. Although chemical

Figure 10.2. Processes involved in bamboo manufacturing

processing is not environmental friendly, it is preferred by many manufacturers as it is a less time consuming process (Fig. 10.2).

In mechanical processing, the crushed bamboo is treated with biological enzymes. This breaks the bamboo into a mushy mass, and individual fibres later combed out. Although expensive, this process is eco-friendly.

10.4.2   Processing

Bamboo fibre being a regenerated cellulosic fibre, it can be processed same way as the viscose materials. But certain precautions have to be taken due to special physical and chemical characteristics of the fibre.

Light sergeant, enzyme desizing, moderate bleaching and semimercerizing should be applied to the bamboo fabric during its dyeing and finishing process. Avoid drastic conditions and use small mechanical tension. Bamboo fabric can be signed in moderate condition. Desizing rate should be over 80%.

Scouring: Pure bamboo normally needs no scouring; sometimes washing with a little alkaline soap may serve the purpose. The scouring process should be made in terms, if fibre blend contains cotton. When pure bamboo fabrics are scoured, the alkali should not be over 10 g/l but can be applied in accordance with the thickness of fabrics.

Bleaching/mercerizing: The processing should be made in terms of the specification and thickness of fabrics. Fabrics of bamboo fibres normally should not need mercerizing due to their sound lustre and poor anti-alkaline properties. However, in some cases, to increase their absorbance capacity to dyestuff, it can be mercerized.

Dyeing: Reactive dyestuffs are used during dyeing process – alkali should not be over 20 g/l; temperature should not be over 100°C. During drying process, low temperature and light tension are applied. In yarn dyeing, alkali should not be over 8 g/l. Chemically manufactured bamboo rayon has some wonderful properties which are adored by conventional and eco-aware designers and consumers involved in towel sector.

Precautions:

1. Bamboo fibres are sensitive to alkaline conditions hence has to be pretreated at lower alkali concentrations after trials. To achieve good whiteness, we need higher concentrations of peroxide than other fibres.
2. Since the fibre is sensitive to alkali, it is preferred to be dyed with cold reactive dyes (lower alkali requirement).
3. During finishing, it should be born in mind that the fibre as such has a soft drape and the finishing agents has to carefully selected which can enhance the natural feel unless otherwise required.
4. It has been found that the natural antibacterial properties are reduced during wet processing. If necessary, they may be supported by anti-microbial treatments.

10.4.3   Specialities of bamboo fibre

It has a basic round surface which makes it very smooth. Bamboo fabric is softer than the softest cotton, and it has a natural sheen like silk or cashmere. Bamboo drapes like silk or satin yet is less expensive and more durable. Bamboo/organic cotton blends are also extremely soft but heavier in weight.

Bamboo is durable and long wearing fabric. It is very sturdy and does not spoil easily, even in the most severe conditions or environments.

Many people are allergic to natural fibres, such as wool. Bamboo has had the smallest number of allergic reactions with people barely having any. It is the best choice for people having allergy or skin problems. Bamboo fibre absorbs and evaporates sweat very quickly. Fabrics made from bamboo fibre are highly breathable in hot weather and also keep the wearer warmer in cold season. Bamboo is naturally cool to the touch. The cross-section of the bamboo fibre is filled with various micro-gaps and micro-holes leading to much better moisture absorption and ventilation.

Pure bamboo fabric has a 99.8% bacterial kill rate. Most of the varieties of bed sheets do not possess these qualities but bamboo bed sheets possess specific antimicrobial and antibacterial properties, making them more odour and bacteria resistant. This is because bamboo itself is highly resistant to the growth of microorganisms and bacteria. With sweat, dead skin cells, oil and many other things coming in contact with our bed sheets every day, this reasoning makes the bamboo bed sheets with antimicrobial properties all the more appealing.

Bamboo naturally provides added protection against the sun's harmful UV rays. Due to its high moisture absorption property, bamboo fabric results in the enhancement of antistatic property.

Bamboo is an environment friendly product. As a regenerated cellulose fibre, bamboo fibre was 100% made from bamboo through high-tech process. They are all 3 to 4-year-old new bamboo of good character and ideal temper. The whole distilling and producing process in our plant is green process without any pollution. It produces natural and eco-friendly fibre without any chemical additive. As a natural cellulose fibre, it can be 100% bio-degraded in soil by microorganism and sunshine. The decomposition process does not cause any pollution environment. Bamboo fibre is praised as 'the natural, green and eco-friendly new-type textile material of 21st century'.

Bamboo bed sheets are highly long lasting and do not demand any extra care. They can easily be machine washed and dried. The threads of the bamboo bed sheets are extremely soft and actually become softer the more they are washed and the more they are used. This makes bamboo bed sheets an ideal choice for those individuals who prefer bed sheets that have a longer life. The cost differences range from 20% to 40% more expensive than cotton. Bamboo bed sheets are slightly expensive than the other typical varieties of bed sheets but their ability to last for a longer duration makes it highly cost-effective in the long run. The cotton bed sheets begin to feel rougher as they are used but for the bamboo bed sheets, it is completely opposite.

10.4.4   Advantages

Apart from above characteristics, bamboo also has the following benefits:

1. The fabric is light and strong.
2. It has the ability to take colours well.
3. It is an anti-fungal and hypoallergenic fabric– even after 50 washes.
4. It dries quickly – about twice as fast as most other fabrics.
5. Perfect travelling clothes – the wrinkles fall right out, so there is no need to iron.
6. Bamboo is less than half the price as compare to cashmere.

10.4.5   Disadvantages

1. Bamboo loses 60% of its strength when wet, so it may not be suitable for industrial uniforms and washings.

2. It is not ideal for screen printing.

3. Expensive as compared to cotton.

10.4.6   Applications of bamboo fibre

Bamboo fabric is used in home furnishings such as sheets, comforters, duvet covers, pillow shams, pillows, bed skirts, table cloth, drapes, blinds, sofa slipcovers and bathroom textiles such as hand towels, wash cloths, body towels, tub mats, bath rugs, decorative lid covers, shower curtains, etc.

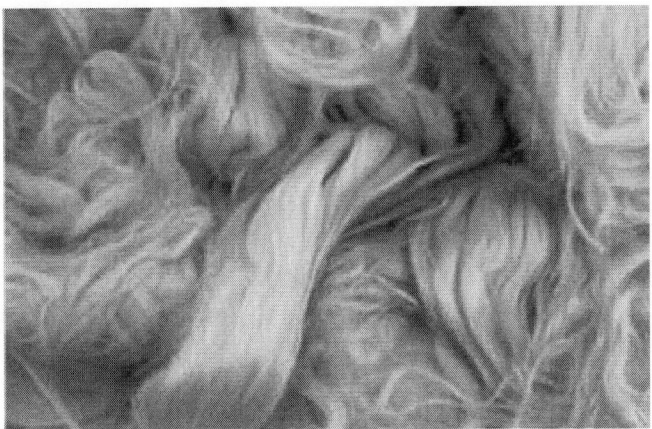

Figure 10.3. Ingeo fibre

10.5   Ingeo fibre

World's first and only man-made fibre based on nature and based on 100% renewable resources such as maize oil. Manufactured by NatureWorks LLC, Ingeo for the first time gives the appeal and aesthetics of a natural fibre combined with the performance of a high-tech synthetic fibre. Ingeo is available in various counts as micro denier to high counts suitable for various applications such as finest fabrics to fine fabrics (Fig. 10.3).

Special properties:

1. Good moisture management and wicking properties.

2. Exceptional odour absorption.

3. It has natural resilience and good shape retention for long time.

4. Ingeo filament fabrics have a subtle lustre and drape with a natural hand, offering a beautiful new material to stimulate creativity.

5. They show good soil release property which offer easy washing cycles at the same time excellent after wash appearance.

6. Its quick drying property makes the washing still easier.
7. It has a soft fluffy feel which retains even after ironing and pressing.
8. It has heat insulation properties.

Because of the above properties, the Ingeo fibres are best used for making shirts. It is also used for sportswear, trousers and under garments.

10.6   Modacrylic fibre (Teklan)

It is a copolymer made with equal weight proportions of vinylidene chloride and acrylonitrile and some small but critical quantities of other ingredients. The fabric made from this is flame proof passing some standards like British Standard 3120.

Some special properties of this fibre/fabric include flame proof, good abrasion resistance, resistance to photodegradation, resistant to microbial attack, low shrinkage in boiling water, etc. Used for making T-shirts and children's clothing.

10.7   Dralon fibre

It is a type of acrylonitrile fibre, made with acrylonitrile copolymer and small proportion of some other monomer by Bayer and company initially. It is available as filaments only. It has good elasticity with 95% recovery with 2% stretch. It is also heat and chemical resistant and dimensionally stable. Because of the elasticity, it is used for making domestic gloves, suits, socks, etc. But with the advent of elastane fibres, the importance of these fibres has been diminished (Fig. 10.4).

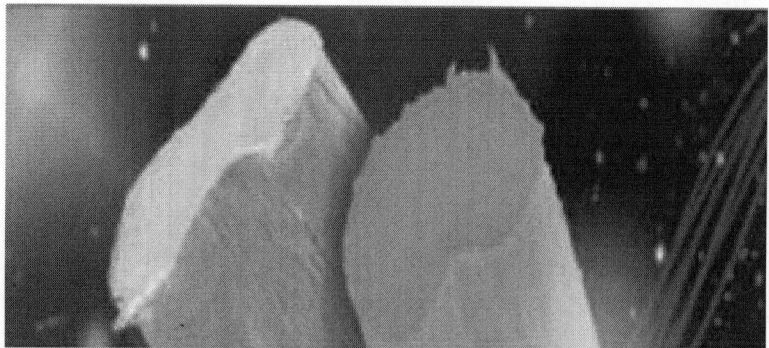

Figure 10.4. Microscopic view of Dralon fibre

10.8   Nitrile fibre

Nitrile is a man-made fibre containing at least 85% of a long chain polymer of vinylidene dinitrile, where the vinylidene dinitrile content is no less than every other unit in the polymer chain. The fibre is also called Dravan and Travis.

Special properties of the fibre are as follows:

1. Good handle.
2. Resistance to sunlight.
3. Moisture regain between 2% and 3% even though fully synthetic.
4. Good biological resistance to mildew, moth, etc.
5. Good dimensional stability.

Because they are shrink resistant, it is used in making sweaters.

10.9   Anidex fibre

A man-made fibre in which the fibre forming substance is a long synthetic polymer chain composed of at least 50% by weight of one or more esters of a monohydric alcohol and acrylic acid, $[-(CH2=CHCOOH)-]x$. First commercial Anidex fibre was started in 1970 by Rohm and Haas Company. It gives permanent stretch recovery properties to the garment. It retains dimensional stability to the garment even after repeated washing. Good wear comfort and excellent resistance to gas fading, oxidation, sunlight, oil and chlorine bleach. Used in athletic wear, blouses, dresses, hosiery, jackets, linings, shirts, slacks, suits and sweaters.

10.10   Hollow fibres

These fibres are developed for improving wearing properties of synthetic fibres like polyester and nylon which otherwise is uncomfortable to wear. During melt spinning of polyester and nylon, special spinnerets are used to inject air into the filament and staple fibres. This imparts high bulk, higher moisture absorption, wicking properties and keep warmth better compared to its parent fibre. These fibres are having better aesthetic qualities, good feel, softer and fuller handle, etc. Because of its fuller handle and keeping warmth, these fibres are best suited for sweaters, jackets and tights.

11

The Yarn

Any fibre to be used in woven of knitted fabric has to be converted to yarn. We will mainly take the cotton system to explain regarding the yarn. The end product of the cotton fibre-to-yarn conversion system is a spun yarn or a staple-fibre yarn, which is suitable for making numerous end products from knit apparels to woven fabrics, from towels to sheets and from carpets to industrial fabrics.

The yarn quality requirements for different fabric manufacturer differ. Processor may not have a big say in this, but when a grey yarn quality requirement of a fabric manufacturer is provided by the spinner the same way a dyed yarn quality requirement of a manufacturer has to be provided by a processor. To achieve this, one should see that his requirement for yarn for dyeing is provided by the spinner. Hence processor should have the basic knowledge of yarn characteristics – quality, winding, package, count, twist, etc.

Fibres are normally spun into yarns with the exception of nonwovens. A selection of typical yarn structures is Fig. 11.1. The so-called 'spun yarns' are yarns made from staple fibres (e.g., cotton and cut man-made fibres). All other yarns are made from man-made fibres. Plied yarns consist of two or more parallel-oriented yarns; twisted yarns consist of at least two twisted yarns.

The knitter may have more detailed criteria of yarn quality. These may include:

- A yarn that can unwind smoothly and conform readily to bending and looping while running through the needles and sinkers of the knitting machine. This translates to flexibility and pliability.
- A yarn that sheds low fly in and around the knitting machine. This translates to low hairiness and low fibre fragment content.
- A yarn that leads to a fabric of soft hand and comfortable feeling. This translates to low twist, low bending stiffness and yarn fluffiness or bulkiness.
- A yarn that has better pilling resistance. This translates to good surface integrity.

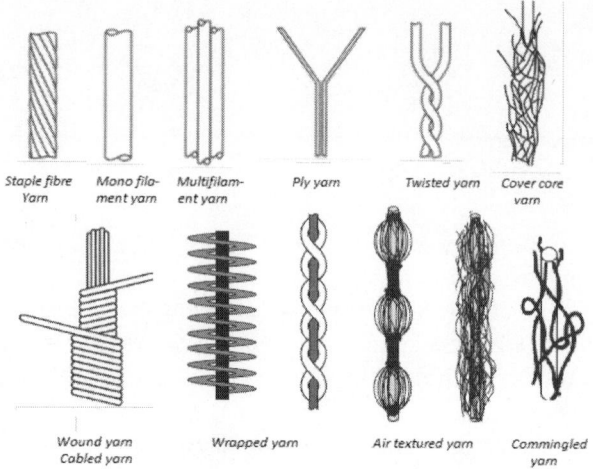

Figure 11.1. Selected yarn structures

The weaver may have a different set of yarn quality criteria:

- A yarn that can withstand stresses and potential deformation imposed by the weaving process. This translates to strength, flexibility and low strength irregularity.
- A yarn that has a good surface integrity. This translates to low hairiness and high abrasion resistance.
- A yarn that can produce defect-free fabric. This translates to high evenness, low imperfection and minimum contamination.

11.1   Classification of yarns

Figure 11.2 summarizes the classification of yarns.

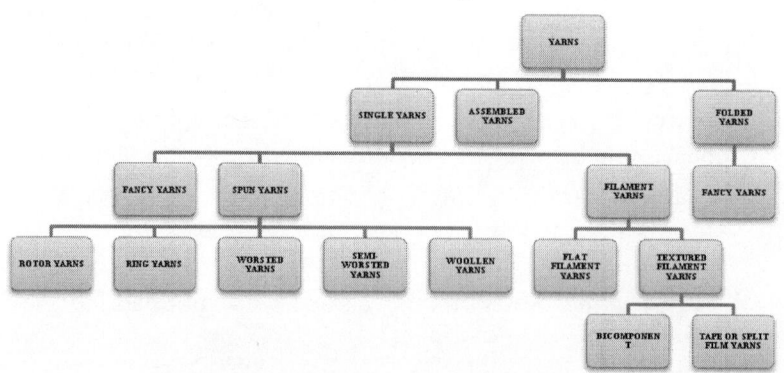

Figure 11.2. Classification of yarns

11.2   Cotton spinning process

At present, the short-staple (also called cotton or three-roller) spinning process is the most common spinning method worldwide. The name three-roller spinning comes from the arrangement of the rollers in the drafting zone in the most commonly used spinning machine, the ring spinning frame. This spinning principle is suitable for all fibre types with lengths up to 40 mm. It is very flexible with regard to the properties and applications of the produced yarns. Yarns manufactured by ring spinning are processed into wovens, hosiery, knits and braidings in the areas of apparel, home textiles and technical textiles. Yarn properties are parameters to describe the yarn, such as fineness or titre, elongation, hairiness, tenacity, twist and volume. It takes several processing steps to manufacture fibres into a yarn. Fig. 11.3 gives an overview of the various processing steps.

From the bale of raw cotton or cut synthetic fibres to the final yarn. Depending on the desired yarn properties and the fibre material, various different machine sequences and spinning principles are applied to reach a compromise between optimum yarn properties and cost-saving manufacture.

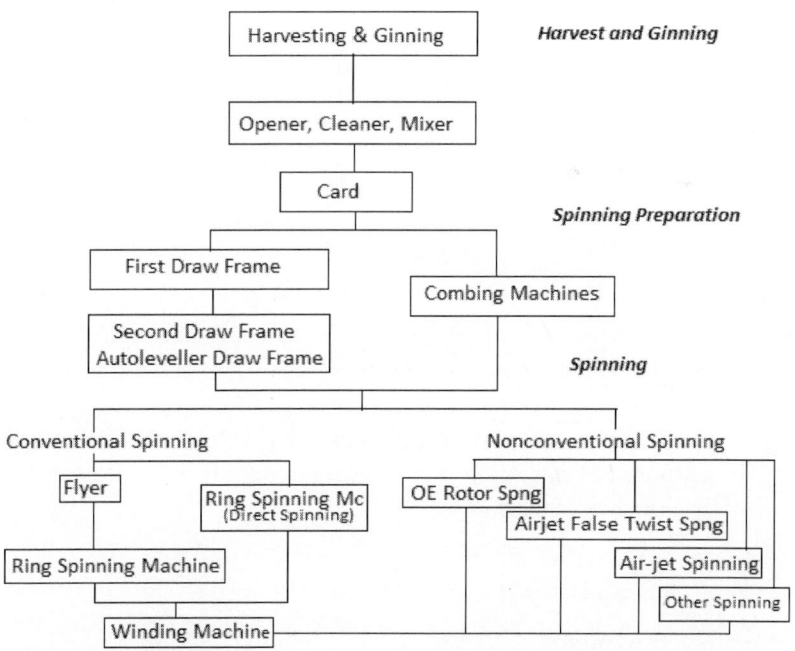

Figure 11.3. Overview of short staple spinning

Table 11.1 shows the main function of each machine during processing. The generic term mixing means a thorough blending of nonhomogeneous fibres of the same type as well as a quantitatively defined combining of different fibre types. Opening is the disentanglement of compressed fibre packages into single fibres. Cleaning is the removal of particles such as wood, leaf or seed coat fragments (cotton) or fibre knots and neps that cannot be opened. Parallelizing leads to an orientation of the fibres in one direction.

Table 11.1. Functions of various machines during yarn production

Processing steps	Mixing	Opening	Cleaning	Parallelization	Sliver formation	Drafting
Harvest and ginning	Yes	Yes	Yes			
Opener, cleaner, mixer	Yes	Yes	Yes			
Card	Yes	Yes	Yes	Yes	Yes	
Draw frame	Yes	Yes		Yes	Yes	Yes
Combing machine		Yes	Yes	Yes	Yes	Yes
Flyer machine				Yes	Yes	Yes
Ring spinning machine				Yes		Yes
OE rotor spinning machine	Yes	Yes	Yes	Yes		
Air-jet spinning machine				Yes		Yes

Note: OE, open end.

Drawing is the drafting of an oriented fibre web or sliver in a drafting field. The purpose of drawing is to provide an optimally straightened, parallel and uniform fibre orientation.

11.2.1   Bale feeding

Bale feeding is normally done by hopper feeders or by top feeders. In bale feeding, the compressed bale of cotton is broken down into smaller tufts of fibre. Of the two feeding systems, top feeders are more prevalent today when processing cotton fibres. Cotton bales are selected for desired properties and arranged in a row formation, two or three bales wide. This is referred to as a lay down and normally consists of 40–80 bales. The top feeders move across the top of the lay down from end-to-end plucking a very thin layer of fibres from each bale. Those tufts of fibres are then transported through air ducts to the opening/cleaning/blending equipment.

11.2.2 Fibre opening and cleaning

Cotton fibres then pass through a succession of machines designed to gradually and progressively open the dense tufts of fibres and reduce tuft size. Less intensive opening action will proceed to action that is more intensive. This gradual reduction and opening of the tufts help preserve fibre quality, create less fibre entanglements (referred to as neps) and promote better cleaning and blending. Impurities of concern in cotton are plant leaf, stem, seed particles, neps, short fibres less than 0.5 inch (12.7 mm) and immature or not fully developed fibres. These initial opening and cleaning machines utilize large beaters covered with either spikes or coarse saw-tooth wire. These machines will be followed by beaters with finer and finer wire surfaces. Initially larger particles of waste will be removed followed by smaller particles removed with more intensive action. Care is taken to preserve fibre quality.

With the working principle *free beat*, the flocks are caught in free fall and accelerated by the cleaning elements. The opening is created by the interaction of forces of acceleration and inertia. Contaminating particles resulting from the effect of centrifugal and gravity forces can be separated by grids. There are machines with one roller (single-roller cleaner) or with two rollers (double-roller cleaner).

With the *restricted beat*, the flocks are 'squeezed' between two feed rollers or between one feed plate and one feed roller during the operation of the

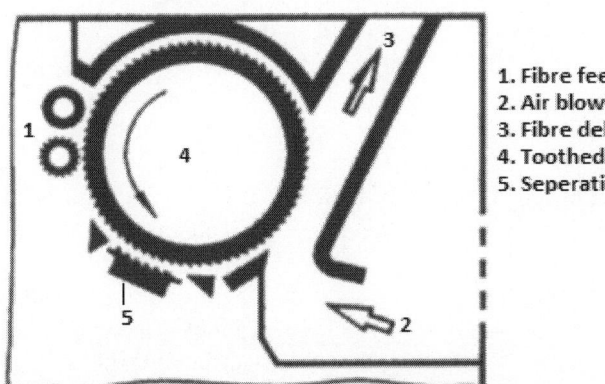

1. Fibre feed
2. Air blowing
3. Fibre delivery
4. Toothed drum
5. Seperation knife

Figure 11.4. Principle of saw tooth cleaner

beaters. Depending on the degree of opening, the beaters are nose beaters, pins or saw teeth (see Fig. 11.4). Because of the squeezed position of the fibres, the effect of opening with the restricted beat is more intense but also more aggressive compared to the free beat. This may lead to fibre damage if the flocks have not been sufficiently preopened.

11.2.3   Fibre blending (mixing)

The consistency in yarn quality depends heavily on the homogeneity of the material composition. The objective of mixing is to optimize the homogeneity of the material mixture by combining several bales. Further objectives of mixing are

- decrease in irregularities in bales of different origins,
- economic processing,
- recycling of comber waste and other offal,
- improved properties of the final product, and
- reduction of raw material costs.

The expression mixing is divided into the two functions: blending (dosage) and mixing thoroughly (mixing).

A popular machine for blending fibres from bales of the same fibre type is the cell mixer. Fibre stock is fed from the top into a series of parallel vertical cells or chambers and removed at the bottom or delivery end of the chambers. Fibres from the different cells are randomly mixed due to each cell delivering its fibres in different time intervals. For this reason, these machines are also called time-delay blenders or blender/reserves.

When blending different fibre types such as cotton and polyester, each type can be weighed by a weigh-pan type hopper feeder. Many times, six to eight of these machines will be parallel to one another, each with the potential of dropping a designated weight of fibre onto a conveyor belt. The combined drops of different fibres are then moved forward and blended by a mixing beater. They then move into a cell mixer for further blending to promote a more even product in subsequent processing. When fibres are blended in this way, it is referred to as an 'intimate blend'. In addition to weigh-pan type machines, there are belt-weighing and chamber-type machines available.

The purpose of mixing thoroughly is to achieve a homogeneous distribution of the various components in the final product. The machines most often used for this task are the mixing chamber and the multiple mixer. With the mixing chamber principle, a very large volume (e.g., 230 m³) is filled

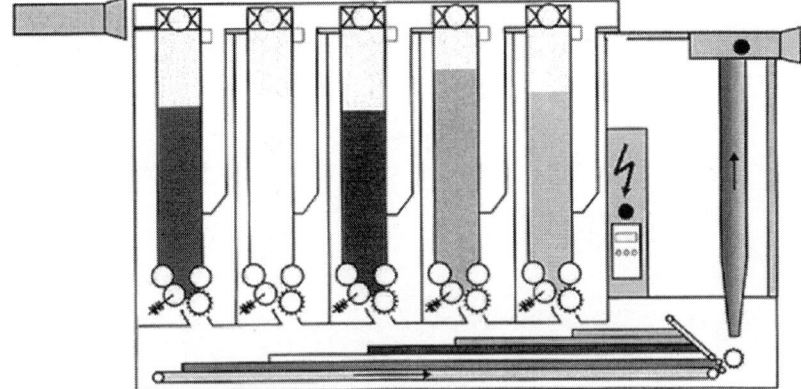

Figure 11.5. Multiple mixer (Rieter)

in horizontal layers and subsequently reduced vertically from one side. This may be done continuously or noncontinuously. The largest mixers are used in mock-worsted spinning. A multiple mixer (see Fig. 11.5) is composed of multiple chambers lined up behind each other. There are two working principles: either flocks that had been fed at the same time are processed at different times or flocks fed at different times are processed at the same time.

11.2.4   Carding

Following are the functions of the carding machine:
- removal of dirt particles and short fibres,
- disentanglement of the fibre flocks into single fibres,
- parallelizing of the fibres,
- mixing thoroughly,
- drafting,
- sliver formation, and sliver delivery and storage in a can.

Opened, cleaned and blended fibres are transported by air ducts to a chute feeder, which prepares the fibres for feeding into the carding machine. The chute will form a uniform mat of fibres about 40″ wide and 3″ thick. The card makes use of a series of saw-tooth, wire covered surfaces that move in close proximity to one another and at different speeds. This speed differential and wire action causes the fibres to be brushed and is designed to remove trash matter, short fibres and neps. The carding action also achieves a total opening of the cotton fibres, which further enhances fibre cleaning. In addition, the fibres are

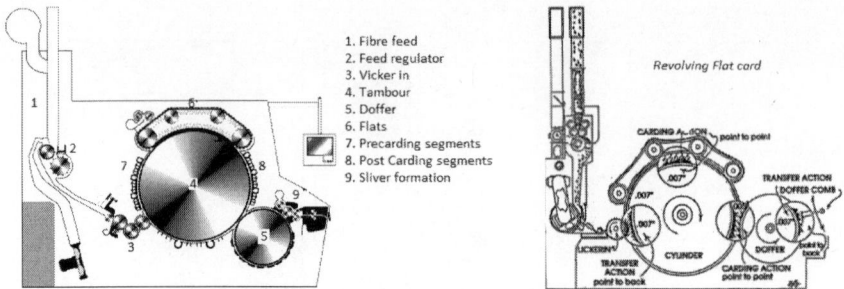

Figure 11.6. Principle of a card (Trutzschier)

also aligned somewhat parallel in the direction of material flow out of the card. Finally, due to the wire surface orientation and speed differences, the fibres are actually 'drafted' or thinned down to a spider web-like structure. The thin web of fibres will be 38–40″ in width and at the front of the carding machine will be condensed into a rope-like strand called a card sliver. Card slivers are then levelled and coiled into large cylindrical cans. The delivered card slivers will be routinely monitored and checked for correct weight per unit length and degree of uniformity. Variations in the card sliver will lead to eventual variations in the final yarn product (Fig. 6).

11.2.5   Drawing

Carding is usually followed by a one or more drawing machines, the functions of which are:

- doubling of multiple slivers and drafting for mixing and homogenizing,
- removal of dust from the slivers,
- production of homogeneously mixed slivers from slivers of different materials, for example, four cotton slivers and two slivers of chemical fibres for the production of a yarn of cotton/chemical fibres 67/33, and
- additional control for sliver homogenization.

Multiple cans of card sliver are then fed into drawing machines. Strands of sliver going into the drawing machines are typically called 'doublings' or 'ends'. The number of doublings or ends is normally six to eight. The ends of sliver can be of different fibre types, creating what is called a draw blend. For example, four cotton slivers and four polyester slivers of the same linear density would eventually form a 50 cotton/50 polyester blended yarn.

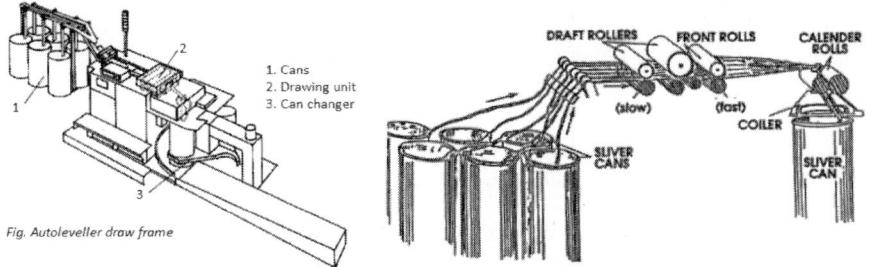

Figure 11.7. Drawing frame

One or more slivers could contain dyed fibre, and the result would be a heather blend for an eventual heather yarn such as used in fleece wear and underwear.

On the drawing frames, these multiple ends of sliver are fed into a roller-drafting system. A series of pairs of top and bottom rollers grip the mass of fibres coming through, and with successive increases in roller speeds, the fibre mass is reduced or drafted. Therefore, only one sliver is levelled and delivered from each drawing machine. The roller action helps to further align the fibres, and the multiple ends fed create further blending of fibre variables and/or different fibre types (Fig. 11.7).

There may be multiple processes of drawing. Drawing before combing is called predrawing. When not combing, first process drawing is called breaker drawing, and a second process drawing is called finisher drawing. As noted on an earlier flow chart, there are several variations of drawing steps according to the method of yarn spinning.

11.2.6   Combing

For combing which is done for production of combed yarn, multiple ends of drawn sliver are wound onto a large spool in a process called lap winding. Each lap will typically contain 20–48 slivers and is fed to an individual combing position on the comber machine. A rotating wire-covered cylinder called a half-lap will 'comb' through a short segment of fibres from the lap. These combed fibres are then separated from the comber lap and pieced together with previously combed fibres. The combing action removes short fibres, neps and trash particles while promoting fibre parallelization and blending. The items removed in combing as waste are referred to as 'comber noils'. There are different degrees of combing as related to the per cent noil removed (Fig. 11.8).

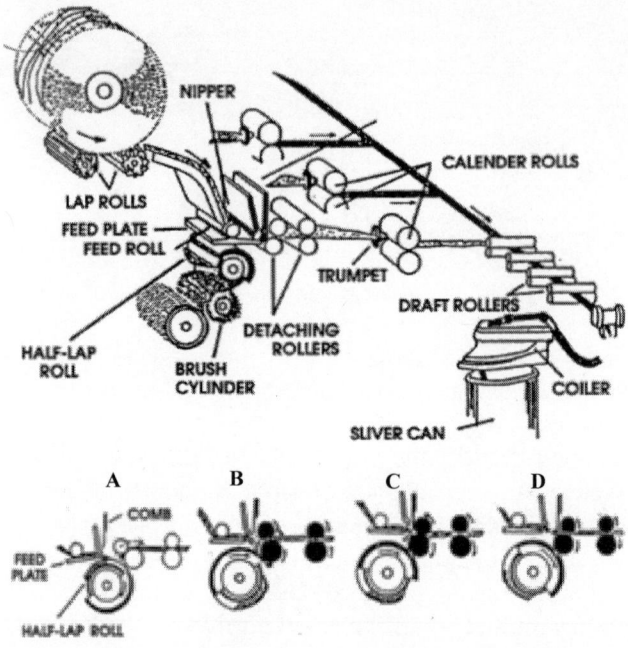

Figure 11.8. Combing

After combing, there should be more drawing processes. Normally, one or two drawing processes will follow combing in order to improve the short-term variation created by the combing and piecing action.

11.2.7   Roving

Roving is a process utilized only in the production of ring spun yarns. Drawn carded or combed sliver is fed into a set of paired top and bottom drafting rolls. The mass of fibres in the sliver is thinned to a cross section about the size of a pencil. Because of this small cross section of fibres, aprons (small flexible rubber belts) are used for better fibre control. Because of such a thin strand (roving), there is not a sufficient amount of inter-fibre cohesion to give the roving adequate strength for transporting to the ring-spinning process. Therefore, a small amount of twist is added as the roving is delivered from the machine by a rotating device called a flyer. The rotating flyer aids in twist insertion and the laying of the roving onto a roving bobbin. These bobbins are in turn transported to ring spinning, where they become the input material for making ring spun yarn (Fig. 11.9).

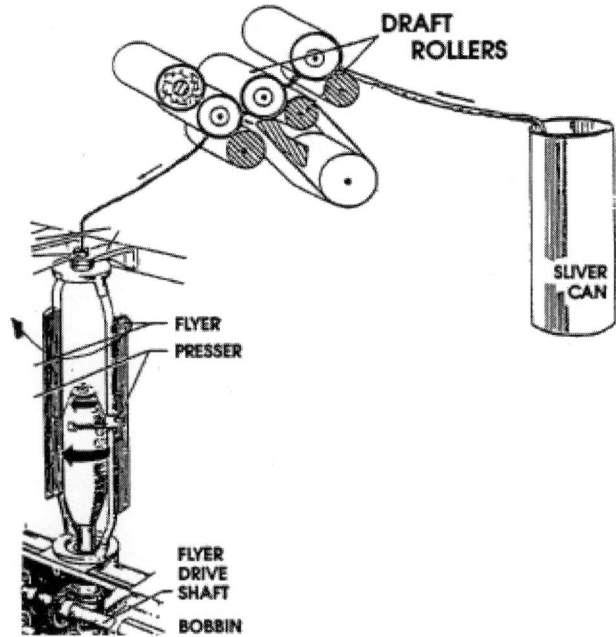

Figure 11.9. Roving

11.2.8   Conventional spinning

11.2.8.1   Ring spinning

In ring spinning, roving is fed into a roller-drafting system with aprons, which has the responsibility of drafting (thinning down) the fibre mass to that which is desired for a given yarn size (count). As the fibre stream leaves the front or delivery roll of the drafting system, it forms into a triangular shape. Fibres on the fringe of the triangle tend not to be totally attached to the yarn core and thus increase the yarn hairiness and potential shedding of fibres. Compact spinning is a relatively innovation where an additional attachment to the delivery roll of the ring-spinning machine causes the yarn to be 'compacted'. This add-on unit helps create a ring spun yarn with greater strength, reduced hairiness, greater yarn elongation, increased lustre and the possibility of reducing yarn twist resulting in an improved fabric hand.

From the front roll of the spinning machine, the yarn is directed downward and then through a small 'c-shaped' steel clamp called a traveller. The traveller is free to rotate around a metal ring, hence the term 'ring' spinning. The package (ring bobbin) that the yarn is wound onto is located inside of the metal ring. The ring bobbin is situated on a rotating spindle, which can

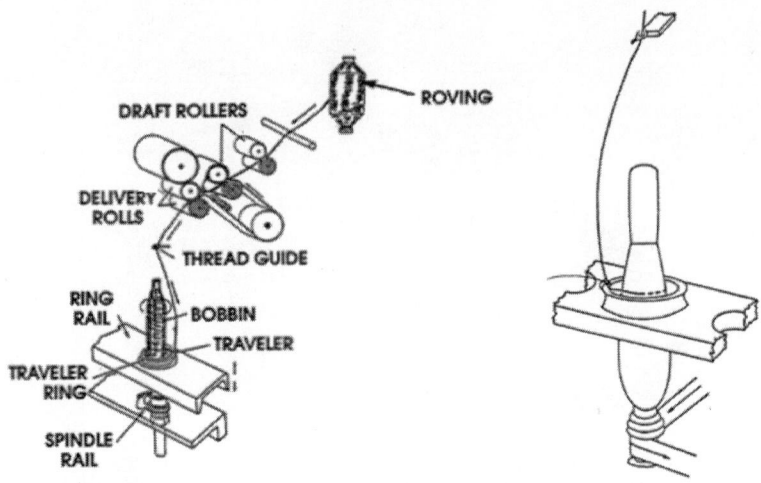

Figure 11.10. Ring spinning

rotate clockwise to produce Z-twisted yarn or counterclockwise to produce S-twisted yarn. As the yarn is taken up on the spinning bobbin, the yarn itself pulls the traveller around the ring at speeds up to 8000 feet per minute. The twist in the fibres created by the rotation of the bobbin and traveller migrates all the way up to the front roll of the drafting system. This twist gives the yarn its final strength values. The amount of yarn on each spinning bobbin is only a few ounces in weight, which necessitates a subsequent winding operation where numerous spinning bobbins will be individually wound onto a much larger yarn package (Fig. 11.10).

Ring spun yarn properties can be altered according to customer requirements or to meet a specific need mainly by changing the level of yarn twist. Higher twisted yarns will tend to be stronger and more abrasion resistant, but harder, less flexible and have more torque.

11.2.8.2   Flyer spinning

In the first step, the drawn sliver is predrafted in the flyer spinning frame (draw ratio of 1:5 to 1:50), while it is also twisted slightly to prevent wrong drafts. This safety twist has to be small enough to still allow drafting of the flyer yarn to the final yarn titre in the drafting field of the ring-spinning machine. Fig. 11.11 schematically shows a flyer spinning frame. The sliver is first drafted in the flyer-drafting field, which is often designed as a three-roller-two-apron-drafting unit. From the drafting field, the drawn sliver is transported over the flyer top into the flyer leg, which it exits at the bottom. A finger guide leads the sliver to the bobbin surface. With this mechanism, the

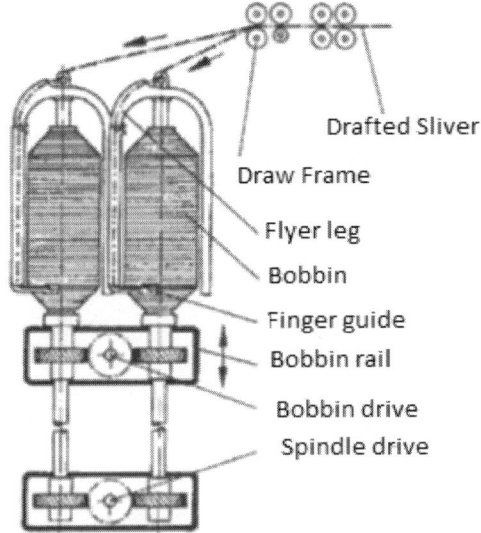

Drafted Sliver

Draw Frame

Flyer leg

Bobbin

Finger guide

Bobbin rail

Bobbin drive

Spindle drive

Figure 11.11. Schematic view of a flyer spinning frame

sliver obtains one twist with each revolution of the flyer. The winding itself is caused by a lead of the bobbin against the flyer top. The vertical movement necessary for the winding is accomplished by the bobbin. Because the flyer operates at a constant feed speed, the vertical movement and the revolutions per minute have to be adjusted continuously according to the bobbin diameter. The limit of revolutions per minute for the flyer is about 1300–1500 rpm and depends on the maximum feed speed.

For a bobbin change, the flyer legs have to be removed, which makes automation difficult. Several more disadvantages associated with the flyer spinning principle include that the stopping of the entire machine is necessary if a sliver breaks. For decades, people have been working on eliminating the flyer process and spinning drawn slivers directly on the ring spinning frame (so-called direct spinning).

11.2.8.3   Compact spinning

This relatively new development reduces the size of the spinning triangle to a minimum. This is achieved through a condensing of the fibres after the main draft by using a perforated roller in combination with a suction unit. The hairiness of the yarn is thus reduced, and the tenacity is higher when compared to ring spun yarns. The yarn evenness is also improved. See Fig. 11.12 for a comparison of ring and compact spinning, as well as the principles of the three systems currently on the market.

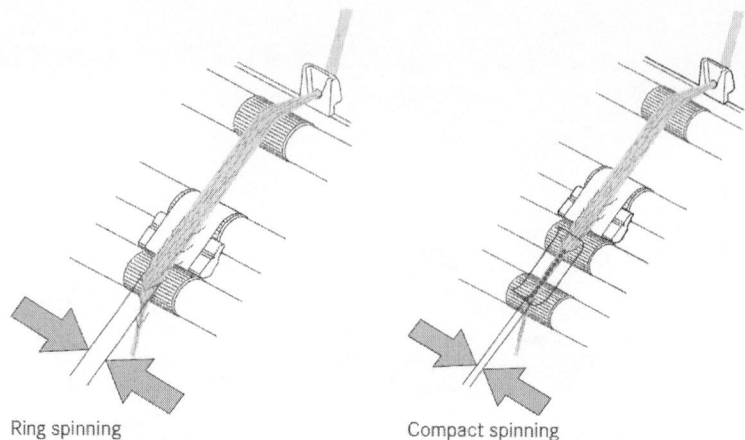

Ring spinning   Compact spinning

Figure 11.12. Principles of ring and compact spinning

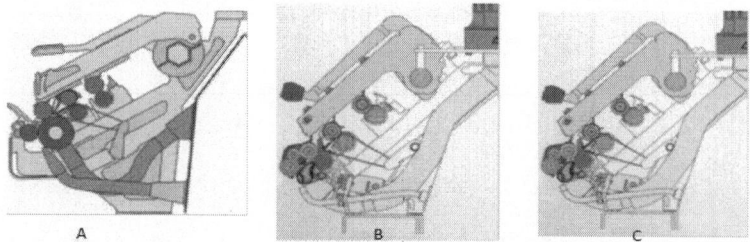

A   B   C

Figure 11.13. Three types of compact spinning. (A) Compact spinning rieter. (B) EliTe, Suessen. (C) Air-Com-Tex, Zinser

There are different types of compact spinning unit available in the market (see Fig. 11.13).

11.2.9   Nonconventional spinning

11.2.9.1   OE spinning

In this spinning (sometimes called rotor spinning) process, sliver (not roving) is fed through a feed roll to a rapidly revolving comber (opener) roll, which is covered with either saw-tooth wire or short pins (Fig. 11.14). The comber roll opens, separates individual fibres from the end of the sliver and removes trash particles. These opened and separated fibres go through a tapered transport channel and then into the groove of a rapidly revolving rotor (up to 150,000 revolutions per minute). Fibres are impinged into the rotor groove on top of each other to create the needed mass for the desired yarn size. The fibres are twisted together as the rotor revolves, while the just formed yarn is

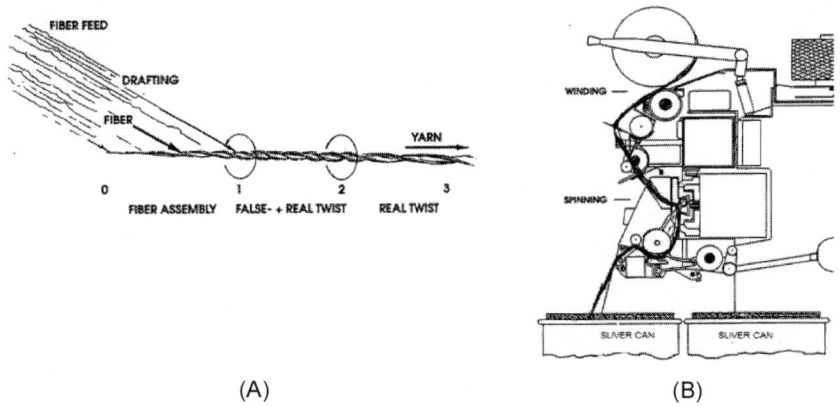

Figure 11.14. (A) Principle of OE spinning. (B) Schematic diagram of OE spinning

removed from the spin box. Instead of being twisted together, some fibres are wrapped around the waist of the yarn and do not contribute to yarn strength. These 'wrapper' fibres are unique to OE yarns. The speed of the delivery roll divided by the speed of the feed roll is what determines the yarn size (count). The amount of twist in the yarn will be determined by the speed ratio between the rotor and the delivery roll. After the yarn leaves the delivery roll, it is normally passed through some types of mechanism designed to 'clear' (remove) objectionable defects from the yarn. The yarn is then waxed and

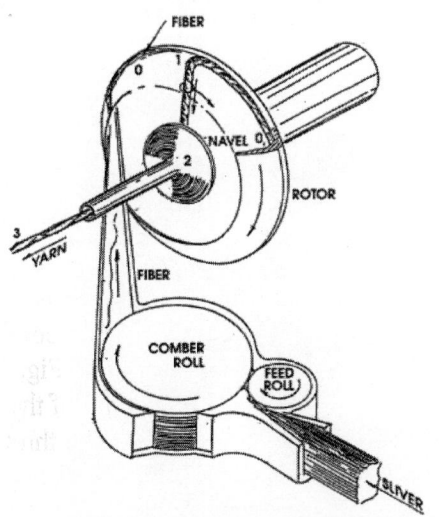

Figure 11.15. Typical OE machine

wound onto a large package that will be used in knitting and weaving (requiring no wax). Productivity of each OE spinning position is about 8–10 times greater than a typical ring-spinning position.

Open-end yarn characteristics can be altered by changing the navel, which is located at the centre of the rotor in the spin box. As the yarn leaves the rotor via the navel, it can be made softer and bulkier by using an aggressive (higher surface friction) type of navel. A stronger and smoother yarn with more torque can be attained by using a less aggressive, smoother navel. OE yarns with ring-spun-like characteristics can be produced by using computer-aided systems, which help introduce additional variations in the yarn structure. Different types of rotor grooves can also alter yarn properties (Fig. 11.15).

11.2.9.2   Air-jet spinning

There are two types of air-jet spinning, MJS and MVS vortex. In either case, finisher sliver is fed to a high-speed drafting system, consisting of three pairs of rolls and an apron. The amount of draft determines yarn mass (size). After leaving the front roll, the fibres are introduced to an air jet or vortex to create a yarn (Fig. 11.16).

In MJS spinning, introduced by Murata Machinery Ltd, Japan and hence called Murata Jet Spinning (MJS), the finisher drawn sliver is subjected to an air vortices located in a pair of air-jet nozzles. This initial air vortex in the first nozzle adds twist to the leading ends of fibres, while trailing ends are held by the front roll. This vortex also helps to separate a small ribbon of fibres from the main core fibres. This ribbon of fibres wraps around the parallel core fibres in a spiral configuration and provides sufficient fibre-to-fibre

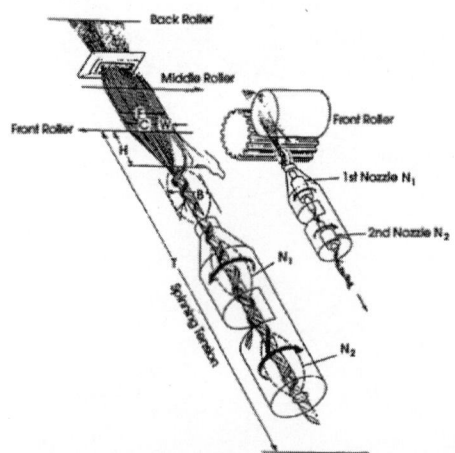

Figure 11.16. Formation and structure of air-jet spun yarn

cohesion for needed yarn strength. Unlike the OE yarn wrapper fibres, the binder fibres that wrap around the air-jet yarn are critical to yarn strength, because they hold the internal bundle tightly together. They are more organized than those of an OE yarn. The second air nozzle creates another vortex, which imparts false twist to the yarn bundle with air revolving in the opposite direction to that of the first nozzle. As the yarn exits the second jet, the false twist is removed and core fibre twist is reduced to zero. As in OE spinning, the air-jet yarn passes through a yarn clearer and is then wound onto a large package. At this point, the package is ready to be the yarn supply for knit and woven fabrics. Productivity for air-jet spinning is 15–20 times higher per position compared to ring spinning.

11.2.9.3   Vortex spinning (MVS)

In this most recent modification of air-jet spinning, introduced by Murata Machinery Ltd and called Murata Vortex Spinning (MVS), the fibres leave the roller drafting system and enter a single nozzle where an air vortex is created by three air inlets (Fig. 11.17). The air vortex creates a circular movement of the fibres and causes the fibre ends to flare or curve outward just before entering the vortex. These flared fibres are then wrapped around the central core of fibres by the air vortex. Fibre ends from the flared fibres give the vortex yarn characteristics and appearance of a ring spun yarn. Shorter

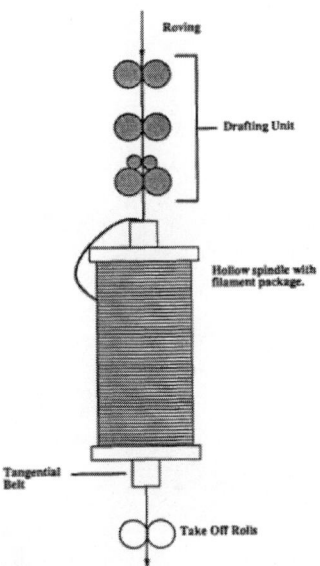

Figure 11.17. Vortex (wrap) spinning principle

fibres in the fibre flow are removed from the main yarn bundle and end up as process waste. This newer system can produce 100% cotton yarns, which is not normally done successfully on conventional air-jet machines.

11.2.9.4   Air false-twist wrap spinning

Another type nonconventional spinning of importance is air false-twist wrap spinning which has gained more and more importance (Fig. 11.18). The feed sliver is first drawn to the final yarn titre via a drafting unit. Two air nozzles are located immediately at the exit of the drafting field. The twisting nozzle N_2 provides a tangential air stream that causes a false twist in the yarn that propagates down to the exit to the drafting zone. This twist is only temporary and disappears on its own after the false-twist zone. The nozzle N_1 provides an air stream through tangential bore holes diagonal to the yarn transport. This setup is designed to orient single fibres that protrude from the surface of the false-twisted fibre assembly perpendicular to the transport direction. After the release of the false twist by the yarn transport, the core of the yarn has lost its twist again. The fibres spread out in and wind around the core during the release of the false twist. The normal force acting on the yarn core by this wrapping mechanism causes the stabilization of the yarn by friction. In the air-friction spinning method, very fine yarn titres can be spun (up to 10 tex) at good uniformity. Favourable yarn properties can be achieved with fine, strong and uniform fibre raw material. Therefore, this method is especially suited for chemical fibres or mixtures (cotton percentage <50%). The processing of 100% cotton fibres has recently become possible. The future will

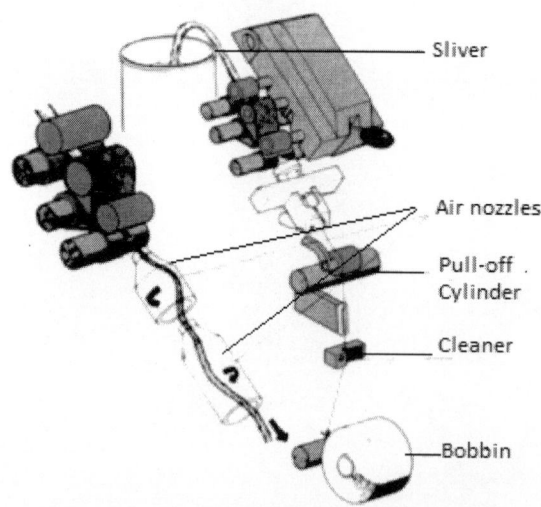

Figure 11.18. Principle of air false-twist wrap spinning

show if staple fibre yarns of 100% cotton will be economical for industrial production.

11.2.10   Winding

While winding of yarn packages is a part of OE and air-jet spinning systems, it is a separate operation when producing ring spun yarns. Yarn is unwound from individual ring bobbins, passed through a 'clearer' mechanism to remove undesired defects, and then wound onto large packages. Due to more torque in ring spun yarns, the yarn packages are many times conditioned by heat and moisture under pressure to relax the torque created by the twist in the yarn.

11.2.11   Twisting (plying)

Sometimes individual cotton spun yarns are plied (twisted together) to introduce different yarn characteristics. Plying of spun yarns can accomplish the following:

- Add to or increase the strength of single strand yarns.
- Yield a balanced yarn with no twist liveliness.
- Utilize multi-strands of fine yarns to produce a thick strand.
- Produce a smoother yarn.
- Produce a yarn with a more uniform diameter.
- Introduce novelty effects.
- Add colour.
- Introduce different fibre yarns.
- Combine spun and filament yarns.

Normally, yarns are ply-twisted in the opposite twist direction from that which is found in the singles yarn. If single yarn is twisted in the Z direction, then the ply-twist will be in the S direction, thus reducing yarn torque (liveliness). For example, if two Ne 40's yarns are plied, the resultant yarn number would be denoted as 40/2. This yarn size would be equivalent to a 20/1 yarn.

Single spun yarn uses TM, and ply uses TM and TPI.

Filament yarn TPI = TM (count)1/2

Generally accepted twist for cotton yarns

Knitting TM = 2.2–3.5

Filling (weft) TM = 3.5–4.2

Warp TM = 4.2–5.0

Voile and crepe TM = 5.0–7.0

11.3   Spun yarn characteristics by different spinning system

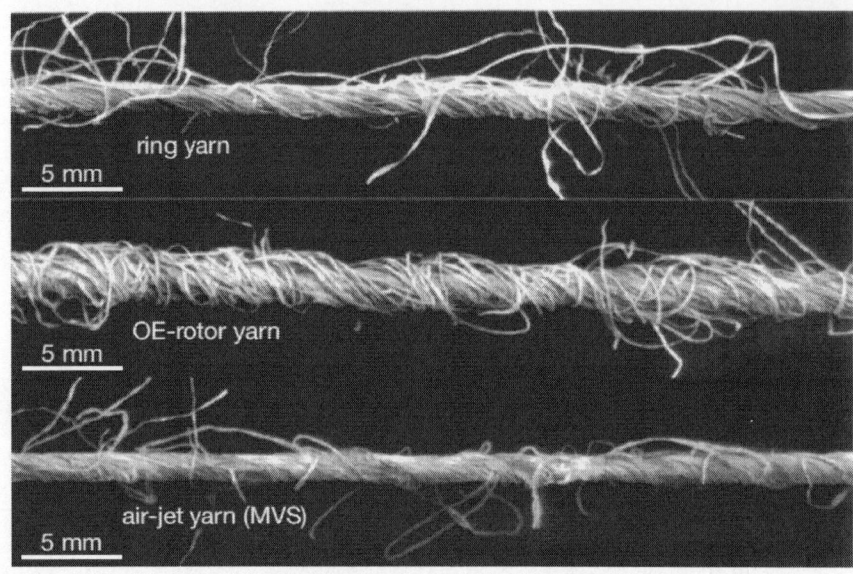

Figure 11.19. Appearance of different spun yarns

Table 11.2. Spun yarn characteristics by different spinning system

Ring spun	OE (rotor)	Conventional MJS to air-jet	MVS/vortex to air-jet
Strongest yarn	More even	Fewer process	More ring like
Finest yarn	Higher strength uniformity	Very high productivity	Lower hairiness
Softest yarn	Higher production rate	Harsher and stiffer hand	Limited count range
Z and S twist	Z twist only	Less pilling	Good hand
Lowest productivity	Lower cost	Fewer imperfections	Higher productivity
Most uneven	Fewer imperfection	Weaker yarn	More toque
Most expensive	Harsher hand (feel)		Dyes darker
More hairy, generally	Not as strong	Good evenness	
High linear density	Medium linear density	Low linear density	Low linear density

Ring spun	OE (rotor)	Conventional MJS to air-jet	MVS/vortex to air-jet
More torque	Fewer process		Lower cost
Low spinning speed	High spinning speed	Very high spinning speed	Very high spinning speed
Widest range of yarn counts	Limited count – coarser yarns	Limited to mostly medium counts	More waste in spinning

11.4   Yarn counts

There are two entirely different yarn counts

- Yarn count by length, i.e., variable length/constant weight
- Yarn count by weight, i.e., variable weights/constant length

11.4.1   Type of counts

1. Yarn count by length: For cotton, flax, sisal, jute, wool and silk yarns
2. Metric yarn count (Nm): Metric yarn count (Nm) indicates the number of metres yarn weighing 1 g. Thus, Nm = length/weight = m/g
3. English yarn counts (Ne): This indicates how many units of a determined number of yards (1 yard = 0.9144 m) in 1 lb (1 lb = 453.59 g); normally number of skeins at 840 yards weighing 1 lb for cotton and number of skeins at 560 yard lengths weighing 1 lb, etc. Thus, 1 Ne = number of units of standard number of yards per pound.

Depending on the type of fibre, English yarn counts are subdivided as follows:

(a) Cotton (NeB) = 840 yards/1 lb = 768.1 m/453.54 g

(b) Wool (worsted) (NeK) = 560 yards/1 lb = 512.06 m/453.54 g

(c) Linen (NeL) = 300 yards/1 lb = 274.32 m/453.54 g

There is a French number (Nf) which is the number of skeins of 1000 m weighing 500 g, which is not much in use.

4. Yarn count by weight:

Decimal count (dtex) = g/10,000 m

Denier count (Td or den) = g/9000 m

Tex (tex) = g/1000 m

Kilotex (Ktex) = kg/1000 m

Conversion formulas

Nm = Ne × 1.693 = Nf × 2

Ne = Nm × 0.591 = Nf × 1.18

Nf = Nm × 0.5 = Ne × 0.947

Tx = 1000/Nm = 1 × 10^{-3} g/m = D/9

Nm = 1000/Tx = 9000/D

11.5   Basic structural features of spun yarn

11.5.1   Yarn density

In a yarn structure, fibres represent the main component. The other component is air pockets created by the technology forming the structure. Accordingly, the yarn bulk density should be determined by the packing fraction, Φ, as defined by the following equation:

$$\Phi = V_f/V_y \tag{1}$$

where V_f is the volume of the fibres in yarn and V_y is the volume of the yarn (fibres plus air).

The packing fraction is an indication of the air spaces enclosed by the fibres. For example, a packing fraction of 0.5 indicates that there is as much space taken by air as by fibre. Most spun yarns have packing fraction well above 0.5.

11.5.2   Yarn bulk integrity

Yarn bulk integrity is determined by the fibre arrangement in the yarn structure. Fibre arrangement is expected to have significant effects on many yarn and fabric characteristics, including yarn liveliness, fabric dimensional stability, yarn appearance, yarn strength and fabric cover. The bulk integrity of a spun yarn largely reflects the impact of the spinning process on yarn structure. In general, different spinning techniques provide different forms of bulk integrity through providing different fibre arrangements. Obviously, the simplest fibre arrangement can be found in a continuous filament yarn where fibres (or continuous filaments) are typically arranged in parallel and straight form. As shown in Fig. 11.20, a slight deviation from this arrangement can be caused by slightly twisting the filaments or through deliberate distortion in the filament orientation as in the texturizing process.

In spun yarns, fibre arrangement is quite different from the simple arrangement discussed above. The discrete nature of staple fibres makes it impossible to fully control the fibre flow in such a way that can produce a

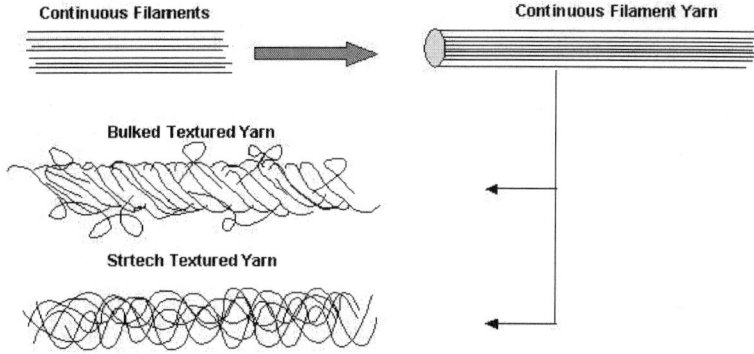

Figure 11.20. Fibre arrangement in continuous filament yarns

well-defined fibre arrangement. For this reason, a spun yarn typically exhibits some irregularities along the yarn axis. In addition, no spun yarn can be free of fibre ends protruding from its surface as shown in Fig. 11.21. Different spinning systems produce different forms of bulk integrity or fibre arrangements.

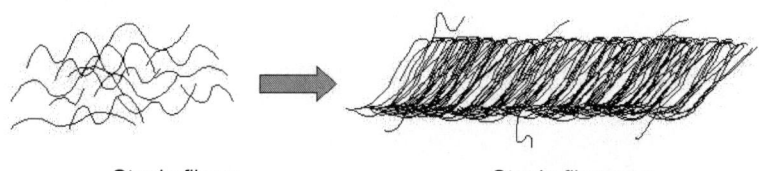

Staple fibres   Staple fibre yarn
Figure 11.21. Fibre arrangement in staple fibre yarn

The general features of fibre arrangement produced by four different spinning systems are shown in Fig. 11.22.

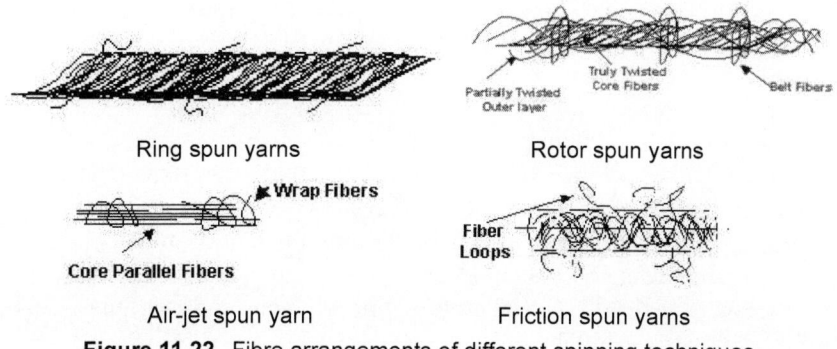

Figure 11.22. Fibre arrangements of different spinning techniques

In ring spun yarn, the fibres show true twist. The fibre axis follows an irregular helical path around the yarn axis. Fibres are crossing each other at different points along their axes creating different fibre segments between points of entanglement.

Rotor spun yarn can be seen as a three-layer structure – truly twisted core fibres, partially twisted outer layer and belt fibres.

In air-jet spun yarn, one can see a core of parallel fibres and wrap fibres holding the core fibre.

Friction spun yarn are truly twisted fibres with fibre loops.

11.5.3   Yarn fineness or count

In practice, yarn fineness is typically described by terms such as yarn count, yarn number or yarn size.

In general, two yarn count systems are commonly used: (a) the direct system and (b) the indirect system.

11.5.3.1   Direct count system

In a direct system, yarn count is the mass of a unit length of yarn. One of the universally used direct systems is known as the 'tex'. This is defined by the mass in grams of 1 km of yarn. For intermediate heavy products, such as slivers, the 'kilotex' is commonly used. This is the mass in kilograms per kilometre (or equivalently, grams per metre). A more common direct system for slivers is the grains/yard, where a grain is 1/7000 lb. For continuous filament yarns, the denier system is used; this is the weight in grams of 9000 m.

11.5.3.2   Indirect count system

In an indirect system, the yarn number or count is expressed in 'units of length' per 'unit of weight'.

For cotton yarns including slivers, the 'English' or 'cotton' count is used to express yarn fineness. The unit of length in an 'English' count system is the hank, 840 yards, and the unit of weight is 1 lb. Normally, yarn count is determined by determining the mass of 120 yards of yarn. For example, if the weight of a 120-yard yarn is 0.004 lb, the English or cotton count will be $120/(840 \times 0.004)$, or 35.7. In symbols, this is commonly written as 35.7Ne. Another indirect system is the 'metric' system commonly used in Europe. In this system, the unit length is kilometre, and the unit of mass is kilogram. The conversion from one system to another is given in the Handbook for Textile Processors Series IX – 'Laboratory Test Methods and Formulas, Tables, Useful Informations etc.' by the same author.

11.5.4   Plied yarns

We can make a plied yarn by twisting two single yarns together (see Fig. 11.23). The resultant yarn count is calculated as follows:

Using a direct count system,

$$\text{Count plied} = \text{Count 1} + \text{Count 2}$$

$$\text{e.g. Tex plied} = \text{tex 1} + \text{tex 2} \qquad \text{....(2)}$$

In indirect system,

$$1/\text{count plied} = 1/\text{count 1} + 1/\text{count 2}$$

$$1/\text{neplied} = 1/\text{Ne1} + 1/\text{Ne2} \qquad \text{.....(3)}$$

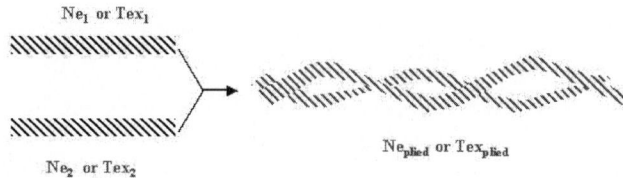

Figure 11.23. Plied yarns

Yarn may be plied for different purposes: (1) Introduce different fibre yarns. (2) Combine spun and filament yarns. (3) Add to or increase the strength of single strand yarns. (4) Utilize multi-strands of fine yarns to produce a thick strand. (5) Produce a smoother yarn. (6) Produce a yarn with uniform diameter. (7) Introduce textured or novelty yarns. (8) Add colour interest (Fig. 11.24).

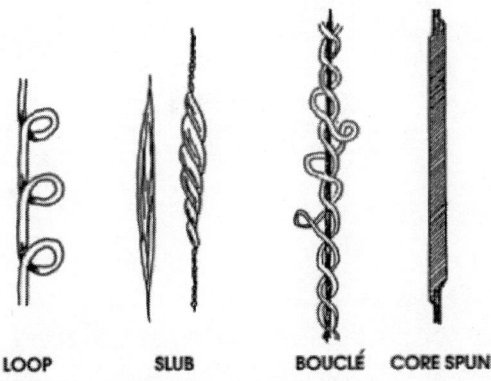

LOOP   SLUB   BOUCLÉ   CORE SPUN

Figure 11.24. Novelty yarns produced by plying

Characteristics of plied yarns: (1) Thicker and heavier. (2) Coarse. (3) Differ in count. (4) Less flexible than single yarns. (5) Affect drapability quality of fabric. (6) May be constructed with no twist at all. (7) May be highly twisted. (8) May differ in tension and direction of twist.

11.5.5   Count variation: CV_{count}, %

High-count variation can result in many quality problems including: high yarn irregularity, variation in fabric weight and variation in dye uptake or barré. Count variation is defined by:

$$CV_{count} = \sigma 100/\mu \qquad .....(4)$$

where CV_{count} is the coefficient of variation of yarn count, σ is the standard deviation and μ is the mean value of count.

11.5.6   Twist direction

Twist may be performed in the following two directions (see Fig. 11.25).

S direction: A single yarn has 'S' twist if, when it is held in the vertical position, the fibres inclined to the axis of the yarn conform in direction of slope to the central portion of the letter S.

Z direction: A single yarn has 'Z' twist if, when it is held in the vertical position, the fibres inclined to the axis of the yarn conform in direction of slope to the central portion of the letter Z.

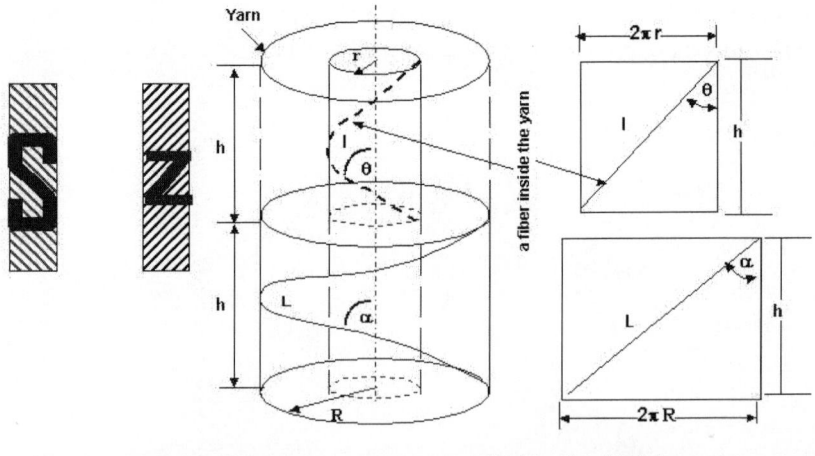

(a) Twist direction   (b) Idealized twist

Figure 11.25. Twist direction and twist level of idealized yarn geometry

11.5.7   Twist level

The amount of twist in the yarn is commonly expressed by the number of turns per unit length. In Fig. 11.25, the yarn is assumed to be built-up of a series of superimposed concentric layers of different radii in each of which the fibres follow a uniform helical path so that its distance from the centre remains constant. Based on this model, the length of one turn of twist, h, is given by:

$$h = 1/T \qquad\qquad ......(5)$$

where T is the twist level expressed in the number of turns per unit length.

Using the opened-out surface of the ideal yarn (see the above figure), we can derive the following basic relationships:

$$L^2 = h^2 + (2\pi R)^2$$

$$\tan \alpha = 2\pi R/h = 2\pi RT \qquad\qquad ..... (6)$$

where R is the yarn radius, L is the length of the fibre in the yarn outer layer and α is the twist angle.

Using a general concentric layer of the yarn at a radius r, similar relationships for the fibre can be obtained.

$$l^2 = h^2 + (2\pi r)^2$$

$$\tan \theta = 2\pi r/h = 2\pi rT \qquad\qquad ..... (7)$$

where 2 is the helical angle of the fibre layer, and l is the fibre length. The range of the angle 2 is from zero at the centre of the yarn to ω at the yarn outer layer. This means that fibres at the centre are straight, and the helix angle reaches a maximum ω at the yarn outer layer.

11.5.8   Twist factor

According to the above equation the twist angle α is a function of the twist level, T, and the yarn radius, R. The twist factor is a measure of twist, which accounts for the yarn radius as well as the twist level. Refer to Fig. 11.25, the linear density of the yarn (mass/unit length) is given by:

$$\text{Mass/Length} = \pi R^2 \rho$$

Thus,

$$R = [(\text{Mass/Length})/\pi\rho]^{1/2} \qquad \text{..... (8)}$$

From equations (6) and (8),

$$\tan \alpha = K_0 T \times [(\text{Mass/Unit length})/\rho]^{1/2} \qquad \text{..... (9)}$$

where K_0 is a constant.

Equation (9) shows the relationship between the twist angle, the linear density and the volumetric density. In theory, the twist factor (or twist multiplier – TM), is defined by the following equation:

$$TM = \tan \alpha \, (\rho) 1/2/K_0 = T \, (\text{Mass/Unit length})^{1/2} \qquad \text{.... (10)}$$

or

$$TM = TPC \, (\text{tex})^{1/2}$$

or

$$TM = \frac{TPI}{(Ne)^{1/2}} \qquad \text{...... (11)}$$

where TPC is turns per centimetre and TPI is turns per inch.

In practice, equation (11) is commonly used to determine the TM of yarn for a given yarn count and a given twist level. It simply indicates that the TM is an expression of the twist level adjusted for yarn count.

Normally, the Z on Z twist will result in a contraction of the plied yarn, while the S on Z twist will result in an increase in length. This amount of contraction or expansion will depend on the amount of twist inserted. When the yarn is woven or knitted into a fabric, the direction of twist influences the appearance of fabric. When a cloth is woven with the warp threads in alternate bands of S and Z twist, a subdued stripe effect is observed in the finished cloth due to the difference in the way the incident light is reflected from the two sets of yarns.

In twill fabric, the direction of twist in the yarn largely determines the predominance of twill effect. For right-handed twill, the best contrasting effect will be obtained when a yarn with Z twist is used; on the contrary, a left-handed twist will produce a fabric having a flat appearance. In some cases, yarns with opposite twist directions are used to produce special surface texture effects in crepe fabrics. Twist direction will also have a great influence on fabric stability, which may be described by the amount of skew or 'torque' in the fabric. This problem often exists in cotton single jersey knit where knitted wales and courses are angularly displaced from the ideal perpendicular angle. One of the solutions to solve this problem is to coordinate the direction of twist with the direction of machine rotation. With other factors being similar, yarn of Z twist is found to give less skew with machines rotating counterclockwise. Fabrics coming off the needles of a counterclockwise rotating machine have courses with left-hand skew, and yarns with Z twist yield right-hand wale skew. Thus, the two effects offset each other to yield less net skew. Clockwise rotating machines yield less skew with S twist.

The relationship between yarn strength and twist level is well recognized among textile technologists and engineers. This relationship is generally illustrated in Fig. 11.26. Initially, as the twist level (number of turns per unit length) increases, yarn strength will also increase. This effect holds only up to a certain point beyond which further increase in twist causes the yarn to become weaker. Thus, one should expect a point of twist at which yarn strength is at its maximum value. This point is known as the 'optimum twist'.

The various effects of degree of yarn twist are diameter or fineness, contraction, softness or hardness (hand), bending behaviour, absorbency, covering power, permeability, tensile strength, elastic performance/extension and recovery, resistance to creases and abrasion, pilling behaviour, lustre, etc. Effects of degree of yarn twist on the fabrics are hand, appearance, texture, drapability qualities, performance expectations, durability, serviceability.

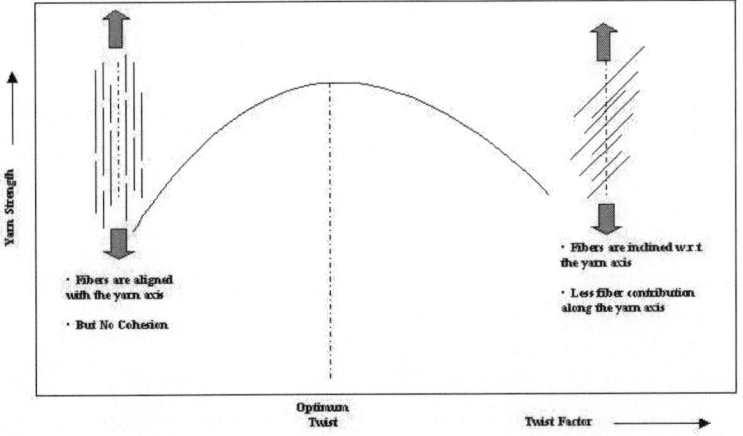

Figure 11.26. Effect of twist on spun yarn strength

11.5.9   Yarn diameter

In certain applications, yarn fineness expressed in diameter or thickness provides more useful information. For example, determining the structural features of a fabric (e.g., cover factor, yarn crimp, etc.) requires a prior knowledge of yarn diameter. It is important, therefore, to measure yarn diameter or to provide an estimate of its value. Theoretically, equation (8), introduced earlier, provides a general expression of yarn radius as a function of the linear density and the volumetric density of the yarn. For direct count system (say, tex), this general relationship will be as follows:

$$d = k_1 \, [\text{tex}/\rho]^{1/2} \qquad\qquad ....(12)$$

For indirect systems (say, cotton count), the general expression of yarn diameter is as follows:

$$d = 1/[k_2 \times P \times \text{Ne}]^{1/2} \qquad\qquad .....(13)$$

The above expressions indicate that the value of yarn diameter mainly depends on the linear density or yarn count, tex or Ne, and the volumetric density of yarn. As indicated earlier, volumetric density describes the degree of compactness of fibres in the yarn structure. This means that yarn twist will have a significant effect on yarn diameter.

Table 11.3. Formula for yarn diameter

Yarn Type	Expression	Units
Ring Spun	$d = \dfrac{1}{28\sqrt{Ne}}$	Inch
Ring Spun	$d = -0.10284 + \dfrac{1.592}{\sqrt{Ne}}$	Mm
Rotor Spun	$d = -0.16155 + \dfrac{1.951}{\sqrt{Ne}}$	Mm
MJS Air-Jet Yarn	$d = -0.09298 + \dfrac{1.5872}{\sqrt{Ne}}$	Mm

As indicated above, yarn diameter is used to estimate fabric structural parameters such as width and cover factor. Since thousands of ends or wales are presented side-by-side in the woven or the knit fabrics, a slight change in yarn diameter can result in a substantial change in the overall cover factor of fabric. Factors affecting yarn diameter are essentially those that affect yarn density or fibre compactness.

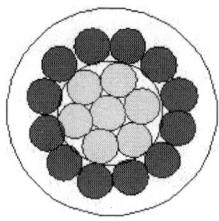
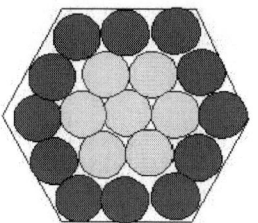

Open-packed structure   Closed-packed structure

Figure 11.27. Theoretical fibre packing in the spun yarn

For a given yarn count and at the same twist factor, the larger the fibre length, the higher the yarn density and the smaller the yarn diameter. In theory, fibre compactness may be characterized by two main categories of fibre arrangement in the yarn cross section: (a) the open-packed structure and (b) the closed-packed structure. These are illustrated in Fig. 11.27. In practice, fibre packing may deviate largely from these idealized forms due to many facts like noncircularity of fibres, dimensional variability, the relaxation and coherence of fibres in the yarn structure and the effect of twist.

11.5.10   Yarn strength

In practice, the strength of staple fibre yarn is commonly described using the following parameters:

- Skein strength
- Count-strength product (CSP)
- Single end strength
- Strength irregularity ($CV_{strength}$, %)

11.5.10.1   Yarn skein strength

The skein strength is typically measured by winding a 120-yard skein on a wrap reel. The yarn is then removed from the reel and tested in the form of several revolutions of parallel threads using a pendulum tester at a constant rate of traverse. When the specimen is subjected to tensile loading, all threads will resist the loading until a break occurs in one of the threads (the weakest point). The remaining unbroken threads will then support the skein until a second thread breaks. This process continues through a succession of thread breaks until a total failure occurs. The parameter obtained from this test is called the skein or lea strength expressed in pounds.

11.5.10.2   Count-strength product

A yarn count test in which the same test specimen is weighted to determine the cotton count. The CSP provides a strength measure commonly known as the skein break factor (Ne, lb). In practice, this measure is used more commonly than the absolute value of skein strength. Typical values of skein break factor for different yarns are given in Table 11.4.

Table 11.4. Typical values of skein break factor (Ne, lb) (based on US Crop Data, 1990, 1991)

Yarn Type/Ne	Range*	Mean Value (lb.Ne)
Ring-Spun [Carded] Ne = 20's Ne = 36's	 1774-3129 1584-3096	 2227 2045
Ring-Spun[Combed] Ne = 22's Ne = 36's Ne = 50's	 2886-4009 2653-3737 2433-3587	 3430 3223 3042
Open-End [Carded] Ne = 10's Ne = 22's Ne = 30's	 1979-2898 1663-2669 1507-2415	 2347 2005 1814

11.5.10.3   Single end strength

The single-end strength represents a more fundamental parameter than the skein strength. Using modern tensile testers (e.g., Uster TensoRapid), strength parameters can be obtained at a constant rate of extension of 5 m/min and a gauge length of 50 cm. These parameters include: breaking load, breaking elongation, load-elongation (or stress-strain) curve, yarn tenacity, yield stress and strain, specific work of rupture and tensile modulus.

11.5.10.4   Strength irregularity (CV$_{strength}$, %)

Similar to count variability, strength irregularity is commonly defined by the coefficient of variation of yarn strength:

$$CV_{strength}, \% = \rho \times 100/\mu$$

where ρ and μ are the standard deviation and the mean of yarn strength, respectively.

In any spinning technique, yarn strength represents a crucial parameter, which determines the performance of spinning. During yarn preparation for weaving, the yarn is subject to continuous tension as a result of the repeated winding and unwinding necessary for weaving preparation. This tension should be within the elastic boundaries of the yarn to avoid permanent deformation. During dyeing or sizing, the yarn is subjected to chemical treatments that can alter its mechanical behaviour.

During the weaving process, thousands of yarns are simultaneously subject to continuous cyclic loading, which is a necessity for the interlacing actions required to make cloth. Weaving peak tension may reach levels exceeding 35% and in knitting about 30% of the average breaking force of the yarn. When a maximum tension coincides with a minimum strength point of the yarn, failure of yarn to withstand the tension will occur. This failure may result in an end breakage and a complete stop of the weaving process. A yarn break during knitting will have an adverse effect not only on the machine efficiency but also on the fabric quality.

11.5.11   Yarn evenness

In practice, the concept of limiting irregularity can be used to estimate the partial effect of process-added variability on the overall irregularity. In this regard, the uster evenness tester can provide the so-called 'irregularity index' defined by the following equation:

$$I = \frac{CV_{eff}}{CV_{lim}}$$

where CV_{eff} is the effective or measured irregularity.

The limiting irregularity, CV_{limit}, %, is simply defined by

$$CV_{lim} = (apprx.) = 100/n^{1/2}$$

where n is the average number of fibres per yarn cross section estimated by

$$n = \frac{tex_{yarn}}{tex_{fibre}}$$

11.6   Flax spinning

After cotton, flax is the most used plant fibre in spinning yarns. Hence, we give a short explanation of spinning process of flax fibres. Even though many spinning systems are used for flax spinning, the most important ones are wet and dry spinning. The principal difference between the spinning of flax fibres and the spinning of other fibres is in wet spinning. Tow is both dry and wet spun, and line is nearly always wet spun. Dry spinning is similar to the semi-worsted method of yarn production and uses the same preparation and spinning machinery. This system is used both for 100% flax yarns and for blends with other fibres. On the contrary, wet spinning requires specific yarn preparation and spinning frames.

Figure 11.28 shows the flow chart of dry and wet spinning of flax yarn.

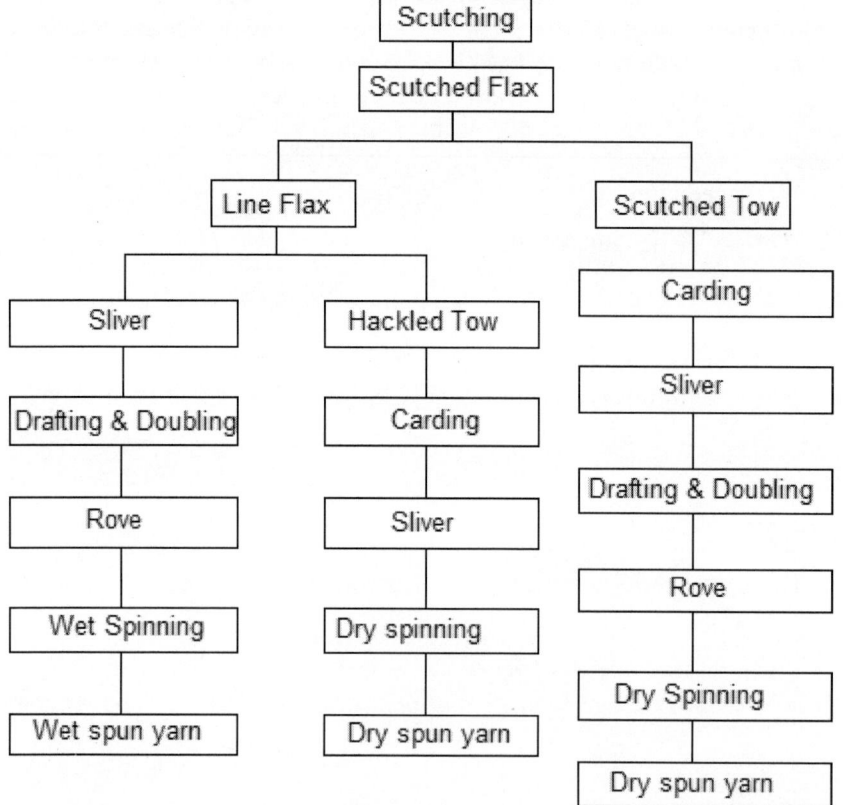

Figure 11.28. Flax spinning flow chart

Earlier, under fibre section, we have explained the process up to scutch-ing in the fibre manufacturing. In scutching process, retting and drying enable the fibre bundles to be easily separated from the shiv. However, the pectins cementing the fibres to each other within the bundles would not have been completely removed, unless the straw is over-retted. The removal of pectins can be accomplished by placing the fibres in warm water at 60°C. The pect-ing softens and enables the primary fibres to slide against each other when longitudinal force is applied, as it is in the drafting zone of a wet spinning frame. The line fibres are usually longer and used for spinning finer yarns. The shorter fibres are termed as Tow.

11.6.1   Sliver formation from line flax

Next process for line flax is hackling which is equivalent to combing in cot-ton spinning. This enables the fibres to straighten, disentangle and parallelize and at the same time separates the fibres within the fibre bundles. Hackling produces hackled tow as a by-product which is uncombed flax fibres, which is normally spun by dry spinning process. After hackling, the combed line fibres are placed manually on a slanted apron in which is a series of gills. As they proceed through the gills, they are formed into a sliver which passes through two calender rollers and is then either coiled into cans for further processing on draw frames or pressure baled for delivery to spinners. Even though the processes are similar but the machines are specially designed to handle flax fibres.

Even though the grey spinning is common, these days there are more and more preference for degummed bleached spinning. The fibre in the rove form is degummed and bleached for the further removal of gums, lignin and pectins from the fibres. This process has many advantages:

1. It is possible to spin finer yarns by this route.
2. It facilitates yarn dyeing, as it is difficult to degum and bleach the grey yarn and dye.
3. The yarn produced by this route is more consistent in the technical quality.

Nowadays, bleached spun yarns are more preferred by customers be-cause of the bettermoval of the extraneous matter, which is more difficult in the spun or woven/knitted form. Degumming is usually done by boiling the rove in boiling sodium carbonate or sodium hydroxide solution which removes the noncellulosic substances. It is further bleaches using sodium hy-pochlorite, sodium chlorite or hydrogen peroxide on perforated cylindrical packages done on roving frame. Hydrogen peroxide and sodium hypochlorite

in better preferred as they are more environmental friendly and they give a
better whiteness.

11.6.2   Spinning

The bleached or grey rove is directly spun by wet, dry or semi-wet spinning.

11.6.2.1   Wet spinning

Many wet spinning machine designs are available one of which is shown in
Fig. 11.29.

Wet spinning frame used for flax has a trough of warm water heated to
about 60°C placed before the drafting zone which helps the fibre to swell and
soften the gums and allow the fibres to slide over each other so that the drafting becomes easier and more efficient. After the drafting, the spun yarn can
be directly used for weaving, knitting, etc.

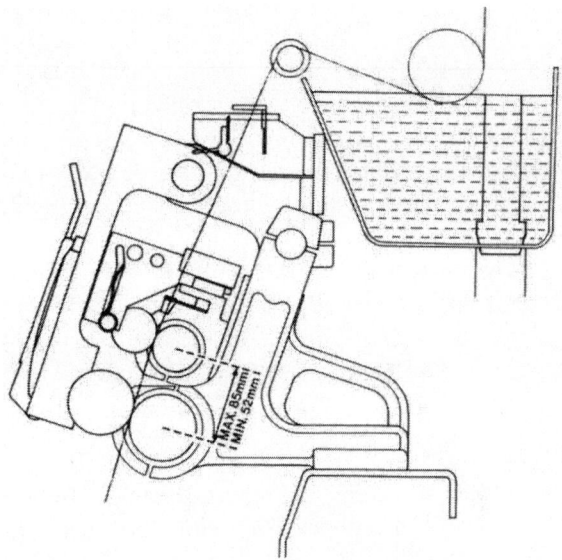

Figure 11.29. Wet spinning machine

11.6.2.2   Dry spinning

Dry spinning machinery is similar to that used on the semi-worsted system
for wool. Dry spinning is not only used to produce 100% flax yarns but also
for yarns blended with other fibres, for example, cotton, wool, polyester,
acrylic, etc. Counts produced are normally from 2.5s metric (450 Tex, 4 lea)

to 9s metric (115 Tex, 15s lea) but depending on the fineness of the flax or of the other fibres in the blend, and finer counts of up to 12s metric (82 Tex, 20 lea) can be achieved. The principal uses of dry spun flax yarns spun from scotched tow are for twines, ropes and cordage and industrial fabrics. Semi-wet spinning is a normal dry spinning with the addition of a dip roller which transfers water from a trough to the surface of the yarn. Semi-wet spinning is used principally to produce yarns for sewing threads and sometimes for apparels, but manufacturers prefer wet spinning to semi-wet spinning as the cost is almost the same.

12

Fabric

In this book, we discuss only two types of fabric production: weaving and knitting.

12.1 Weaving

Weaving is the most widely used method for making cloth. It is simple, inexpensive, suitable for high-quality fabrics and adaptable to special effects. Weaving is the interlacing of yarns in a regular order to create a fabric. The operation is performed in a machine called a *loom*. Two sets of yarns are interlaced, almost always at right angles to each other. One, called the *warp*, runs lengthwise in the loom; the other, called the *filling*, *weft* or *woof*, runs crosswise. Woven fabric is normally much longer in the warp direction than it is wide, that is, in the weft direction. Fig. 12.1 gives an overview of the entire manufacturing process of woven fabrics. Warp and weft may be processed raw white or as coloured yarns. For the production of the warp threads, two alternative methods are used.

12.1.1 Weaving preparation

12.1.1.1 Winding

Ring spun yarn which cannot be directly used in the weaving department is wound from the feed onto a bobbin, which is usually a conical or cylindrical tube made of cardboard, plastic or (seldom) metal.

Winding has following purposes:

1. enlarging the yarn package for economic reasons and handling (e.g., rewinding from the cops of the ring spinning frame onto cross bobbins),
2. improvement of the draw-off properties,
3. improvement of yarn quality by cleaning out thick and thin places of a yarn,

4. improvement of cops building for uniform dyeing with package dyeing, and

5. winding after package dyeing and winding of dyed hanks.

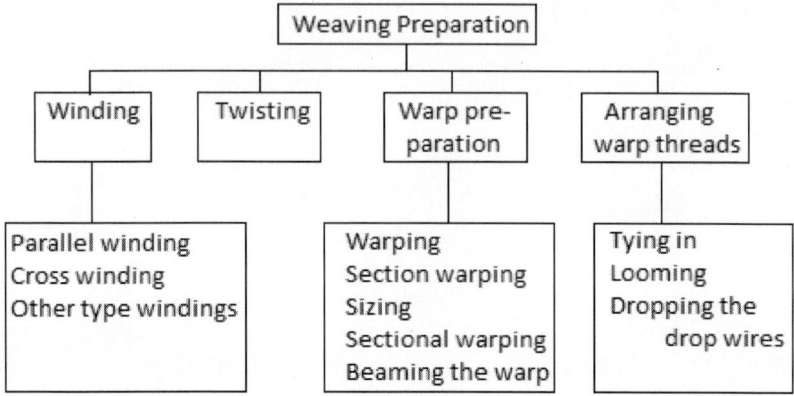

Figure 12.1. Basic structure of the preparation process for weaving

Cross-wound bobbins that may be produced with random winding or precision winding are the most commonly used (Fig. 12.2).

Mainly there are two types of winding: random winding and precision winding. Bobbins with random winding have a more stable construction than bobbins with precision winding. The surface drive of the random winding is much less expensive than the spindle drive of the precision winding. The essential disadvantage of the random winding is the evolution of images at bobbin diameters with integer i. Such bobbins are not suitable for dyeing. Furthermore, layers of threads in the areas of the images may slip off while drawing off the yarns and produce filament breaks. These consequences of the image turns can be reduced by such methods as overlapping the basic side traverse frequency with a wobbling frequency, by slippage between the bobbin and the surface drive, by periodic lifting of the bobbin, by stroke displacement, or by variation of the bobbin speed (step precision).

12.1.1.2   Twisting

In twisting process, two or more of same twist or different twists (S or Z twist) are plied together. The twisting has following purposes:

1. increase the tenacity (especially with staple fibre yarns),

2. adjust the elongation properties,

3. reduce yarn unevenness, and

4. obtain a desired surface structure and colour effects.

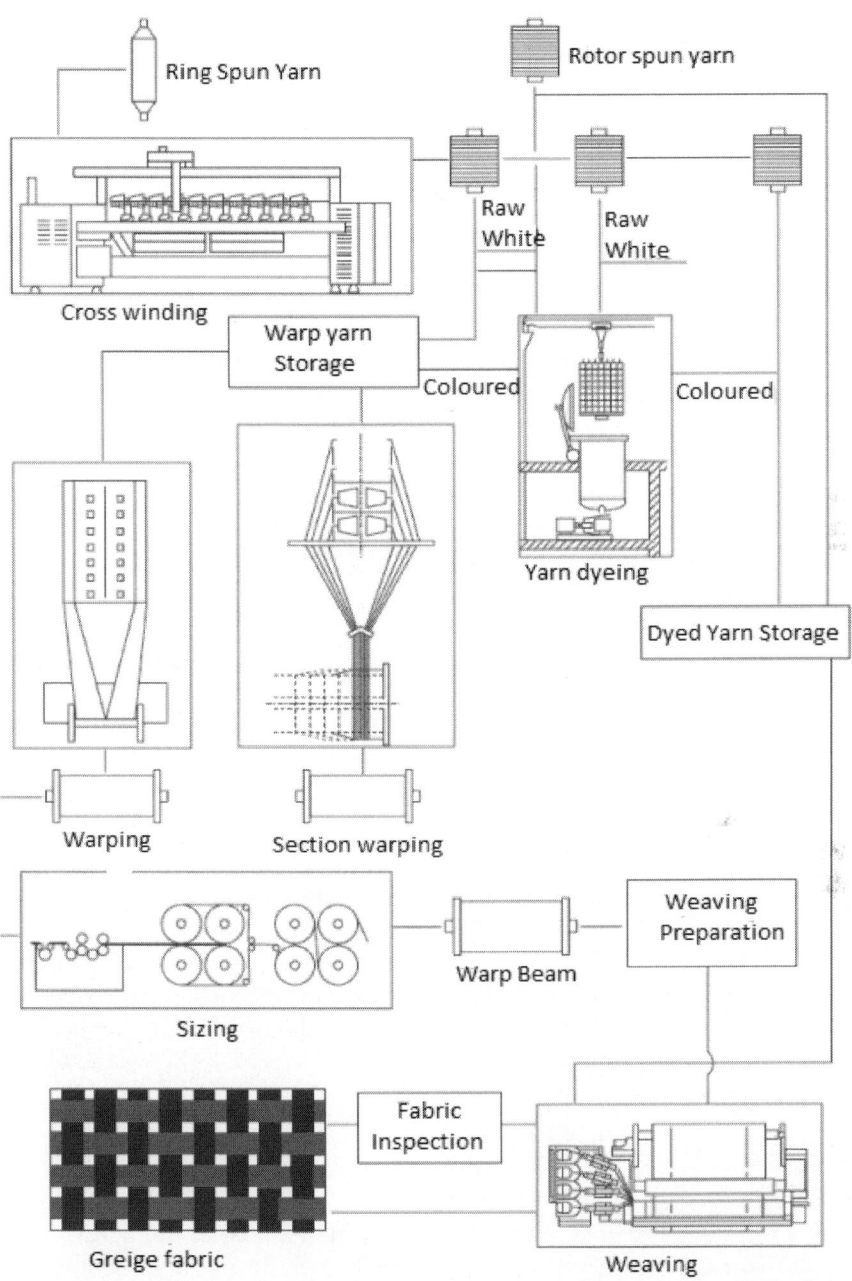

Figure 12.2. Manufacturing steps for the production of woven fabrics

The twist direction is usually opposite to the previous twist direction. Twisting in the same direction results in very hard and stiff ply-yarns with low elongation values. Because of the warp in weaving, the mechanical stresses on the yarns are large, so relatively high-tenacity and highly twisted ply-yarns are used.

12.1.2   Preparation of warp beam

For the economic production of a fault-free woven fabric, a system of warp threads with a defined number of threads of equal length and tension has to be fed to the weaving loom. Various feeding principles are distinguished: weaving from the creel, direct beaming of the warp, warping and section warping. All of these principles use a bobbin creel from which the warp threads are drawn. Weaving from the creel or direct beaming of the warp is used less frequently, as the total number of warp threads is restricted by the capacity of the bobbin creel. With warping or section warping, the warp beam is obtained after several production steps, and only a small number of bobbins are required.

The warp yarns may be coated with a temporary sizing for protection against damage during the operation. The process of applying this coating by taking yarn from a large rack, called a creel, passing it through comb guides and through a bath of starch, and winding it on a warp beam, is called beaming or slashing.

12.1.3   Warping

For warping, a portion of the total number of threads is wound on a warp beam across the entire width (partial warp beam). Several warp beams are rewound together onto the final warp beam. This procedure is called assembling. It is often combined with a sizing process. This procedure is preferably used for large lots with simple designs because the number of patterns produced by the warp is restricted.

12.1.4   Section warping

With section warping, narrow warp sections of the desired length are wound parallel to each other onto a warping drum with the final thread density. The slide-off of the warp sections is prevented by a conical arrangement. Many sections are deposited next to each other, until the whole width of the warping drum is filled. After section warping, the warp threads are rewound from the warping drum onto the warp beam. The potential for patterns is unlimited,

Figure 12.3. A sectional warping machine

which makes this procedure suitable for smaller lots and for complicated warp designs (Fig. 12.3).

12.1.5   Sizing

Sizing is a process by which the warp threads are coated with a size for better efficiency during weaving. Sizing has the function of enhancing the mechanical properties and the loading capacity of the thread without reducing the thread elasticity. It also serves to reduce yarn hairiness. The sizing procedures can be divided according to groups of slashing products:

1. Hot-melt sizing: Application of a melt-liquid, water-free bonding agent.
2. Cold sizing: Application of a small amount of bonding agent at room temperature.
3. Dry sizing: Application of a melt-liquid, water-free product without a sticky component.
4. Wet sizing: Traditional wet sizing. Application of a water-soluble size followed by drying.
5. Solvent sizing: Dissolved size remains on the warp after evaporation of the solvent.
6. Warp waxing: Application of a liquid, water-containing product without a sticky component.

12.1.6   Weaving

Warp yarns are fed from large reels called *creels* or *beams*. Typically, these hold about 4500 separate pieces of yarn, each about 500 yards (450 m) long. The filling yarns are fed from bobbins, called quills, carried in shuttles (hollow projectiles) that are moved back and forth across the warp yarns, passing over some and under others. The shuttle is designed so that the yarn it carries can unwind freely as the shuttle moves. Each length of yarn, fed from the shuttle as it moves across the loom, is called a *pick*. The yarn folds over itself at the end of each pick and forms another pick as the shuttle returns. When the yarn in a particular shuttle is exhausted, current production looms have automatic devices that exchange the empty quill with a full one.

Looms perform the following functions: (1) raising selected warp yarns, or *ends*, with suitable *harnesses*, consisting of frames of *heddles*, with taut vertical wires and eyelets, or strips with openings in the middle. There is one heddle for each end that is threaded through the eyelet. The heddles guide and separate the warp yarns, raising some of them to make room for the shuttle during the pick. This action is called *shedding*, and the space between the warp yarns is called the *shed*. Simple weaves require only two harnesses; complex weave patterns may require as many as 403. (2) *Picking*, laying a length of the filling or weft yarn between warp yarns from the shuttle (a hollow projectile that holds weft yarn inside) as it moves across the shed. (3) *Battening* or *beating in*, forcing the filling yarn from the pick against the

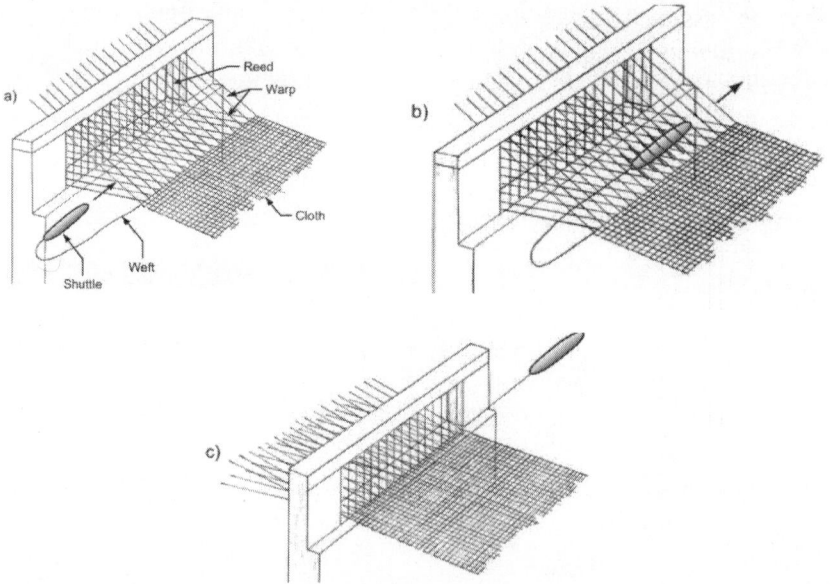

Figure 12.4. Diagrammatic representation of a typical loom in operation (weaving): (a) shedding, raising some warp yarns to make room for the shuttle, (b) picking, laying the weft (filler) yarn across and between warp yarns and (c) beating in, pushing the reed against the last filler yarn against the woven cloth.

just-formed cloth next to the previous pick. This step is necessary because the shuttle requires some space in its movement across the loom and it is not possible to deposit the pick closely against the previous picks. Battening is done with the *reed*, which is a grating of parallel vertical wires between the warp yarns. (4) *Taking up*, winding the cloth, as it is formed, onto a take up reel, the *cloth beam*. (5) As the cloth is taken up, warp yarn is released from the warp beam. This action is called *letting off*. Fig. 12.4 illustrates major loom operations.

In the simplest weaving arrangement, alternate warp yarns are over or under the shuttle as it moves in one direction, and the warp yarn positions are reversed for the return stroke of the shuttle. This weave can be made on a loom with only two harnesses. In other arrangements, several warp yarns may be moved upward or downward together, or several filling picks may take place before the warp yarns change position. In still other cases, the warp yarns are raised or lowered with respect to the picks in some prede-termined sequence, creating a pattern in the appearance of the weave. These patterns may affect the feel and strength of the woven fabric. Such weaves may require looms with five or more harnesses.

Three basic weave patterns: (a) *Plain weave*, also called *taffeta*: Fill-ing yarns pass over and under alternate warp yarns. Other plain weaves are *broadcloth, muslin, batiste, percale, seersucker, organdy, voile* and *tweed*. (b) *Twill weave*: Filling yarns pass over two warp yarns and under a third, and repeat the sequence for the width of the fabric. The next filling yarn re-peats the sequence but shifts one warp yarn sideways, creating a diagonal pattern. *Herringbone, serge, jersey, foulard, gabardine, worsted cheviot* and *drill* are twill weaves. (c) *Satin* weave: Filler yarns pass over a number of warp yarns, four in this illustration and under the fifth. *Damask, sateen* and *crepe satin* are satin weaves. Exposed yarns reflect light and give the weave its sheen (Fig. 12.5).

Jacquard-type looms are looms with an automatic, selective method for *shedding*, the lifting of certain warp yarns for each cycle of the loom. The

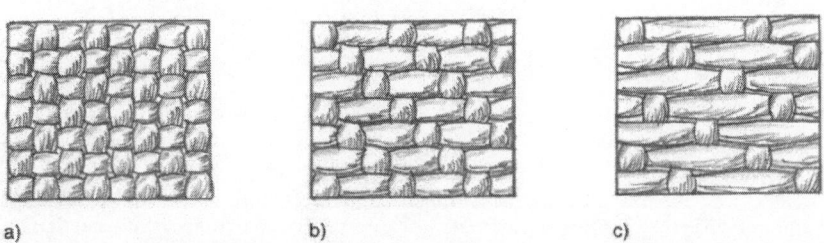

a)   b)   c)

Figure 12.5. Weave patterns: (a) Plain weave, (b) Twill weave, (c) Satin weave

mechanism permits the use of continuously varying shedding patterns to create corresponding patterns in the woven cloth. Complex patterns, including pictures, can be woven into the cloth. The original Jacquard process used a series of perforated cards to control the operation. Needle-like components, connected to hooks that controlled the heddles, passed through holes in the cards and raised the warp yarns. Each heddle moved independently of the others. Where there were no holes, the needles did not move through and the heddles were not raised. The cards were moved with each cycle of the loom, creating a variable weaving pattern corresponding to the hole patterns in the cards. Current Jacquard looms use sophisticated electronic means to control the pattern of shedding. Jacquard loom weaving is used in making upholstery and drapery fabrics, in table linens and in some garments. Damask, brocade, brocatelle, matelasse and tapestry fabrics are made on Jacquard looms.

Because of space and size limitations, the amount of filling yarn that can be carried in a shuttle is also limited. Bobbins in shuttles must be replaced when the yarn is exhausted. As noted above, this operation has been automated. Automatic bobbin loaders sense when the filling yarn is exhausted, remove empty bobbins and insert full ones when the shuttle is momentarily stationary. The operation does not reduce the speed of the loom.

Shuttle looms have its limitations. Many current production looms do not use shuttles. In some looms, air or water streams propel the end of the filling yarn for each pick. In others, dummy shuttles pull the filling yarn but do not

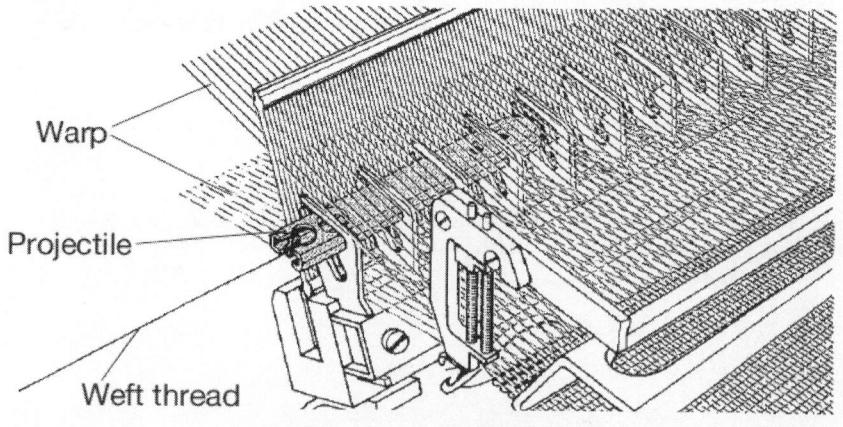

Figure 12.6. Projectile loom

carry a bobbin. These are high production machines, e.g., projectile looms, rapier looms, air/water jet looms, etc.

12.1.6.1   Projectile looms

With this weft insertion principle, the weft thread is pinched in the projectile and then shot through the shed. After insertion, the thread is tensioned tightly and cut off the external weft bobbin. The projectile is transported back outside the shed. Therefore, one weaving loom operates with multiple circulating projectiles (Fig. 12.6).

12.1.6.2   Rapier looms

The rapier method uses an arm or tape-like machine element that grasps the filling yarn and pulls it across the web of warp yarns (Fig. 12.7). One arm usually feeds the yarn halfway across the loom and an arm on the other side grasps the end of the filling yarn and pulls it the rest of the way across. Actually, a gripper head catches the yarn end from the feed bobbin and transports it through the shed. In the middle of the fabric, the yarn is transferred to the opposing second gripper. The weft insertion is controlled at every single moment of the process. Therefore, this principle is very flexible and extremely suitable for delicate materials. Rapiers are available in rigid and flexible forms (Fig. 12.8).

Figure 12.7. A rapier loom (Dornier)

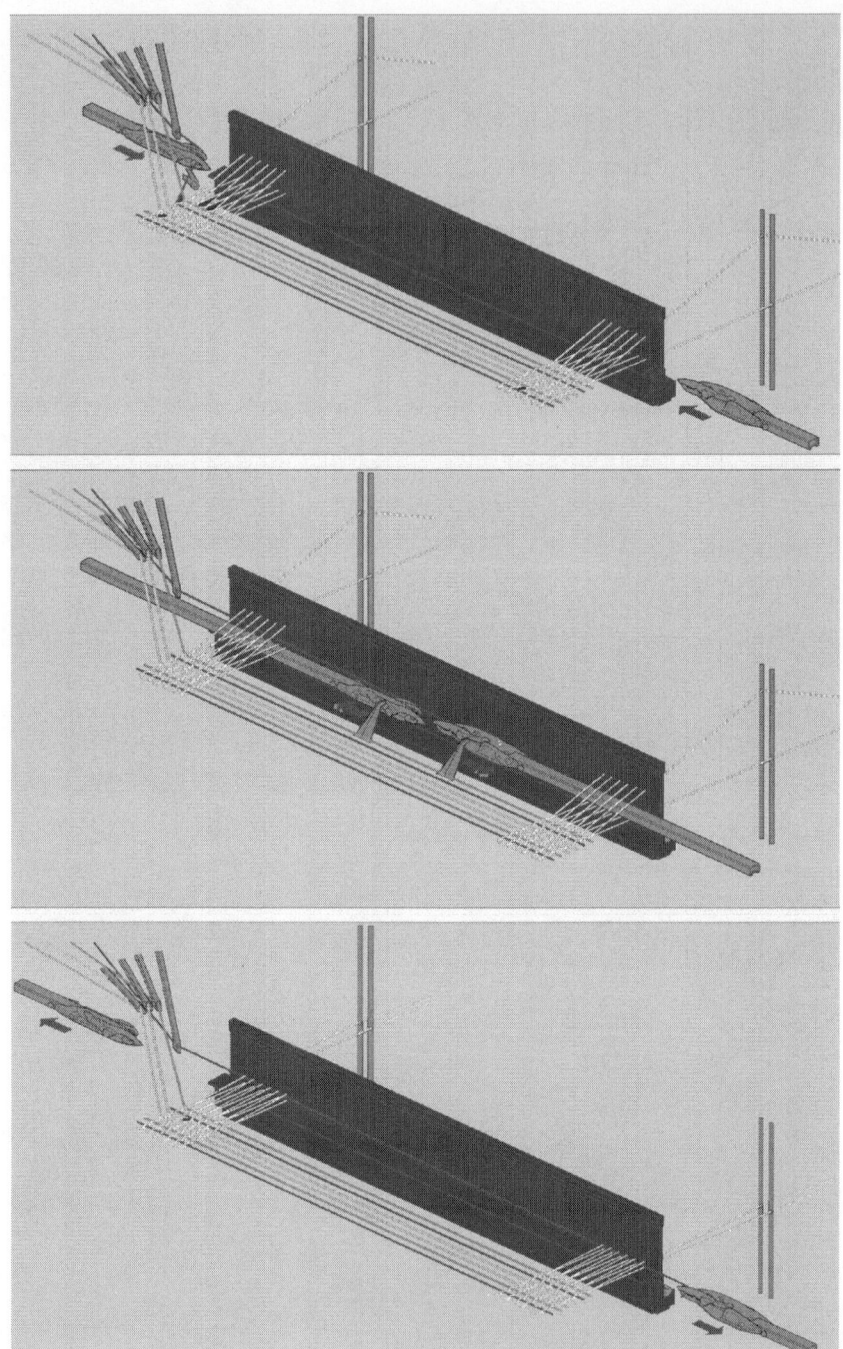

Figure 12.8. Weft insertion action in a rapier loom

12.1.6.3   Jet loom

Newer looms simply propel the end of the filling yarn across the loom by inertia. All these arrangements provide quieter operation, reduced wear, elimination of the need to protectively coat the warp yarn and increased weaving production (Fig. 12.9).

Jet weaving looms are divided into air-jet and water-jet weaving looms according to the medium of weft insertion. The weft thread is transported through the shed with the help of air or water jets. Secondary jets located in the shed assist the main jet in transporting the thread. This technology can

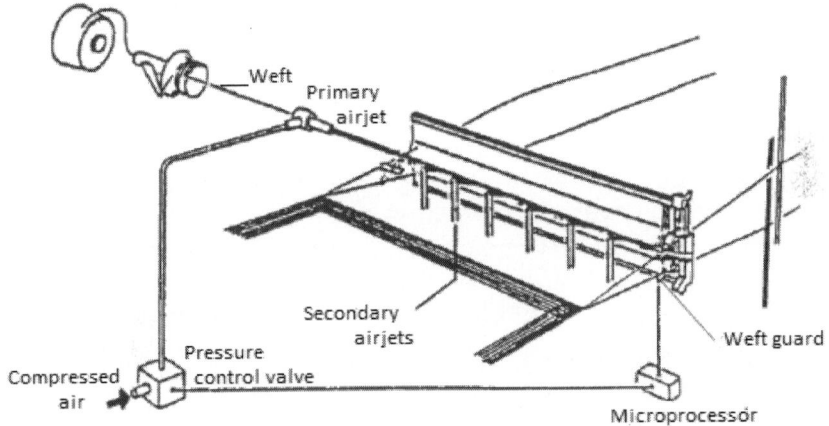

Figure 12.9. Weft insertion principle of jet looms

only be used for lightweight fabrics. Jet looms have the highest weft insertion speeds of all looms, but they are limited with regard to suitable weft materials.

12.1.6.4   Multiphase loom

In multiphase weft insertion systems, several weft threads are inserted at the same time. In the wave-shed weaving machine, the sheds are oriented wave-like across the whole warp width. In each wave, a small weft cop is drawn off as with the shuttle loom. Production machines have already been sold by various European loom manufacturers (Fig. 12.10).

However, this principle has not succeeded because of a lack of quality and flexibility. An automated system to fix weft breaks is another challenge. Sequential multiphase looms provide sequential sheds in the warp direction, each for one weft insertion. This principle has been known for a long time. Despite the high production speed (up to 5000 m/min), it was no commercial success, the main reasons being the low flexibility regarding patterns and the difficulties when fixing a weft yarn break.

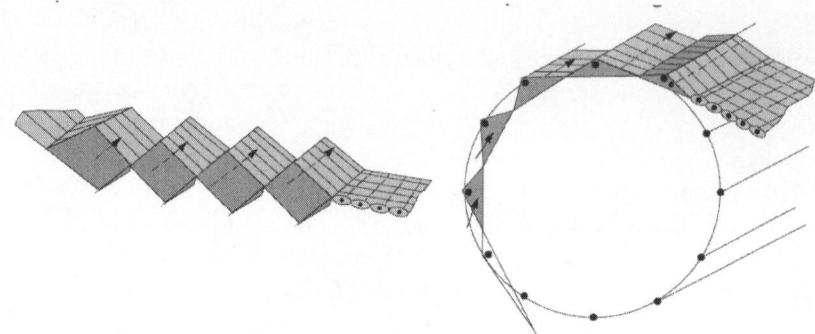

Figure 12.10. Weft insertion principle in a multiphase weaving loom

Is usually a plain weave in which either the filler or the warp yarn is drawn from the fabric to form loops between the intersecting yarns. The loops provide a thickness to the cloth. Turkish towelling is made from pile weaves with the loops uncut. Velvet is a pile fabric, but the loops are cut. In another method, special looms weave two fabrics face-to-face simultaneously. They are connected together by pile yarns. When the pile yarns are cut, two fabrics result, each with a pile. The process is less costly than weaving individual fabrics with a pile which must then be cut. Woven rugs and carpets are pile fabrics.

12.1.7   Weave diagram

The modern weaver or a person handling the woven fabric should have the basic knowledge of woven designs and how it is represented on paper. The weave diagram is a graphical sketch of the weave pattern on graph paper. Each square represents a potential interlacing point of warp and weft threads. It has been standardized to represent the warp threads in a vertical direction and the weft threads in a horizontal direction. In the diagram, a square is filled in if the warp thread lies above the weft thread at this crossover point.

12.1.8   Weave symbol

Weave patterns can be displayed with weave symbols. These are codes composed of repeated patterns of weave, number of warp threads up, number of warp threads down, number of threads and shift counter. The international standard for the construction of the weave symbol is defined in textile standards (e.g., DIN ISO 9354). The first number describes the basic weave pattern, where 1 stands for plain weave, 2 for twill and 3 for satin weave. Fig. 12.11 shows an example.

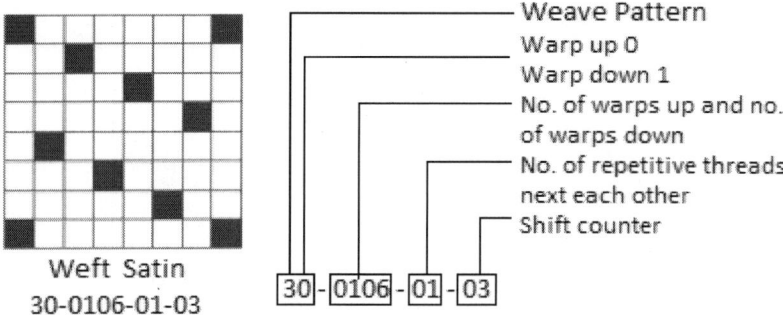

Figure 12.11. Weave symbol

12.1.9   Basic weave patterns

Most of the woven designs are derived from three basic weaves, namely plain weave, twill weave and atlas or satin weave.

12.1.9.1   Plain weave

The plain weave is the most basic weave but tightest crossing of the warp and weft threads. The plain weave and its variations can be produced with only two shafts, as the threads or groups of threads alternate in tying up. For very tightly woven fabrics with a high density of threads, four, six or more shafts are used. This pattern is also called tabby or linen weave. The basic plain weave and some variations are shown in Fig. 12.12.

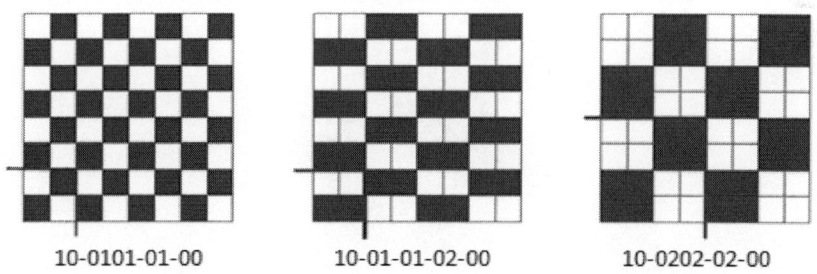

Figure 12.12. Basic plain weave and its variations

12.1.9.2   Twill weave

Twill weaves are characterized by a diagonal seam. Based on the direction of the seam, Z and S twills are distinguished. The seam is caused by shifting the first warp thread or group of threads to the upper right or the upper left, respectively. The magnitude of the shift is defined by the shift counter. Fig. 12.13 shows basic twill weave patterns.

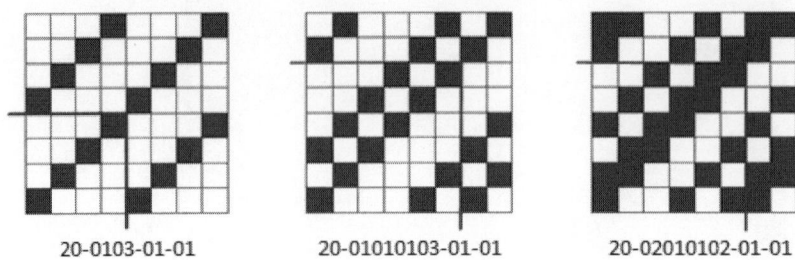

20-0103-01-01   20-01010103-01-01   20-02010102-01-01

Figure 12.13. Basic twill weave patterns

Classification of twill weave

Twill can be classified in different ways.

In *warp-way twill*, weave warp float run in the warp direction. Formula number of every yarn is same, i.e., all warp ends interlace in the same way but displacing the interlacing points of each end by one pick relative to that of the previous end. In this case, any sign or colour in the square of graph or design paper represents warp up, and empty square represents weft up (Fig. 12.14).

In *weft-way twill*, weave weft float run in the weft direction. Formula number of every yarn is same, i.e., all weft yarn interlace in the same way

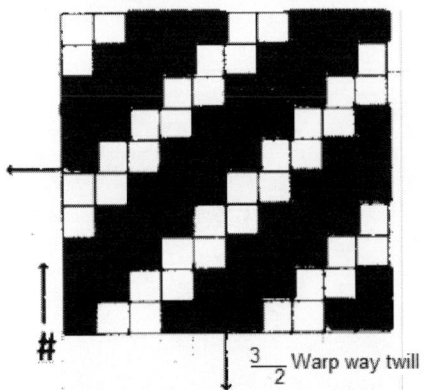

Figure 12.14. Warp-way twill

but displacing the interlacing points of each pick by one end relative to that of the previous pick. In this case, any sign or colour in the square of graph or design paper represents weft up, and empty square represents warp up. This is exceptional than other normal system (Fig. 12.15).

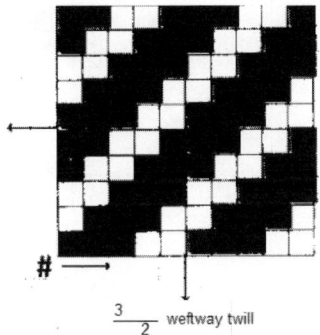

$\frac{3}{2}$ weftway twill

Figure 12.15. Weft-way twill

1. Based on the direction of the twill line on the face of the fabric
    The lines created on the face of the fabric are called 'twill lines' or di-
    agonal lines or wales. When the cloth is held in the position in which
    it was woven, the diagonal lines will be seen to run either from the
    lower left corner to the upper right corner or from the lower right to
    the upper left corner.

    S-twill or left-handed twill weave: When the twill runs from the lower
    right to the upper left corner, the twill is known as a left- hand twill.
    It is produced by down ward displacement of the interlacing points,
    if the starting point is bottom left corner or upward displacement of
    the interlacing points, if the starting point is bottom right corner. For
    example, it is expressed by the formula 1/3, number S, where S indi-
    cates the direction of twill line (Fig. 12.16).

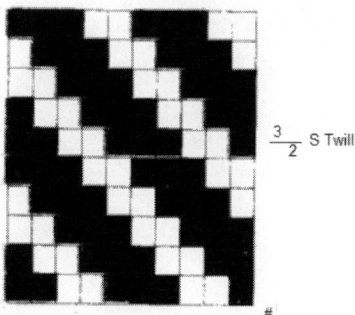

$\frac{3}{2}$ S Twill

Figure 12.16. S-twill weave

    Z-twill or right-handed twill weave: When the diagonal line runs
    from the lower left corner to the upper right corner, the twill is known

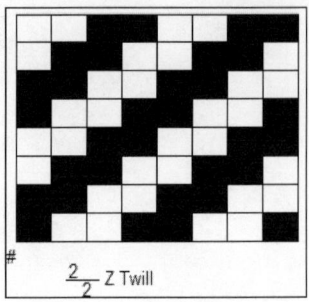

Figure 12.17. Z-twill weave

as a right-hand twill (e.g., 3/2 Z, etc.). About 85% of all twill-woven fabrics are right-hand twills. It is produced by upward displacement of the interlacing points. For example, it is expressed by the formula number 2/2 Z, where Z indicates the direction of twill line. Fig. 12.17 shows one of the Z twill with drafting and lifting plan.

2. Based on the face yarn (warp or weft)

   Warp-faced twill weave: Warp-faced twills have a predominance of warp yarns on the face of the fabric (e.g., 2/1, 3/1, 3/2, 4/2, etc.). The top digit of the fraction line is higher than the bottom one, so it is called warp-face twill. Since warp yarns are made with higher twist, these fabrics are stronger and more resistant to abrasion and pilling.

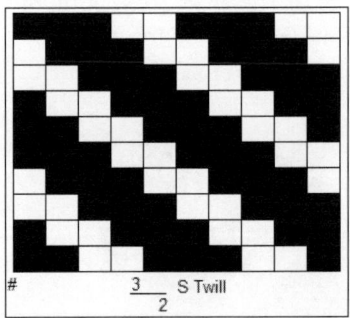

Figure 12.18. Warp-faced twill weave

Weft-faced twill weave: Weft-faced twills have a predominance of weft yarns on the surface of the fabric. Examples of weft face twill weave are 2/3, 3/4, 1/2, 2/4 Z, etc.

The top digit of the fraction 3 line is smaller than the bottom one, so it is called weft-face twill. Weft yarns are generally weaker than the

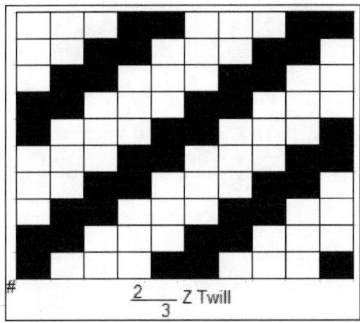

Figure 12.19. Weft-faced twill weave

warp yarns, so that relatively few weft-faced twills are made. Fig. 12.19 shows the weave plan of weft-face twill.

Double face twill weave: Even-sided twills expose an equal amount of warp and weft yarn on each side of a fabric. They are also known as reversible twills because they look same on both sides, although the direction of the twill line differs. Better quality weft yarns are used in these fabrics as compared with warp-faced twills because both sets of yarn are exposed to wear. They are most often two twills and have the best balance of all the twill weaves.

3. Based on the nature of the produced twill line

    Simple twill weave: There are two types of simple twill, such as simple warp twill and simple weft twill. Each warp end is raised over or lowered under only one pick in the repeat. Examples of simple twills are 1/2, 1/3, 1/4 for weft twills and 2/1, 3/1, 4/1 for warp twills (Fig. 12.20).

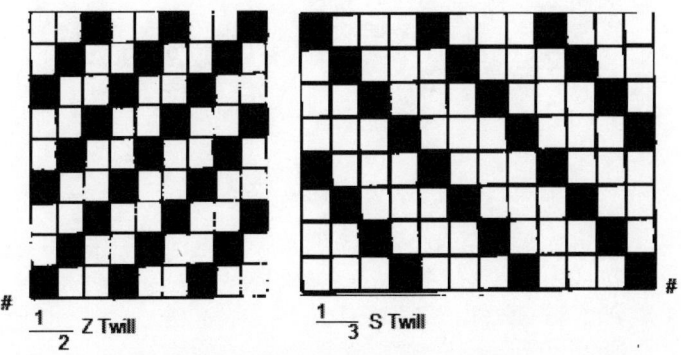

Figure 12.20. Simple twill weaves

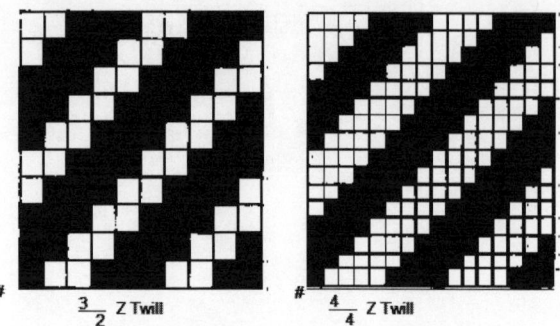

Figure 12.21. Expanded twill weave

Expanded twill weave: Each warp end is raised over or lowered under more than one adjacent pick in the repeat. If the warp and weft twill lines are of equal width, the fabric is double faced. It is represented by the formula number of 2/3, 3/2, 4/4, 2/4, etc. (Fig. 12.21).

Multiple twill weave: In each repeat, there are at least two warp twill lines or two weft twill lines of different width. If the prominency of warp yarn is more then it is called warp-face multiple twill, and if the prominency of weft yarn is more then it is called weft-face multiple

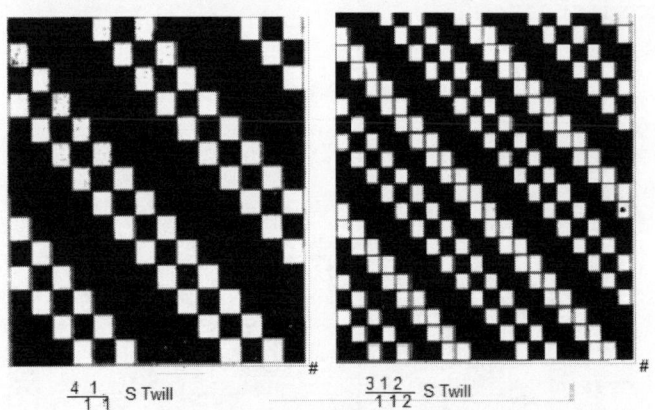

Figure 12.22. Multiple twill weave

twill. If the prominency of both warp and weft yarns are same then it is called double-face multiple twill. Examples are 4 1/1 1, 3 1 2/1 1 2, 1 3 1/1 1 5 etc. (Fig. 12.22).

Derivatives of twill weave

These are twills developed from the basic twill weave, such as

1. Zig-zag or waved or pointed twill
2. Herringbone twill
3. Diamond design
4. Diaper design
5. Broken twill
6. Re-arranged twill or transposed twill
7. Stepped twill
8. Elongated twill weave
9. Combined twill weave or combination of twill
10. Shaded twill weave or shaded design
11. Curved twill weave

1. *Zig-zag or pointed or waved twill weave*

In zig-zag twills, the diagonal line proceeds either to the left or right. Where two lines meet, they create a point, forming a continuous zig-zag effect in the fabric. If one takes a twill weave and reverses the drafting order in the heald shafts regularly after a certain number of ends, the twill lines will run across the width of the fabric in a zig-zag configuration. The reversing of the draft can occur after a repeat or after any number of warp ends. Each reversal produces a point. Selecting the right twill weave when constructing zig-zag or waved effects is of great importance. Short warp or weft floats should be used so as to avoid long floats when the weave is reversed. So zig-zag weave is produced by the combination of S-twill and Z-twill weaves. According to the change of twill direction, there are two types of zig-zag weaves, such as horizontal zig-zag and vertical zig-zag weaves.

Horizontal zig-zag weave: If the direction of twill line change depends on the warp yarn then horizontal zig-zag twill weave is produced. The repeat size of horizontal zig-zag is calculated from the regular or base twill weave (Fig. 12.23). In one system, the number of warp yarn in zig-zag weave is double of the number of warp yarn of base twill and the number of weft yarn is same as base twill weave. For example, if the repeat size of basic regular twill is 4 × 4, then the repeat size of horizontal zig-zag is 8 × 4. In other system, the number of warp yarn in zig-zag weave is two less from double of the number of warp yarn of base twill and the number of weft yarn is same as base twill weave. For example, if the repeat size of basic regular twill is 4 × 4, then the repeat size of horizontal zig-zag is 6 × 4. In general, the direction of twill line is changed after the completion of repeat of regular twill weave and the point is created at the changing time. Normally, warp-way twill is used as regular basic twill. Returning a

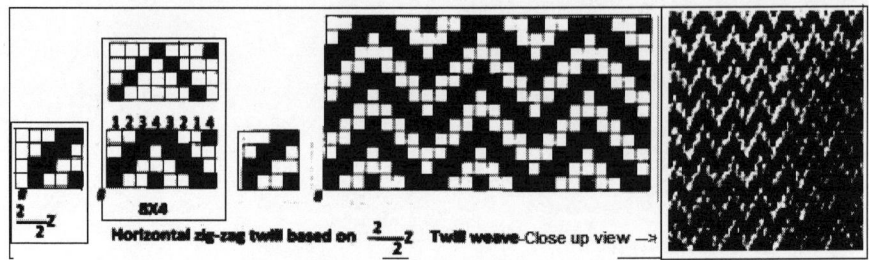

Figure 12.23. Horizontal zig-zag weave

straight draft in the opposite direction will create a pointed draft and does result in a horizontal waved effect if programmed with a peg plan of the twill weave. So pointed or V-draft is used to produce horizontal zig-zag weave. It is possible to produce this weave from any type of regular basic twill weave.

Vertical zig-zag weave: If the direction of twill line is change depends on the weft yarn then vertical zig-zag twill weave is produced. The repeat size of vertical zig-zag is calculated from the regular or base twill weave. In one system, the number of weft yarn in zig-zag weave is double of the number of weft yarn of base twill, and the number of warp yarn is same as base twill weave. For example, if the repeat size of basic regular twill is 5 × 5, then the repeat size of vertical zig-zag is 5 × 10. In another system, the number of weft yarn in zig-zag weave is two less from double of the number of weft yarn of base twill and the number of warp yarn is same as base twill weave. For example, if the repeat size of basic regular twill is 5 × 5, then the repeat size of horizontal zig-zag is 5 × 8. In general, the direction of twill line is changed after the completion of repeat of regular twill weave and the point is created at the changing time. Both weft- and warp-way twills

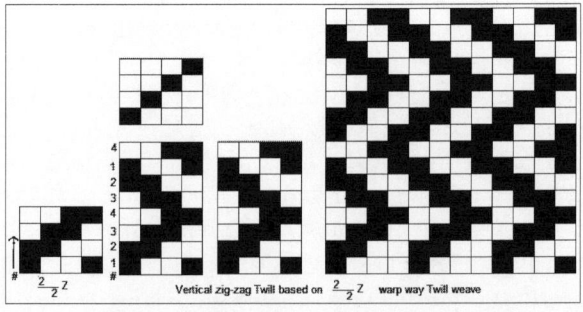

Figure 12.24. Vertical zig-zag weave

are used as regular basic twill. Straight draft is used to produce vertical zig-zag weave. It is possible to produce this weave from any type of regular basic twill weave. Fig. 12.24 shows the weave plan with drafting and lifting plan of some vertical zig-zag weaves.

2. *Herringbone twills*

These weaves are generated by introducing a step into the design after a certain number of ends or picks. At the step, every thread changes from up to down or vice versa. If the original weave is not double faced, this means that, at every step, a warp twill changes into a weft twill or vice versa. It is also produced by the combination of S twill and Z twill like zig-zag weave but it is not create a point. It is also divided into two groups depending on the change of the direction of twill line, such as horizontal herringbone twill weave and vertical herringbone twill weave.

Horizontal herringbone twill: If the direction of twill line is change according to the herringbone principle depends on the warp yarn then horizontal herringbone twill weave is produced (Fig. 12.25). The repeat size of horizontal herringbone is calculated from the regular or base twill weave like as horizontal zig-zag weave. In this case, the number of warp yarn in herringbone weave is double of the number of warp yarn of base twill and the number of weft yarn is same as base twill weave. For example, if the repeat size of basic regular twill is 4 × 4, then the repeat size of horizontal herringbone is 8 × 4. In general, the direction of twill line is changed after the completion of repeat of regular will weave. Normally warp-way twill is used as regular basic twill. Broken draft is used to produce horizontal herringbone weave, if double-face twill weave is used as base twill. When uneven twill such as warp- or weft-face twill is used as base twill, then straight draft is used to produce horizontal herringbone weave.

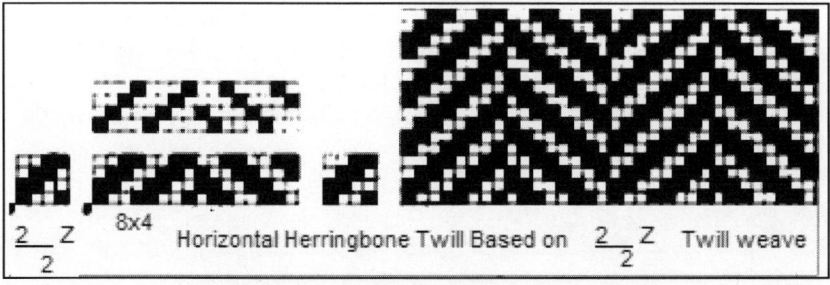

Figure 12.25. Horizontal herringbone twill

Vertical herringbone twill weave: If the direction of twill line is changed according to the herringbone principle depending on the weft yarn then vertical herringbone twill weave is formed (Fig. 12.26). The repeat size of vertical herringbone is calculated from the regular or base twill weave like a vertical zig-zag weave. In this case, the number of weft yarn in herringbone weave is double of the number of weft yarn of base twill. For example, if the repeat size of basic regular twill is 4 × 4, then the repeat size of vertical

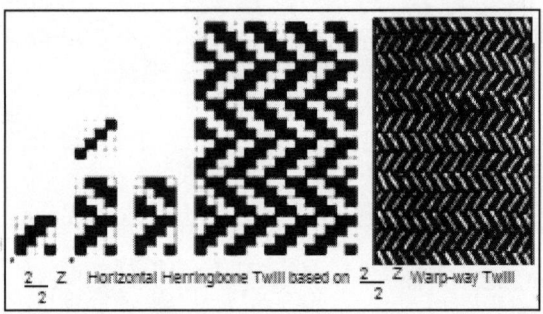

Figure 12.26. Vertical herringbone twill weave

herringbone is 4 × 8. In general, the direction of twill line is changed after the completion of repeat of regular twill weave. Normally, weft-way twill is used as regular basic twill. Straight draft is used to produce vertical herringbone weave from all types of base twills. It is possible to produce this weave from any type of regular basic twill weave. The horizontal stripe effect is produced on the surface of this fabric.

3. *Diamond design twill:*

This is a derivative of twill weave. Diamond design is developed on the basis of pointed principle. It is build-up by the combination of vertical and horizontal zig-zag weaves. The repeat size of diamond design is also calculated from the regular or base twill weave (Fig. 12.27). In this case, the number of both warp and weft yarns in diamond weave are double of the number of warp and weft yarns of base twill, respectively. For example, if the repeat size of basic regular twill is 4 × 4, then the repeat size of diamond design is 8× 8. Diamond is a reversible design. So it may be divided into two equal parts in both vertical and horizontal axis. Pointed or V-drafting system is used to produce diamond design.

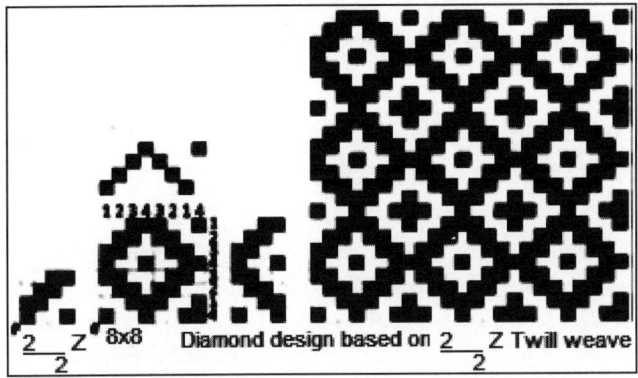

Figure 12.27. Diamond design twill

4. *Diaper design twill*

This derivative of twill weave is developed on the basis of herringbone principle. It is built-up by the combination of vertical and horizontal herringbone weaves. The repeat size of diaper design is also calculated from the regular or base twill weave like in diamond design (Fig. 12.28). In this case, the number of both warp and weft yarns in diaper weave are double of the number of warp and weft yarn of base twill, respectively. For example, if the repeat size of basic regular twill is 4 × 4, then the repeat size of diaper design is 8 × 8. Diaper is not a reversible design like diamond. It may be divided into two parts in diagonal axis. Broken

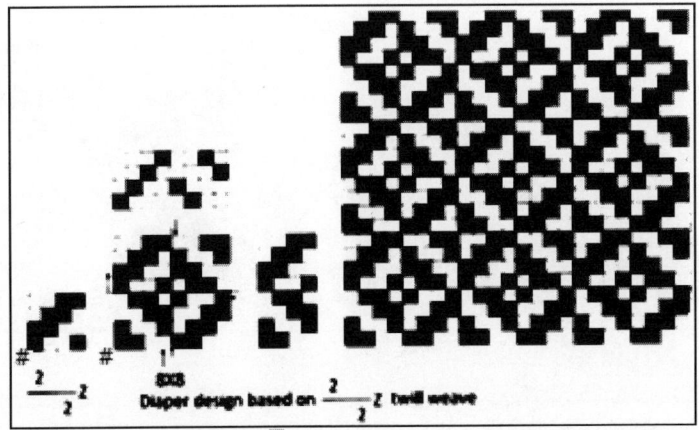

Figure 12.28. Diaper design twill

draft is used to produce diaper design; if double-face twill weave is used as base twill. When uneven twill such as warp- or weft-face twill is used as base twill then straight draft is used to produce this design.

5.   *Broken twill weave*

A broken twill is formed by a break in the continuation of the twill line at predetermined intervals. There is several systems to produce broken twill. Usually, the break will be at the centre of the repeat, with only one reversal, but more complicated breaks can be made. One divides the original weave into two halves and copies the first half unchanged, starting from the first warp end. The second half is copied in reverse order, starting from the last end. Broken twills are also produced by dividing three or more parts. Normally, straight draft is used to produce this weave. The pattern can be broken either in the warp or in the weft direction and no twill line will be generated (Fig. 12.29).

6.   *Re-arranged twill weave or transposed twill weave*

Rearranging a weave means taking single thread or group of threads of the base weave and arranging them in a different order (Fig. 12.30). If the rearrangement does not exceed the repeat of the base weave, then same straight draft can be used. The pattern can be rearranged either in the warp or in the weft direction. By this rearrangement the different types of novelty and attractive designs can be developed in the fabric. The appearance of some rearranged twill is same as broken twill.

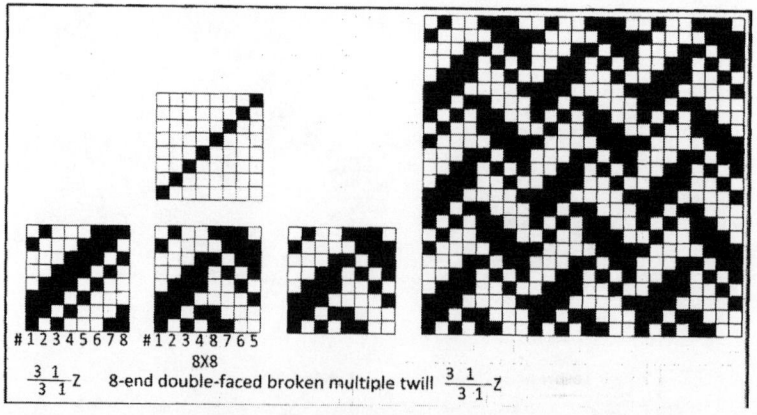

Figure 12.29. Broken twill weave

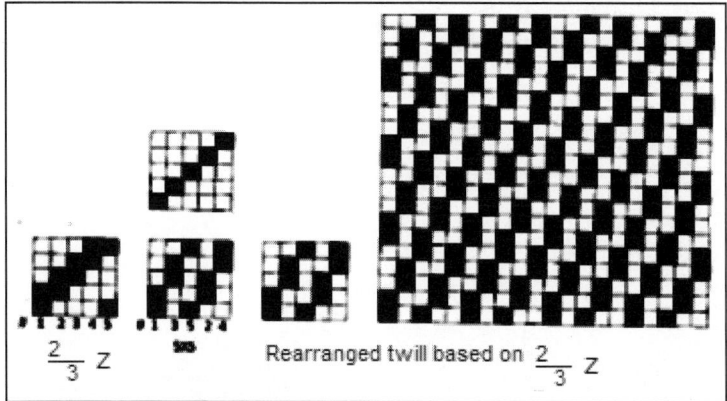

Figure 12.30. Re-arranged twill weave or transposed twill weave

7. *Stepped twill weave*

These weaves are generated by introducing a step into the design after a certain number of ends or picks. At the step, every thread changes from up to down or vice versa. If the original weave is not double-faced, this means that, at every step, a warp twill changes into a weft twill or vice versa. There are three types of step twill weave, such as warp-way step twill, weft-way step twill and both warp- and weft-way step twill weaves.

(a) *Warp-way step twill weave*

There are two types of warp-way step twill. One is created in the same twill direction and another one is created by reversal of the twill direction.

Same twill direction: In the same twill, direction step may be occur after the repeat or any desired number of thread. Fig. 12.31 represents 4-end double-faced twill with 2/2 step after every 4-ends and same twill direction.

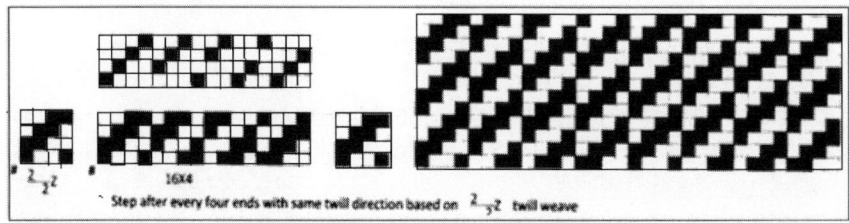

Figure 12.31. Warp-way step twill weave: same twill direction

Reversal of the twill direction: Same as horizontal herringbone twill weave which is discussed above.

(b) *Weft-way step twill weave*

There are also two types of weft-way step twill like as warp-way step twill weave. One is created in the same twill direction and another one is created by reversal of the twill direction.

Same twill direction: In the same twill, direction step may be occur after the repeat or any desired number of thread like as warp-way step twill weave. Fig. 12.32 represents 4-end weft-way step twill weave.

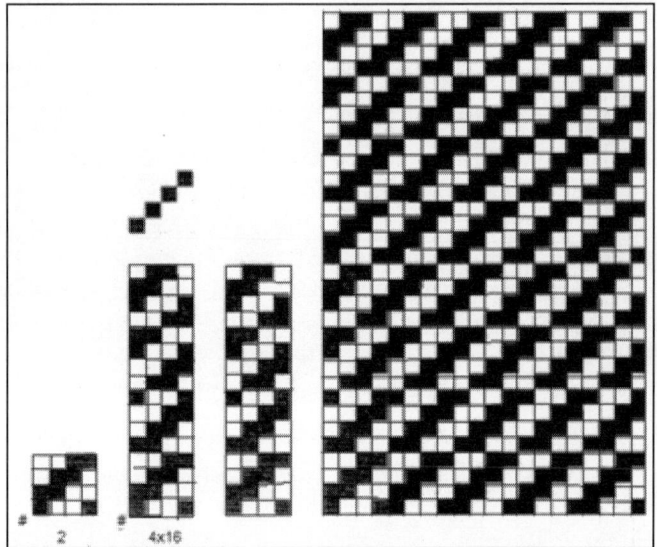

Figure 12.32. Weft-way step twill weave: same twill direction

8. *Elongated twill or steep twill weave*

A peculiar form of twill, known as an elongated or steep twill, is obtained when the warp float of each thread rises two or more picks instead of one pick above the float of the preceding thread. A steep twill can be made by drafting in succession the alternate threads of a regular twill. This is a term applied to a regular twill which has been altered to achieve a steeper or flatter angle. The angle of elongated twill is either below 450 or above 450. These are based on a square sett (same number of ends and picks per 1 inch and an identical yarn count). Any deviation from this will automatically influence the angle. The angle of the twill line is determined by the step number

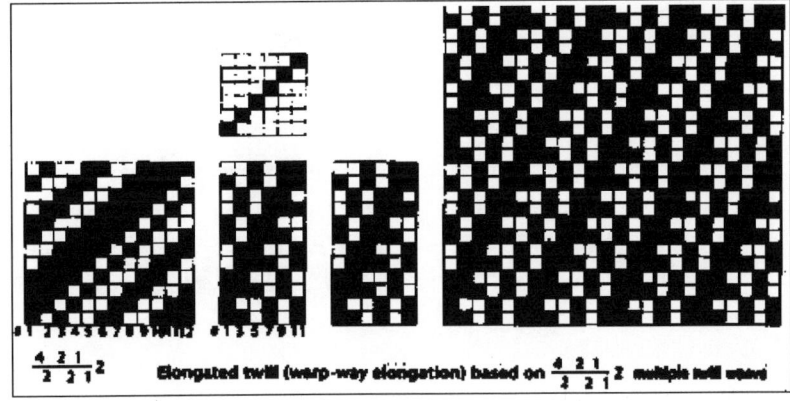

(a)

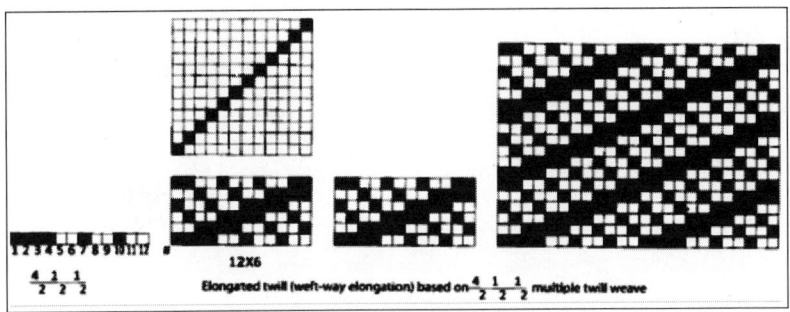

(b)

Figure 12.33. (a) Warp-way elongated twill or steep twill weave. (b) Weft-away elongated twill or steep twill weave

interlacing point to the next. There are two types of elongated twill, such as warp-way elongated twill, i.e., warp-way elongation, and weft-way elongated twill, i.e., weft-way elongation (Fig. 12.33a, b).

9. *Combination twill weave*

For the construction of combined twill, the repeat sizes of two regular base twills play an important role (Fig. 12.34a,b). The repeat size of the combined twill depends on the repeat size of the regular base twill. If the repeat sizes of two base twills are same, then the number of warp yarn in the repeat size of the warp-way combined twill is twice of regular base twill, and the number of weft yarn is same as regular twill. Similarly for the weft-way combined twill, the number of weft yarn in the repeat size is twice of regular base twill and the number of warp yarn is same as regular twill. But if the repeat sizes of the base twills are not same, then it is important to calculate their

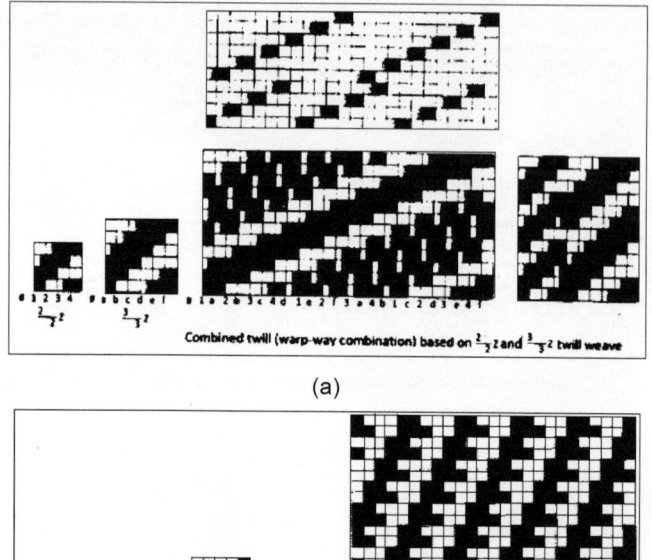

(a)

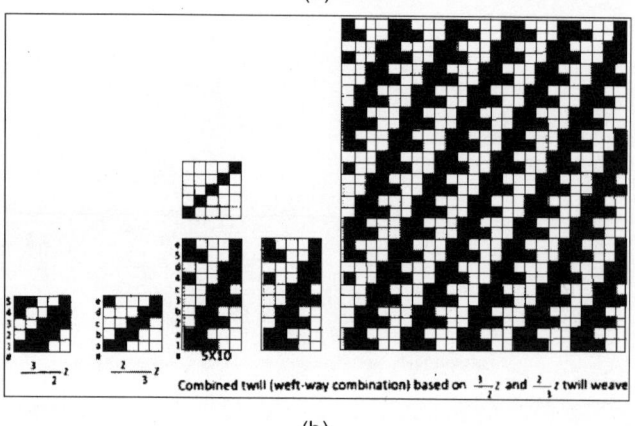

(b)

Figure 12.34. (a) Warp-way combined twill weave. (b) Weft-way combined twill weave

(repeat sizes of the base twills) lowest common multiple (LCM). In this case, the selection of repeat size depends on this LCM value. For warp-way combined twill, the number of warp yarn in the repeat size is twice of LCM value and the number of weft yarn is same as LCM value. Similarly for weft-way combined twill, the number of weft yarn in the repeat size is twice of LCM value and the number of warp yarn is same as LCM value.

10. *Shaded twill weave or shaded design*

Shade effect can be produce in different way on the surface of the fabric. This shade effect can be introduced in any type of cross over or stripe or figure design. There are mainly two types of shading effects, such as single shading and double shading effect. When these shading effects are produce by the use of twill weave, then it is called shaded twill weave. The base twill may be either warp-way or

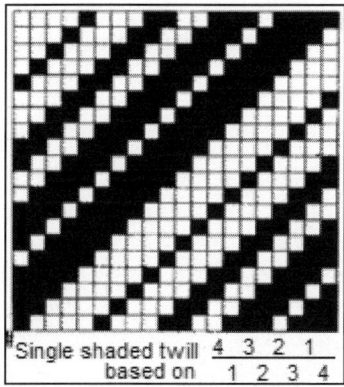

Figure 12.35. Single-shaded twill weave

weft-way twill. So there are two types of shaded twill, such as single-shaded twill and double-shaded twill weave.

Single-shaded twill weave: In this case, the shade effect is gradually decreasing from deep to light by decreasing the number of warp or weft floats and vice versa (Fig. 12.35). These effects are produced from the regular multiple twill. In these multiple twills, the warp and weft floats are arranged in a regular order. Such as $\frac{6\ 5\ 4\ 3\ 2\ 1}{1\ 2\ 3\ 4\ 5\ 6}$, $\frac{5\ 4\ 3\ 2\ 1}{1\ 2\ 3\ 4\ 5}$, $\frac{4\ 3\ 2\ 1}{1\ 2\ 3\ 4}$, $\frac{3\ 2\ 1}{1\ 2\ 3}$, etc.

Double-shaded twill weave: In this case, the shade effect is gradually increasing from light to deep by increasing the number of warp or weft floats and again gradually decreasing from deep to light by decreasing the number of warp or weft floats (Fig. 12.36). These effects are produced from the regular multiple twill. In these multiple

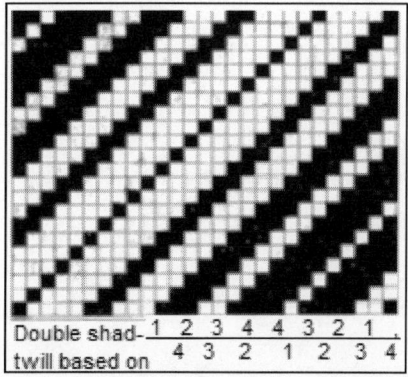

Figure 12.36. Double-shaded twill weave

twills, the warp and weft floats are arranged in a typical order. Such as

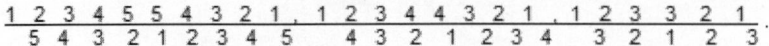

12.1.9.3   Satin or atlas weave

Satin is the third basic weave of the woven fabrics. In basic construction, the satin weave is similar to the twill weave but generally uses from five to as many as twelve harnesses, producing a five to twelve-shaft construction. It differs in appearance from the twill weave because the diagonal of the satin weave is not visible; it is purposely interrupted in order to contribute to the flat, smooth, lustrous surface desired. There is no visible design on the face of the fabric because the yarns that are to be thrown to the surface are greater in number and finer in count than the yarns that form the reverse of the fabric.

Satin weaves have a closed, smooth and dense appearance. The satin weave is characterized by regularly distributed crossing points that do not touch each other. The distance from one interlacing point to another on the next weft line is called shift. In one direction, this distance is always constant (Fig. 12.37).

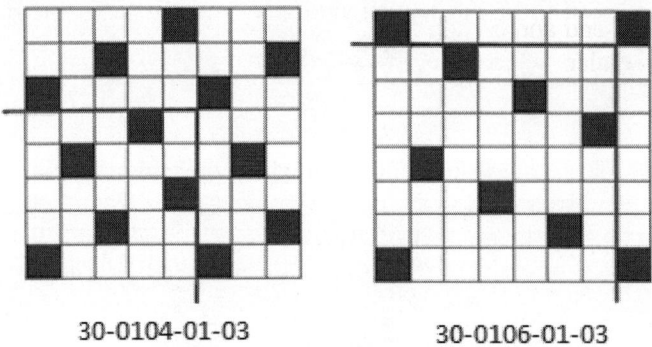

30-0104-01-03   30-0106-01-03

Figure 12.37. Satin or atlas weave

Satin weaves produce a smooth, even and glossy fabric surface. This is due to the interlacing points being covered up by the floats of the neighbouring threads. The smoothness of the fabric surface can be improved by:

1. High thread density

2. Smooth yarn with low twist

3. Filament yarn from man-made fibreSilk yarn

Each end and each pick makes only one intersection, and the intersections are distributed in an orderly manner. They are uniformly separated from each other and nowhere adjacent. Satin is more loose structure fabric, when compared with plain and twill fabrics. Satin is widely used for the foundation of jacquard design.

The difference of warp and weft satins is depending on whether the fabric face shows the warp or the weft. Weft satins are also called sateens. With the most common simple warp satin, each warp end is lowered only on one pick in the repeat while, with the weft satin, it is only raised on one pick. The smallest regular satin weave is the 5-end satin which can be represented either by $\frac{1}{4}$ (2), or by $\frac{4}{1}$ (3) where the figure in the bracket shows the size of the step.

Both warp and weft satins are divided into two groups, such as regular warp satin and irregular warp satin.

Regular weft satin and irregular weft satin

There is a step value or move number for regular warp or weft satin weave but there is no step value for the irregular warp or weft satin weave. In general, 4-end and 6-end satin weaves are irregular, because they have no step value. Other 5-end to 16-end satins are regular, because they have step values.

There is no step value or move number to construct the irregular sateen. So the above mentioned rule is not applicable for the construction of irregular sateen.Only4-end and 6-end sateens are irregular. Fig. 12.38 shows both regular and irregular weft satin constructions.

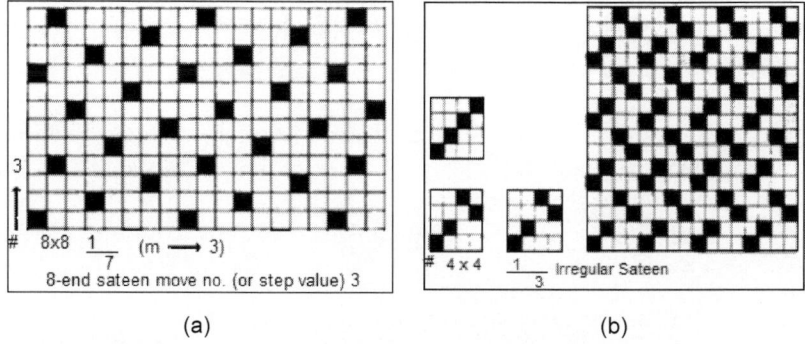

(a)   (b)

Figure 12.38. (a) Regular weft satin. (b) Irregular weft satin

Warp satin is woven so that the warp may be seen on the surface of the fabric. For example, in a 5-end construction, the warp may pass over four weft yarns, and under one pin, a 12-end construction, the warp may pass over 11 weft yarns and under 1. Since the warp lies on the surface and interlaces only 1 weft yarn at a time, the lengths of warp between the weft yarn are called floats. These floats lie compactly on the surface with very little

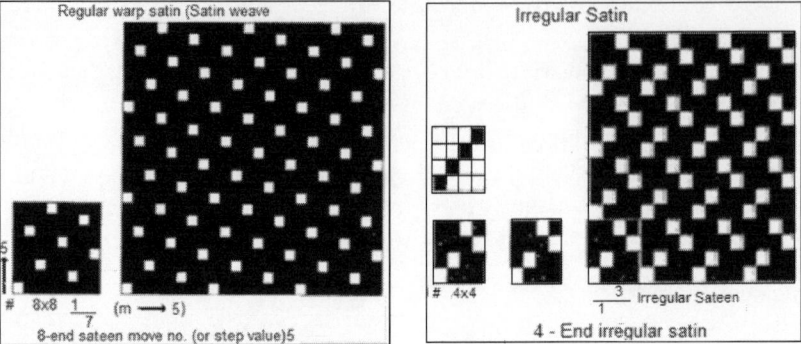

Figure 12.39. Regular and irregular warp satin weaves

interruption from the yarns going at right angles to them. Reflection of light on the.

Derivatives of satin weave

Weaves are produced on the satin base is called derivatives of satin weave. Lot of jacquard designs based on this satin weave. The following are the main types of satin based on its structure.

1. *Crepe weaves*

   Texture is the special feature of crepe weaves. Their surface exhibits an all over, random, small-scale pattern in low relief. There are two ways of obtaining this kind of texture. By using a special 'crepe' or 'oat meal' weave in conjunction with ordinary, normal-twist spun or filament yarns, and high-twist crepe yarns in conjunction with plain or other simple weaves such as twill or satin. The fabric should have a rough irregular surface without any prominent features. This is generally achieved by having approximately equal disposition of warp and weft on the surface of the cloth, and also by avoiding any floats which exceed three.

There are four basic methods of producing crepe weaves:

(a) On a sateen base

(b) By reversing

(c) By superimposing

(d) On a plain weave base

(a) *Sateen base*: Construct a sateen weave and construct a twill weave on the same repeat size. Using the sateen base as the starting point of each lift

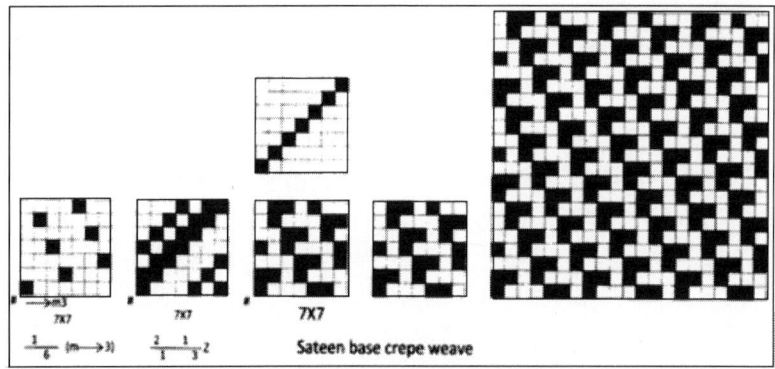

Figure 12.40. Satin base crepe weave

of the twill, rearrange the twill weave on the sateen base. This new weave is called sateen base crepe weave (Fig. 12.40).

(b) *Reversing*: Make a small motif, as at 'a', reverse 'a' by turning it over, so that the warp lifts of the 4th end become the weft lifts of the 5th end and those of the 4th end become the warp lifts of the 5th end; similarly the 6th, 7th and 8th ends are the converse of the 3rd, 2nd and 1st, respectively. And the design is on 8 ends × 4 picks; reverse this by turning it over in the weft direction and using the same technique as described. The final design is thus produced, which is called crepe weave. The method of constructing this weave may lead to a tendency to create grouping of threads, which is generally undesirable in crepe weaves. Fig. 12.41 shows the weave plan with drafting and lifting plan of different crepe fabric. Normally straight drafting system is used to produce this weave (Fig. 12.41).

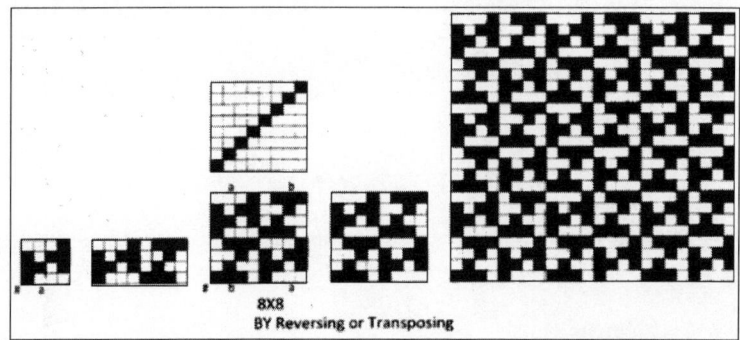

Figure 12.41. Weave plan with drafting and lifting plan of different crepe fabric

(c) *Superimposing*: Construct two different weaves with same repeat size. As sateen weaves are mainly used in this method, there is always a

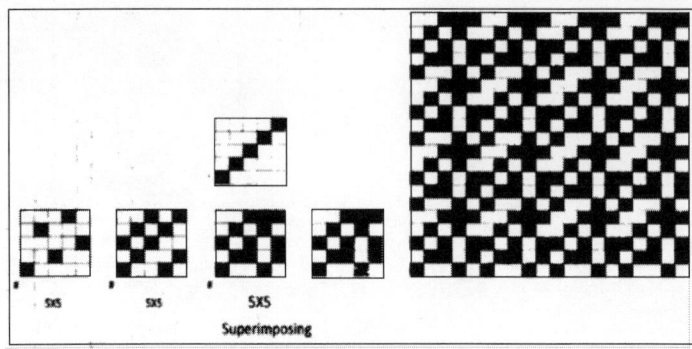

Figure 12.42. Superimposing

predominance of weft over warp. Superimpose one weave on the other to give the final weave (Fig. 12.42).

(d) *Plain base*: Design a sateen on half the number of ends and picks required in the final design – a 6-end sateen will be used for a design to be produced on 12 ends × 12 picks. Expand this weave so that the sateen base appears on alternate ends and picks only, and use this base as the starting point of each lift of at will, in this $\dfrac{1\ 1\ 3\ 1}{3\ 1\ 1\ 1}$ case (Fig. 12.43). On the remaining ends insert alternate ends of plain weave, i.e., all of these ends will weave the same tabby; care should be taken to lift the warp on the picks

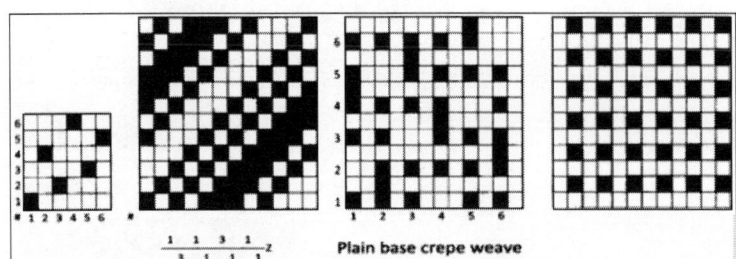

Figure 12.43. Plain base crepe weave

opposite to those on which the sateen base appears. Now combine these to give the final design, which is repeated in the Fig. 12.43.

2. Corkscrew weaves

(a) Odd number cork screw weave: Odd number corkscrew weave is created by rearranging any type of regular twill weave in a sateen order. Both warp and weft face types are available (Fig. 12.44). Warp-face floats

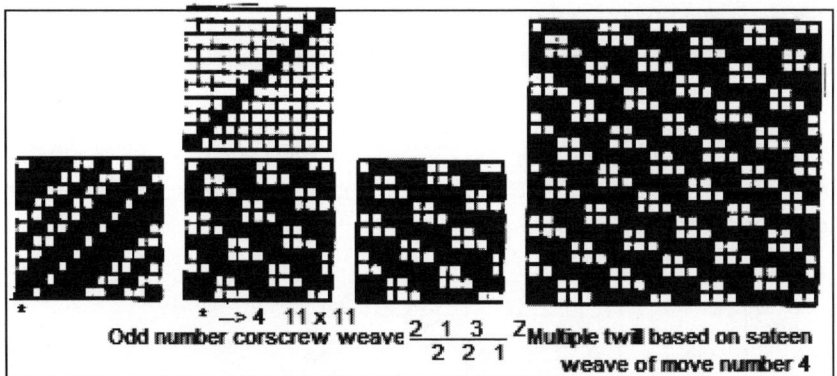

Figure 12.44. Odd number cork screw weave

are one thread longer than weft floats. Same in the case of weft face. They are developed from odd number of ends and picks. The figure show the odd number corkscrew weave from respective regular twill weave with drafting and lifting plan. Straight drafting system is normally used to produce this weave.

(b) Even number corkscrew weave: It is produced from two different regular base twill of the same repeat size. In this case, the number of warp yarn in the repeat size of the resultant corkscrew weave will be the sum of the number of warp yarn of the base twill weave and the number of weft yarn is equal to the base twill. For example, if the repeat size of the base twill is 6 × 6

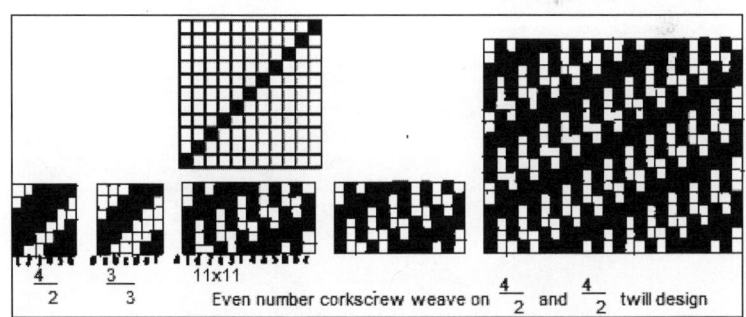

Figure 12.45. Even number cork screw weave

then the repeat size of the resultant even number corkscrew weave is 12 × 6. Straight drafting system is normally used to produce this corkscrew weave. Fig. 12.45 shows the even number corkscrew weave from respective regular twill weave with drafting and lifting plan.

3. Shaded weaves

(a) Single-shaded sateen weave: In this case, numbers of weft satins are developed side by side at first. Then these sateen units are divided into the number of groups. With each group of sateen, one can gradually add warp floats with the interlacing points until it turns into a warp satin. After this, the resultant weave will be a single shaded design. Fig. 12.46 shows the different single-shaded design.

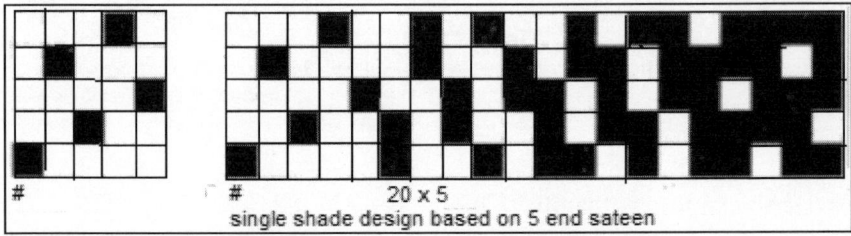

Figure 12.46. Single shade design

Double-shaded twill weave: In this case, the shade effect is gradually increasing from light to deep by increasing the number of warp or weft floats and again gradually decreasing from deep to light by decreasing the number of warp or weft floats (Fig. 12.47). These effects are produced from the regular multiple twill. In these multiple twills, the warp and weft floats are arranged in a typical order, such as

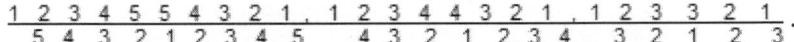

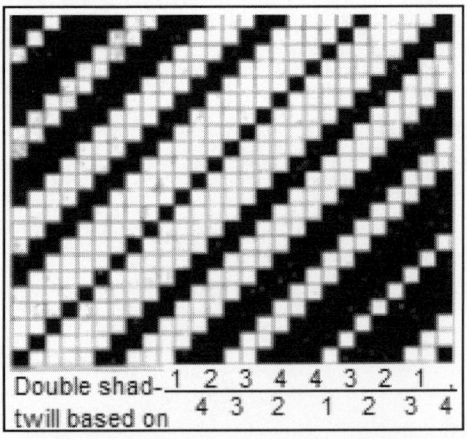

Figure 12.47. Double shade design

12.1.9.4   Huckaback weaves

These weaves are generally applied in the manufacture of non-pile tow-els. Weave construction stages are: (1) Mark out the repeat size, divide into quarters and filling plain weave in two opposite ones. (2) Filling a motif in the other two quarters, which is preferably produced by taking plain weave and adding or removing some lifts, as at second one; care should be taken to ensure that the motif and the plain weave bind together effectively; the final weave is produced by combining first and second one (Fig. 12.48). This weave is characterized by a rough surface, which is produced by floating

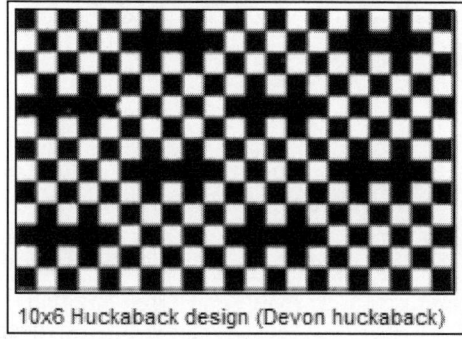

10x6 Huckaback design (Devon huckaback)

Figure 12.48. Huckaback weaves

threads in groups arranged on a plain weave basis. A more balanced hucka-back is produced if the weave-repeat size is twice an odd number (i.e., 2 × 5 = 10; repeat size = 10 ends × 10 picks), but it is by no means impossible to produce the weave on a repeat which is complete on twice an even number of threads.

12.1.9.5   Mock leno weaves

This weave is also referred to as limitation gauze weave. The stages in pro-ducing the weave are: (1) Mark out the repeat size, divide into quarters and fill a small motif in opposite quarters, as in first step. (2) Completely reverse this motif in the two remaining quarters, by substituting warp lifts for weft lifts and vice versa, as in second step. (3) Combine first and sec-ond steps to give the final weave (Fig. 12.49). The main features of this weave are: (a) It is an open perforated weave like as leno fabrics. (b) It is produced in the ordinary way without special leno shafts. The similarity of this weave to the huckaback is quite obvious, but the method of dent-ing is different, as it is necessary to encourage thread grouping. (c) The weave is arranged in groups of equal or unequal sizes. Threads working in

10 x 10 Mock Leno

Figure 12.49. Mock leno weaves

plain weave alternate with threads floating on the face or back of the fabric. The ends from each individual group are whenever possible drawn into the same dent; this bunches the floating ends together and causes a slight gap or opening in the fabric giving an appearance similar to a gauze or leno weave, hence the name 'mock leno'. (d) Even number repeat size is normally used to produce this weave. (e) This weave can also be divided diagonally into two equal parts. (f) The smallest repeat size of this weave is 6 × 6.

12.1.9.6 Honeycomb weave

The term is applied to weaves which resemble honeycomb cells. The cellular formations appear square in the cloth. They are formed by some ends and picks interlacing tighter than others and therefore developing a higher tension. Usually single cloths made by progressively lengthening and shortening both warp and weft floats to form ridges and hollows on a square pattern, to give a cellular appearance. Sometimes called waffle or waffle piqué. There are two types of honeycomb weave, such as ordinary honeycomb and brighton honeycomb.

Ordinary honeycomb weave

A ordinary honeycomb weave can be constructed as follows: (1) Construct $\frac{1}{a}$ – Z twill starting in the bottom left-hand corner, then a similar one running in the opposite direction and starting one square in or one square down from the top left-hand corner, so that there will be a clean intersection of the twill lines, as at first stage. (2) In one of the two diamonds produced, leave a row of stitching points and then lift the remainder of the diamond solid. This

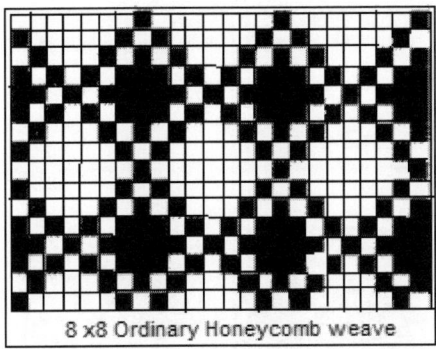

8 x8 Ordinary Honeycomb weave

Figure 12.50. Ordinary honeycomb weave

is the final weave. The main features of ordinary honeycomb weaves are as follows: (a) The characteristic features of this weave are alternate raised and sunk diamond shaped are as which give the effect of a honeycomb. (b) Both sides of the fabric look the same and the surface of the fabric is rough. (c) It has long floats of warp and weft yarns. (d) In the repeat size the number of ends and picks may be equal or unequal and multiple of two. (e) In the larger repeat size a double row of binding has been constructed by using a $\frac{1\ 1}{1\ a}$ twill weave at first stage, so that a firmer structure will be produced.

(f) The long floats in the centre of the diamonds are not equal, and if the fabric is being produced with a square set, this can be detrimental to the appearance of the cloth as they will produce a rectangular pattern instead of a square one. Two methods are available for improving the appearance when this occurs: adjust either the sett or the weave. (g) Pointed drafting system is normally used to produce this weave. This weave is mainly used for making hand towels, glass cloths, dispensed roller towels, bath mats, cellular blankets, etc. (Fig. 12.50).

Brighton honeycomb weave

The construction of brighton honeycomb is $\frac{1}{a}$ more complicated than the ordinary honeycomb. It is constructed as follows: (a) Construct Z a starting in the bottom left-hand corner, and then construct a $\frac{1\ 1}{1\ a}$ S twill, starting with the first warp lifts in the squares to the right and below the square in the top left-hand corner, and indicate the points on the double row of binding which are immediately adjacent to those of intersection that will allow extensive floats in the weft direction, as illustrated in first stage. (b) Using the points indicated in first stage as the extreme lift of the longest float, lift the remaining adjacent ends, as in second stage. (c) Each of these warp floats now form

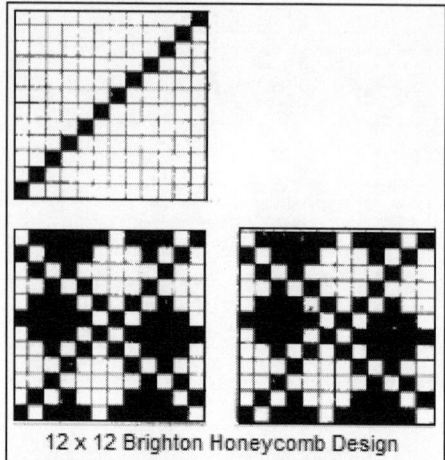

Figure 12.51. Brighton honeycomb design

the centre float of a diamond which can be completed. This is the final weave (Fig. 12.51).

12.1.9.7   *Distorted warp effect weave*

The design is constructed in stages: (a) After indicating the threads in the warp and weft direction which are essential for forming the distortion, fill in plain weave on all the remaining ends and picks, as at first stage. (b) For a warp distortion, lift the preselected warp threads of first stage except where they cross the preselected weft threads, and then lift all remaining ground ends over the preselected weft threads in one group on the first pick and in the other group on the second pick, as at second stage. (c) The completed

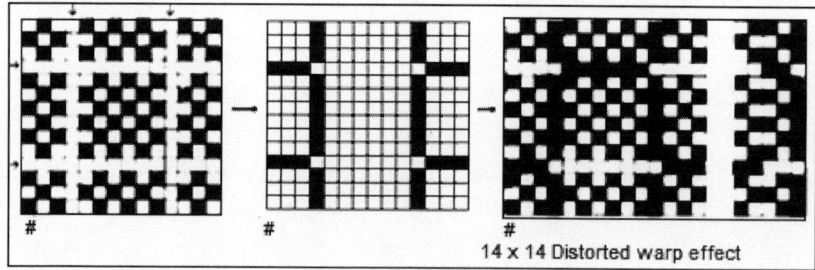

Figure 12.52. Distorted warp effect weave

design, third stage, is then formed by combining first and second stages (Fig. 12.52).

12.1.9.8   Cord weave

The main characteristic of this weave are cords running in warp or weft direction. They have some similarity in appearance to the preceding weft or warp rib weaves but they are not reversible. The end use is mainly for apparel fabrics. Cord effect in fabric can be brought about by using thick yarns and weave-producing cord effects.

Bedford cord weave

Bedford cord is made as follows: (1) Indicate the width of two cords (in this example, each cord has six ends), and then show the outside ends of each cord, known as cutting ends, weaving plain throughout, as at first and second stage. (2) The first pair of picks float under the warp ends in the first cord and weave plain in the second cord. The second pair of picks weave plain in the first cord but float under the warp ends, and thus on the back of the cloth on the second cord. This fourth stage is the final design. Cords running down the piece in the warp direction form the main characteristic of this weave. The face of the cloth is usually plain weave and the corded effect is produced by allowing alternate pairs of weft threads to float on the back of the fabric behind each cord. These threads interweave in plain order with the outside ends of each cord and are known as the cutting ends (sunken line) (Fig. 12.53). There are mainly five types of bedford cords – plain face bedford cord, wadded bedford cord, crepon bedford cord, bedford cords arranged with alternate picks and twill face bedford cord. Fabrics produced with this weave may be

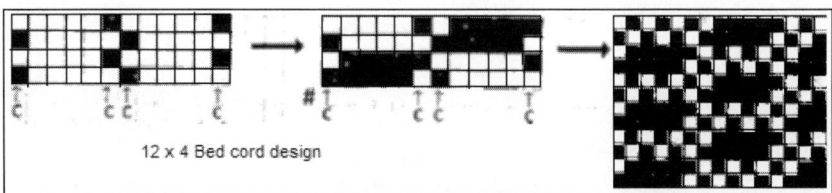

12 x 4 Bed cord design

Figure 12.53. Bedford cord weave

made in medium-weight cotton or spun rayon fabrics for ladies' blouses and dress wear, sportswear and ornamental trimmings. In heavier qualities, it is suitable for soft furnishing when produced with cotton yarns or for trouserings when made of worsted yarns.

Pique weave

A typical pique structure consists of a plain face fabric composed of one series of warp and one series of weft threads, and a series of back or stitching warp threads. Continuous sunken lines or cuts, i.e., cords are run horizontally in the cloth. One cord is produced per repeat. Normally skip drafting system is used to produce this weave. There are four types of pique weave:

1.   Ordinary pique or welt structure/loose back without wadding picks

2.   Weft wadded welts/loose back wadded welt structure

3.   Fast back welt or pique structure

4.   Waved pique structure

1.   *Ordinary pique or welt structure/loose back without wadding pique*: Construction of an ordinary pique is as follows: (a) Indicate the order of the warp thread arrangement, which is always one ground, one stitching end and one ground end, then filling plain weave on the ground ends as at first stage. (b) The stitching warp is lifted over the required number of picks, as determined by the requirements of the final fabric appearance; at second stage a two pick weave is illustrated. (c) The final weave is produced by combining first and second stage. The number of face picks in the width of a cord is varied according to requirements, but usually the number of consecutive picks that are unstitched should not exceed twelve. The order of the warp thread arrangement, which is always one face or ground, one stitching or back end and one ground or face end, in each split of the reed, or in the proportion of two face to one stitching end (Fig. 12.54).

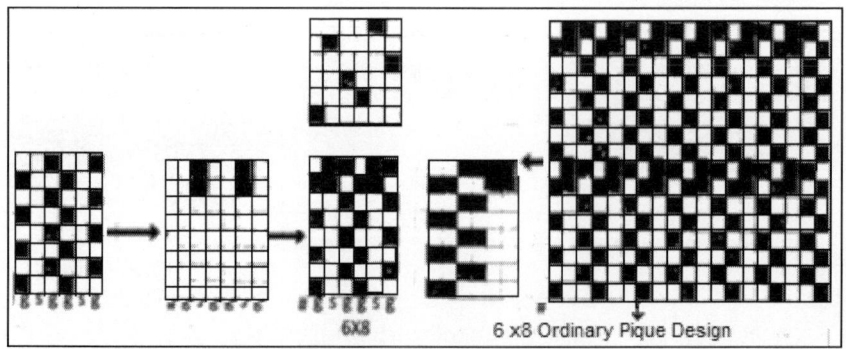

6X8

6 x8 Ordinary Pique Design

Figure 12.54. Ordinary pique

2.   *Wadded welts*: To increase the prominence of the unstitched portions of the cloth, i.e., horizontal cords and to make the cloth more substantial, it is customary to insert wadding picks between the tight back stitching ends and the slack face fabric. Usually the wadding weft is thicker than the ground weft and is inserted two picks at a

Figure 12.55. Wadded welt design with longitudinal section

place, the looms being provided with changing shuttle boxes at one side only. Sometimes, however, the same kind of weft is used for both the face and the wadding, looms with a single box at each side being employed; and, in such a case, one wadding pick at a place may be inserted (Fig. 12.55).

3. *Fast-back welts*: In these designs, the stitching ends are only lifted to form the indentations, the term 'loose-back' being applied to this type of structure. The term 'fast-back' is applied to cloths in which

Figure 12.56. Fast-back welts

the stitching ends are interwoven in plain order with all, or some wadding picks. The reduction of the float length of the stitching ends on the back of the fabric which results from this interlacing helps to produce a more serviceable cloth less liable to accidental damage (Fig. 12.56).

4.   *Waved pique design*: One type of pique weave. A waved pique is a simple modification of the welt structure in which the indentations are not in a horizontal line but are arranged in alternate groups, as the marks in Fig. 12.57 which indicate the lifts of the stitching ends on the face picks. The group of marks does not overlap horizontally, as one commences on a face pick. Immediately following that on which the other has finished.

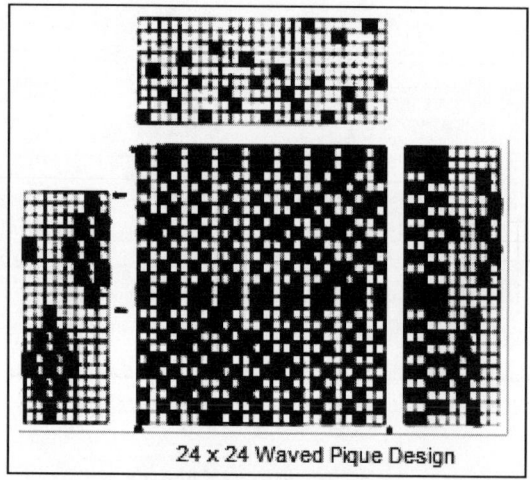

24 x 24 Waved Pique Design

Figure 12.57. Waved pique

Between succeeding groups, two wadding picks are inserted, as indicated by the arrows at the side of first figure. The complete design to correspond with first figure is given at second figure, in which the ends are arranged in the same order as in a welt, while there are ten face picks to two wadding picks. The lifts of the tight stitching ends force the wadding picks first in one direction and then in the other, so that waved lines are formed across the cloth. Fig. 12.57 shows the weave plan with drafting and lifting the plan of a typical waved pique or welt design. This design is used only to a very limited degree, mainly for trimmings and other ornamental uses. It is also used for neckties, ladies light summer holding costume etc.

Sponge weave: Any one of a variety of weave arrangements that groups end and pick together to form a cellular structure and to create a soft spongy effect in the fabric. Examples include spot weaves, diamond effects, honeycombs and sateen-based structures with lifts added. Sponge weave considered as the result of honeycomb effect and also form the cell like honeycomb weave. The characteristics features of this weave are as follows: (a) The number of ends and picks are always equal. (b) 10 × 10 is the smallest repeat size of this weave. (c) Straight drafting system is used to produce this weave. (d) Low twisted and coarser yarns are used to produce this fabric. So the fabric produced by this weave is very soft and absorbent. (e) For the construction of this weave, it is important to calculate the longest float of diamond. This float is depends on the repeat size of the design. The following formula is used to calculate this longest float: Longest float of diamond $= \sqrt{(\text{Number of ends or picks in the repeat} - 1)}$. (f) It is a reversible cloth like honeycomb. (g) Honeycomb weave produce one cell on both sides but in this case number of produced cell on both sides of the weave depends on the number of repeat size. (h) This weave produced on the sateen base. Uses include fancy woollen shawls, bedsheet, towel, counterpanes, drapes,

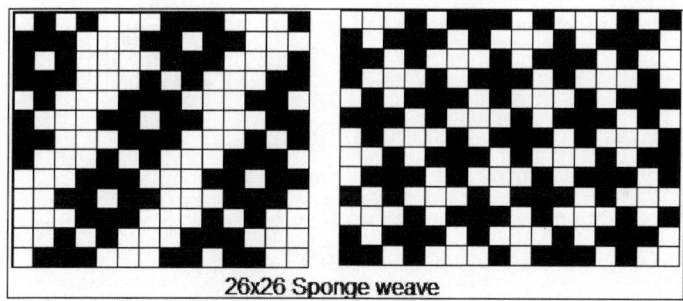

26x26 Sponge weave

Figure 12.58. Sponge weave

bathing wraps and dress fabrics. It is also used as ground of jacquard design (Fig. 12.58).

12.1.9.9   Colour and weave effects

Colour and weave effect are made by mixing colour threads in the warp or weft in the above weave designs. Geometrical designs with two colour warp and weft can be made using simple weaves like 1/1 palin, 2/2 matt or 2/2 twill. The particular design which results depends both on the weave and on the arrangement of the two colours in the warp and weft. These patterns are called colour-and-weave effects.

Colour and weave effects can be of two types – simple colour and weave effects and compound colour and weave effects.

In addition, there are two types of order of colouring – simple order of colouring and compound order of colouring each of which is further two kinds as – regular order of colouring and irregular order of colouring.

Simple order of colouring: In this colouring method, only one ratio of colour is used either for warp or weft, such as

(a) Regular order of colouring, e.g., 4 dark, 4 light, 3 dark, 3 medium, 3 light, etc., same for both warp and weft.

(b) Irregular order of colouring, e.g., 2 dark, 1 light; 3 dark, 2 medium, 1 light, etc., same for both warp and weft.

(c) By arranging the weft in a different order from the warp, for example, 2 and 2 warping crossed with 1 and 1 wefting

Compound order of colouring: In this case, more than one ratio of colour is used either for warp or weft, such as 2 dark–2 light and 4 dark–4 light; 6 dark–6 light and 3 dark–3 light, etc., same for both warp and weft.

In the combination of compound order of colouring, it also may be regular or irregular order like a simple order of colouring.

A convenient classification of the orders of colouring for the threads is as follows:

1. Simple warping and simple wefting

2. Compound warping and simple wefting

3. Simple warping and compound wefting

4. Compound warping and compound wefting

In the above combinations, the order of warping may be the same or different from the order of wefting. To each order of colouring, simple, stripe and check weaves may be applied. The style of pattern which is produced by the combination of each order of colouring with each type of weave is given in Table 12.1.

Table 12.1. Different types of weaves based on order of colouring

Order of colouring	Simple Weave	Stripe Weave	Check Weave
Simple warping and simple wefting	Simple pattern	Stripe pattern	Check pattern
Compond warping and simple wefting	Stripe pattern	Stripe pattern	Check pattern

Order of colouring	Simple Weave	Stripe Weave	Check Weave
Simple warping and compound wefting	Cross-over pattern	Check pattern	Check pattern
Compond warping and compound wefting	Check pattern	Check pattern	Check pattern

Construction principle of colour and weave effect:

Order of colouring and weave structure is fixed for a particular pattern. The stages in producing the pattern are illustrated below:

1. Mark out the repeat size of pattern according to the order of colouring and repeat of the weave.
2. Fill-up the repeat size by particular weave structure with crosses.
3. Indicate order of colouring by shade, the shades indicate the dark yarns.
4. For warp colouring, colour, i.e., shade is put only warp-up position of the particular warp yarn and for weft colouring, colour, i.e., shade is put only weft-up position of the particular weft yarn.
5. This final pattern is produced by combining the colour and weave structure.

Simple colour and weave effects:

In what follows it is assumed that dark and light yarns are used, although any sufficiently contrasting colours are possible. The following designs are the example of simple colour and weave effects.

End and end colouring pattern

The effect of arranging the warp and weft end and end dark and light (i.e., 1 dark:1 light) in a plain weave cloth is shown in the following figure; the shades indicate the dark yarns. The weave and colour arrangement produce the pattern, which consists of fine horizontal lines alternately dark and light.

Continuous line effect

The effect of arranging the warp and weft a 2:2 order of colouring in the twill 2/2 weave cloth is shown in Fig. 12.59; similarly the shade indicate the dark yarns. The weave and colour arrangement produce the pattern, which consists of coarse horizontal lines alternately dark and light.

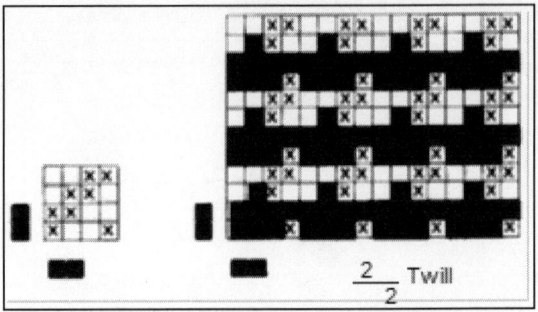

Figure 12.59. Continuous line effect

Hairlines or pinstripe

The effect of arranging the warp and weft a 2:2 order of colouring in the 2/2 Z matt-weave cloth is shown in Fig. 12.60; similarly the shades are indicated by the dark yarns. The weave and colour arrangement produce the pattern, which consists of thick or coarse horizontal lines alternately dark and light like as previous end and end colouring pattern.

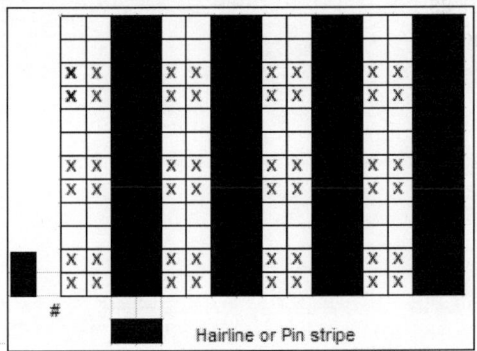

Hairline or Pin stripe

Figure 12.60. Hairlines or pinstriipe

Crows foot pattern

The effect of arranging the warp and weft 2:2 order of colouring in the 1 plain-weave cloth is shown in the following left figure; similarly the shades indicate the dark yarns. The weave and colour arrangement produce the pattern, which is the well-known crows foot design. A similar but larger crows foot pattern results from using a 4:4 colouring with a 2/2 Z matt-weave represent in the following middle figure. The close-up view of a fabric using this weave and colouring is shown in the following right side figure. Other, less

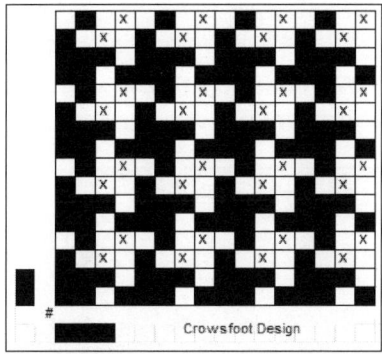

Figure 12.61. Crows foot pattern

useful, pattern result if the footing i.e. the starting point of the 2/2 matt-weave is altered, the order of colouring remaining the same (Fig. 12.61).

Dog's tooth or hound's tooth pattern

The most popular weave for colour-and-weave effects is 2/2 twill. With a 4:4 colouring, arranged as in Fig. 12.62, it gives a distinctive and decorative pattern known as dog's tooth when a relatively fine construction gives a small, and as hound's tooth when a coarser construction gives a larger pattern.

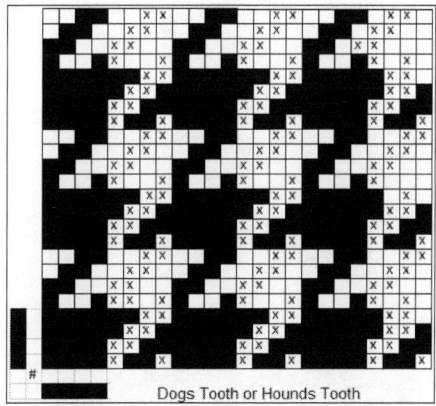

Figure 12.62. Dog's tooth or hound's tooth pattern

Altering the footing of the weave changes the character of the effect produced, but none of the alternatives are as effective or as useful as the one shown.

Shepherd's check pattern

A 6:6 order of colouring with a twill weave gives an effect similar to, but bolder than, dogs tooth. A woollen coating woven in this way from black and white yarns, known as Shepherd's check, is shown in Fig. 12.63.

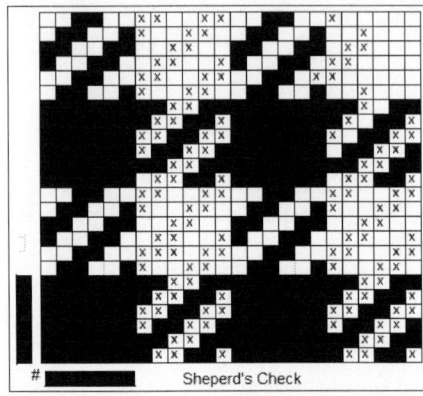

Figure 12.63. Shepherd's check pattern

Bird's eye effect

A useful type of colour-and-weave effect is known as birds eye, defined as a fabric having a pattern of very small and uniform spots, the result of a combination of weave and colour. The development of the pattern and of another pattern of the same type, but having larger spots, is given in Fig. 12.64. Both these patterns use simple fancy weaves. Other fancy weaves used with suitable orders of colouring provide a considerable range of patterns, some of which are distinctive enough to be useful.

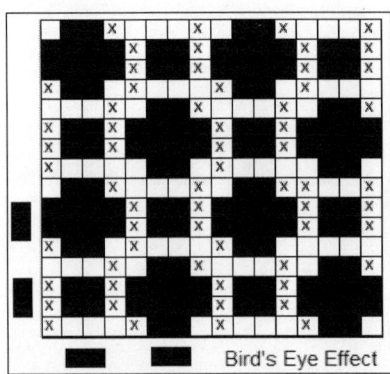

Figure 12.64. Bird's eye effect

Compound colour and weave effects

In this case, different types of check and stripe patterns are produced. Fig. 12.65 shows a stripe pattern, which is produced by simple weave and simple wefting with compound warping. As a weave 2/2 Z matt is used, for warp colouring 1:1 and 2:2 order of colouring and for weft colouring simple 1:1 colour arrangement are used here. The repeat size of this pattern is 32 × 16.

There are many other designs and special fabrics are made by different methods like figuring with extra threads, compound fabrics like tubular cloths, double width cloth, multiply fabrics, stitched double cloths, wadded double cloths, leno or gauze fabrics, quilting, lappet and swivel fabrics, woven pile fabrics, corduroy, velveteen, double cloth, terry, flocked fabrics, carpets, etc., and are not discussed here.

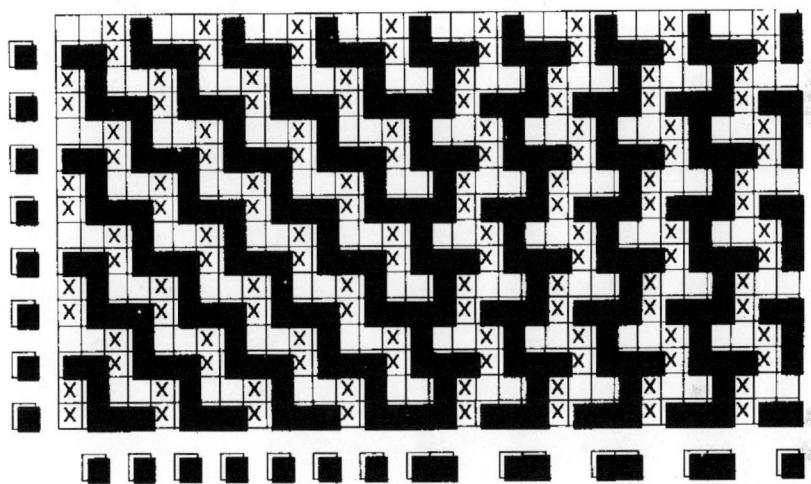

Figure 12.65. Compound colour and weave effects

12.2   Knitting

12.2.1   General

Knitting is fabric or garment making by forming a series of interlocking loops in a continuous yarn or a set of yarns. In production situations, the work is carried out through the movement of hooked needles. (Hand knitting is normally performed with straight needles.) Each row of loops is vertically interlocked with the preceding row. With a sufficient number of loops, the yarn becomes a fabric. Knitted fabrics have the advantage of stretchability, a property not possessed by woven fabrics. Stretching can be in any direction even if the yarn used has little elasticity. Fig. 12.66 illustrates two types of knitted fabric. Mechanized production knitting utilizes a series of needles commonly operated by cams.

There are two basic types of knitting, weft or filler knitting and warp knitting. Weft knitting is somewhat more common. In weft knitting, the courses (crosswise rows of loops) are composed of continuous yarns. Weft knitting can be done by hand or machine but production weft knitting is a machine operation. The individual yarn is fed to one or more needles at a time. In warp knitting, the wales (predominantly vertical columns of loops) are continuous. Separate yarns are fed to each needle. The warp knitting operation is always produced by machine knitted fabrics can be either flat or tubular in form. Warp knits are usually flat; weft or filling knits are most often tubular.

Two types of hooked needles are used in production knitting machines (as shown in Fig. 12.67): (a) the bearded or spring needle and (b) the latch needle. With both designs, the needles draw new loops through the previous loops that they have retained. Once the needle head and new loop have gone through the old loop, the old loop is cast off. The latch needle is most often used. It operates more automatically than the bearded needle which requires other machine elements to present the loop and close the hook.

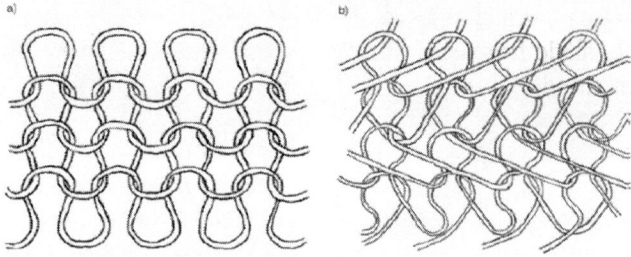

Figure 12.66. Diagrammatic representation of two examples of knit fabrics made by interlocking continuous strands of yarn. (a) A plain knit made on a weft or filler knitting machine. The path of each crosswise yarn is called a course. (b) A single-warp tricot knit.

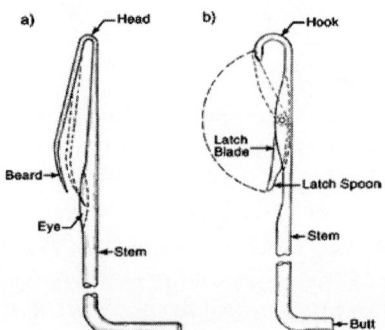

Figure 12.67. Types of hooked needles: (a) bearded or spring needle and (b) latch needle

In weft knitting, one continuous yarn runs crosswise in the fabric and makes up all the loops in one course. Either the needles act in succession or the yarn is fed in succession, so that loop formation and interlocking are not simultaneous. As each needle rises, the needle hook loops over the yarn, which it hooks on the down stroke, and the yarn is held in place by the needle latch. At the bottom of the needle stroke, a previous loop slips off the needle, and the new loop is held in place with the latch. On the next cycle, the loop is released from the latch as the needle rises, another loop is formed and the process is repeated.

Several different stitches can be formed in weft knitting. In the *knit* stitch, the loop is drawn from the back and passed through the front of the preceding loop to the front of the cloth. In the *purl* stitch, the loop is drawn from the front through the back of the preceding loop to the back of the cloth. In the *miss* stitch, no loop is formed. In the *tuck* stitch, two courses on one wale are looped over a third. The stitches, and various combinations of them, make all the patterns of knit and double knit cloth. Distinct patterns can be made from combinations of the knit and purl stitches since the knit tends to advance and the purl to recede. Double knits are made by machine only, using two yarns and two sets of needles. These knits use a variation of the rib and interlock stitches, drawing loops from both directions. Jersey is a common knitted cloth, made from only knit or only purl stitches. Circular weft knitting machines are used to make hosiery, underwear and simulated furs. They can knit shaped garments. Jacquard effects are possible and are now generally controlled electronically. Flat knitting machines can also produce shapes by increasing or decreasing loops. Full-fashioned garments can be made on flat knitting machines.

It is usually accomplished on flat machines but can also be tubular. Warp knitting differs from weft knitting in that each needle has its own yarn. The yarns are fed from a large reel or warp beam as in weaving with a loom. The yarns, then, generally run lengthwise in the fabric. The needles all move together and form parallel rows of loops simultaneously. The loops are interlocked on a zigzag or vertical path. The yarn section is held on one end by the previous loop and at the other end by the yarn guide. The yarn is trapped within the hook of the needle as it descends. With latch needles, the hook is closed as the needle descends. This allows the previous loop to slip off the hook while a new loop is held. If bearded needles are used, a yarn guide, called a *sinker*, positions the yarn across ascending needles and then retracts as the needles descend.

Warp knitting is a versatile process, but standard warp knitting machines make just three basic stitch variations: open loop, closed loop and no loop. Various fabric patterns are created from different combinations of these stitches. One simple pattern produces tricot knit, which consists of a zig-zag pattern of closed loops of parallel wales. Tricot fabrics are run resistant. Other warp-knit patterns are simplex, milanese and raschel. Milanese knitting produces run resistant fabrics with a diagonal rib pattern. Several sets of yarn are used. The Raschel knit is made with latched needles rather than the spring beard needles used for other knits. One or two sets of latch needles are used. Raschel knit fabrics are used frequently for underwear. Warp knitting is used to produce fabric for dresses, lingerie, upholstery and draperies among other products.

12.2.1.1   Different types of knitting machines

There are different types knitting machines which can be classified based on various factors, such as drive, needle, flat or circular, warp or weft, material produced, etc. (Fig. 12.68).

The main classification is individually driven needles and needle bar machines. The former type of machine have needles which are moved individually by cams acting on the needle butt; they are used for producing weft knits and are subdivided into circular knitting machines and flat-bed knitting machines. The needles used can be latch needles or compound needles.

In needle bar machines, the needles move simultaneously, as they are all fixed to the same bar; we distinguish full-fashioned knitting machines and circular loop-wheel machines for the production of weft knit fabrics, which only use spring-beard needles, and warp knitting machines which use spring-beard needles, latch needles and compound needles.

We are mainly dealing with warp and weft knitting machines. The weft knitting machinery may broadly be classified as either straight bar frames, flats or circulars, according to their frame design and needle-bed arrangement.

There are many types of weft knitting machines (Fig. 12.69).

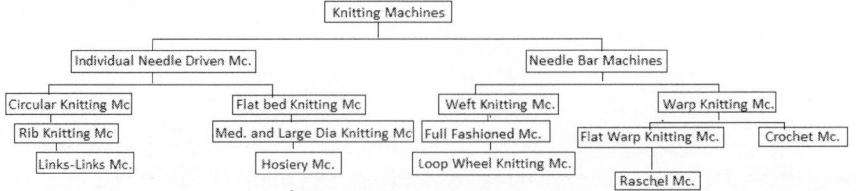

Figure 12.68. Classification of knitting machine

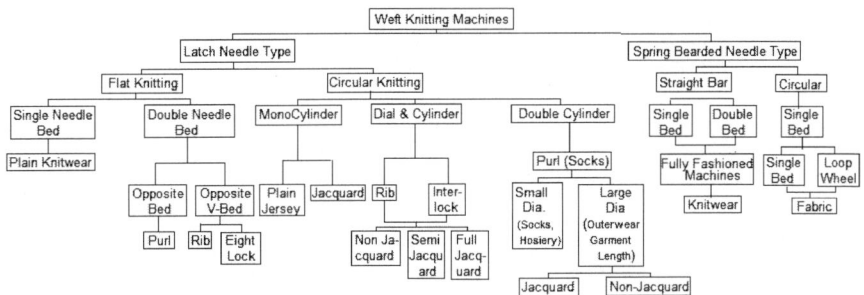

Figure 12.69. Classification of weft knitting machines

12.2.1.2   Needle

In a knitting machine, needle is one of the most important part which forms the loop. There are three most commonly used types of needle: (1) latch needle, (2) spring-beard needle and (3) compound needle (Fig. 12.70).

From a manufacturing point of view, we can have two different types of needle: (1) wire needle and (2) die-cut needle.

Wire needles are made from a steel wire shaped through various machining steps to create a flat profile and form a hook, the section accommodating

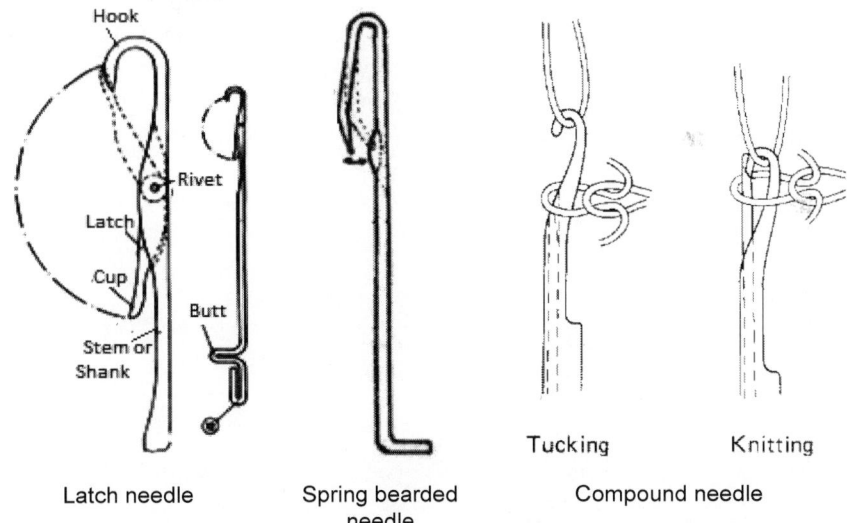

Figure 12.70. Types of needle

the latch and an end butt with tail; sometimes the butt is not obtained with a bending process but by a pressing one (Fig. 12.71).

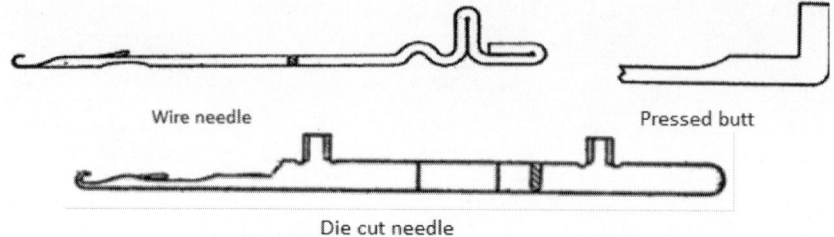

Figure 12.71. Wire and die-cut needles

Die-cut needles are made from a steel plate of the desired thickness, which is die cut so as to create the shape of a butt with or without a tail; the hook and latch fitting are created with a special process.

An important feature of all needles is the gauge; it is directly connected to the strength of the needle which must bear the stress and strain generated during the various technical cycles of the knitting process. The gauge of the needle is directly proportional to the gauge of the machine; the needle must be neither too thick (if so there would not be enough space between a needle and the next one for looping the yarn) nor too fine since in this case the needle, besides being too weak, could compromise the resistance of the binding pattern which would result poorly balanced. Table 12.2 shows the gauge of needles according to their thickness. The values are only a guideline, and it may change as per manufacturers' specifications.

Table 12.2. Various types of gauge of needles based on their thickness

Gauge of the needle	Thickness in mm	Thickness in inches
2.5	2.286	
4	1.778–2.283	0.0700–0.0899
8	1.625–1.750	0.0640–0.0699
10	1.447–1.600	0.0570–0.0639
12	1.168–1.420	0.0460–0.0569
18	0.939–1.170	0.0370–0.0459
24	0.838–0.910	0.0330–0.0369
30	0.762–0.840	0.0300–0.0329
36	0.685–0.740	0.0270–0.0299
42	0.635–0.660	0.0250–0.0269

Gauge of the needle	Thickness in mm	Thickness in inches
48	0.584–0.610	0.0230–0.0249
50	0.533–0.560	0.0210–0.0229
54	0.482–0.500	0.0190–0.0209
60	0.431–0.450	0.0170–0.0189
70	0.343–0.400	0.0136–0.0169
75	0.337–0.340	0.0125–0.0135
80	0.266–0.300	0.0105–0.0124
85	0.245–0.250	0.0100–0.0104

12.2.1.3   Cylinders

Needles are arranged on the cylinder as shown in Fig. 12.72(a). It can go up and down in the position as required during which the loops are formed by the yarns which is guided by the needles. The arrangement of the needles around this cylinder is called a needle bed.

Knitting machine can be equipped with one or two needle beds, according to the model. A needle bed can be flat or circular (see Fig. 12.72b). The needle is free to move inside the grooves in the cylinder. The milled grooves guide the needles during the knitting process. A knitting machine is characterized by the number of needles per inch in the circumference of the cylinder or dial of a knitting machine. The distance between one needle and another is called pitch. It is proportional to the needle gauge or thickness. The space available, which determines the maximum thickness of the yarn (i.e., the yarn

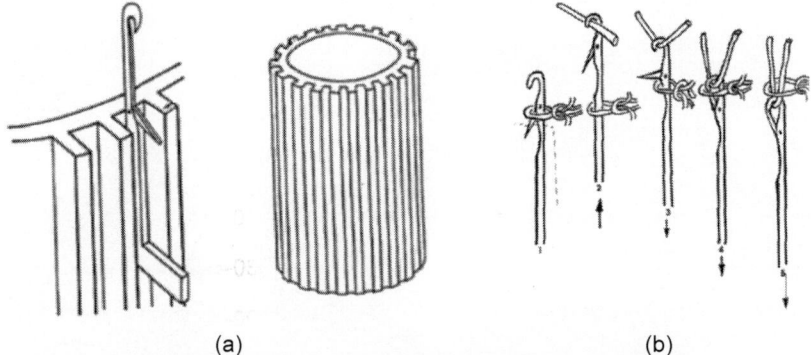

(a)   (b)

Figure 12.72. (a) Arrangement of needles on a cylinder.
(b) Movement of needles for loop formation.

(a)   (b)

Figure 12.73. (a) A flat needle bed. (b) A circular needle bed.

count) that may be knitted, is the gap between the side of the needle and the trick wall as the needle descends to draw a new loop.

The operating width is the maximum working area and varies according to the type of machine: for example, in a flat-bed machine, the operating width is the distance between the first and the last needle, while in circular knitting machines, the operating width is the needle-bed diameter (Fig. 12.74).

As explained earlier, the gauge is the number of needles on a certain length of bed. The English gauge is the number of needles per English inch, that is, to say how 2.54 needle-bed centimetres machine gauge can be calculated by dividing the total number of needles into the length of the needle bed. The figure is rounded to the nearest whole number.

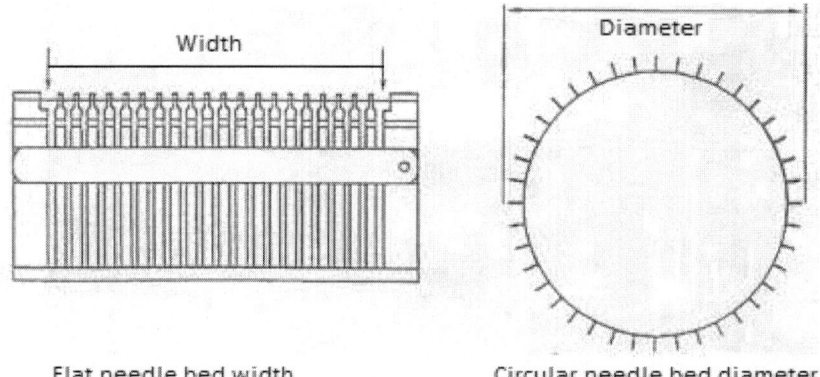

Flat needle bed width   Circular needle bed diameter

Figure 12.74. Types of operating width

For example, a 30-inch diameter circular knitting machine has 1716 needles. The circumference of a circle is πd, where π = 22/7 and d is the diameter. The circumference is therefore 30 × 22/7 = 94.28 inches. The gauge is 1716/94.28 = approximately 18 needles per inch. This may be expressed as 'E 18', E being the number of needles per inch.

Knitting machine can be classified as per diameter or as per number of needle beds.

Classification by diameter may be:

1. Large-diameter circular knitting machines (24–40 inches)

2. Medium-diameter circular knitting machines (8–22 inches)

3. Small-diameter circular knitting machines (3–6 inches)

Classification by needle beds may be:

1. Single-bed circular knitting machines (for jersey and derived stitches: fleece, terry, piquet and floating jacquard)

2. Double-bed circular knitting machines

    (a) Dial-cylinder knitting machines with 90° needle beds (for rib knit and similar: cardigan stitches, milanese rib, bourrelet, rodier, cable, eyelet and jacquard stitches, as well as all interlock and interlock-derived stitches, for example, the pin tuck stitch).

    (b) Double-cylinder knitting machines with 180° needle bed (for stitches based on the links construction)

12.2.1.4   Structure of a circular knitting machine

The main parts of a knitting machine based on (large diameter machine) are machine base, machine core, yarn spool holder, yarn feeding system, fabric take down and winding system and the drive (Fig. 12.75).

Knitting machine has a solid base and an architecture that facilitates access to the machine components for routine operations to be carried out during setting up procedures and production. The 'core' of the machine holds the needle-bed area and all the systems operating during the knitting process. Circular knitting machines can be divided into two basic models: in the first one, the needle beds rotate and the cam frame stands still, while in the other one, the needle beds stand still and the cam frame revolves. Yarn spool holder can be on the machine or separate as shown in Fig. 12.76.

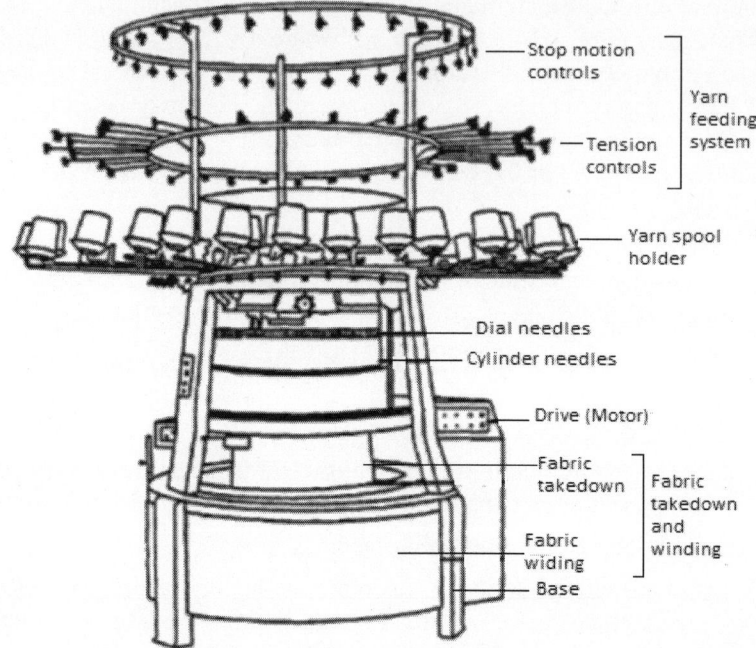

Figure 12.75. Structure of a circular knitting machine

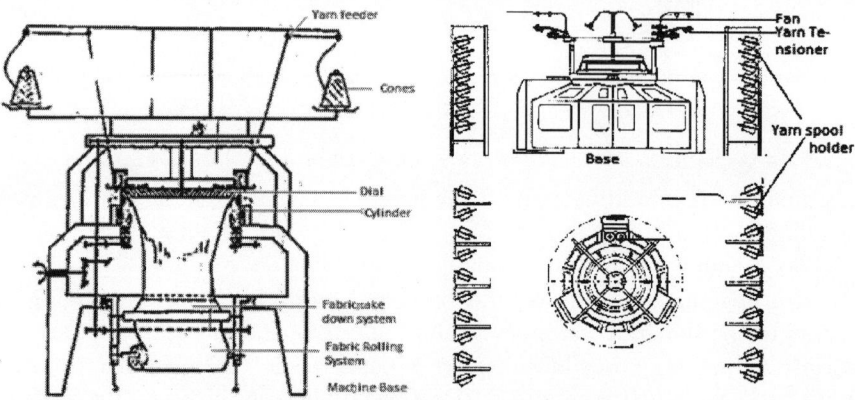

Figure 12.76. Arrangements of yarn spool holders in a
modern circular knitting machine

12.2.1.5   Yarn feeders

There are two types of yarn feeders – 'positive' and 'negative'. A yarn feeder is the negative type when the needle takes the yarn directly from the package during the stitch formation step, and the feeding tension of the yarn cannot be

(a)   (b)

Figure 12.77. (a) Yarn feeding accumulator. (b) Positive belt feeding system.

controlled. This feeding technique can generate differences in the yarn length used for stitch formation. This is due to the variable tension of the yarn since a new spool has a certain diameter which gradually reduces as more yarn is unwound and fed into the machine. In addition, the spool can be too hard or too soft. This problem is solved either by unwinding some yarn from the cone and accumulating in a system called yarn accumulator before the yarn is fed to the needles or by a positive feed systems control the tensions of the yarn fed by means of a drive wheel or a drive belt system and feeds the yarn to the needle with more or less same tension (Fig. 12.77a,b).

12.2.1.6   Thread guides

On circular knitting machines each thread guide corresponds to a feed system. The thread guide is a steel or ceramic plate with a hole for the thread.

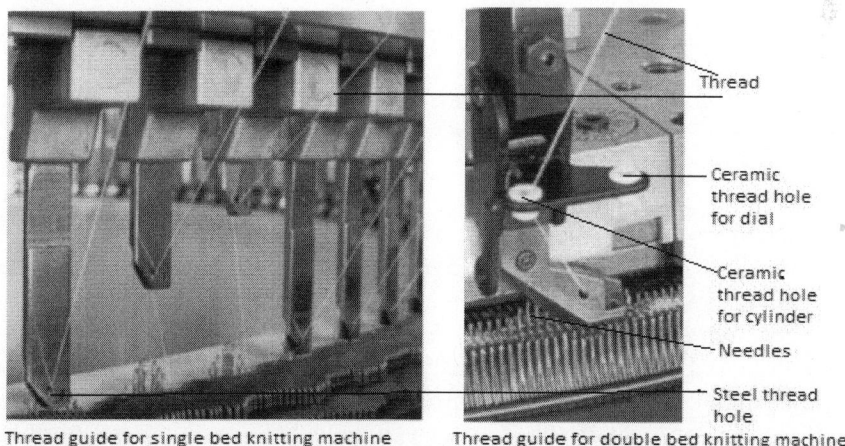

Thread guide for single bed knitting machine   Thread guide for double bed knitting machine

Figure 12.78. Thread guide needles on single and double bed knitting machine

The thread guide is positioned near the hook of the needle and, besides feeding the yarn, it opens and protects the latches. In a single bed machine, each thread guider guides one thread and in a double bed machine one guider will have two holes in which one feeds the thread to needles on the cylinder bed and the other feeds the thread to the dial needles (Fig. 12.78).

Some machines have more thread guides for the same feed system.

12.2.1.7   How the stitch is formed in a knitting machine

The author explains the stitch formation in a single bed circular knitting machine and with latch needles, and in other machines, the stitches are formed in a similar way but may be little more complicated with compound needle.

As explained earlier in a single-bed circular knitting machines, there will be only one series of needles that slide up and down in the grooves which are driven by an important machine part called the cams and causes the stitch formation. They are placed outside the needle bed; each feed system is provided

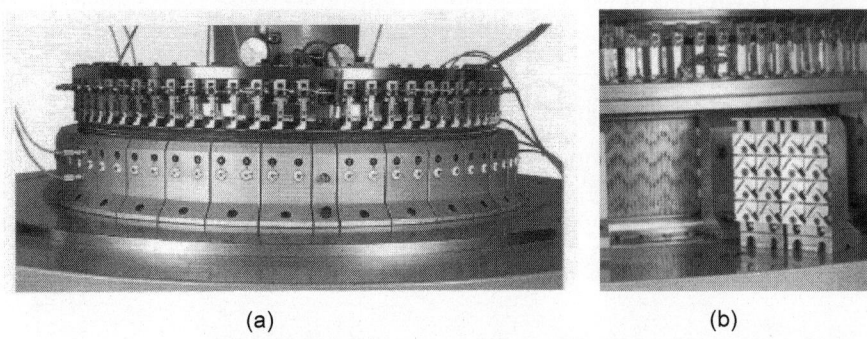

(a)   (b)

Figure 12.79. (a) A knitting head. (b) Cams.

with its own cam group. All the cams are fixed to a bearing structure called 'cam frame'. On single-bed machines, the cam frame is stationary, while the needle bed revolves. Outside the cams, on each feed system, there are special micrometric screws, which adjust the stroke of the lowering cams and determine accurately the length of the yarn fed (Fig. 12.79a,b).

The cams make a track for the needle butts and when the needle butt move in the track the needle end goes up or down as per the design and the stitch requirements.

Figure 12.80 shows the movement of the needle required (left) for the stitch formation and how the cams arrangements achieve this (right).

Knitting cams are attached either individually or in unit form to a cam plate and, depending upon the machine design, are fixed exchangeable or adjustable. At each of at least a raising cam, a stitch cam and an up throw cam

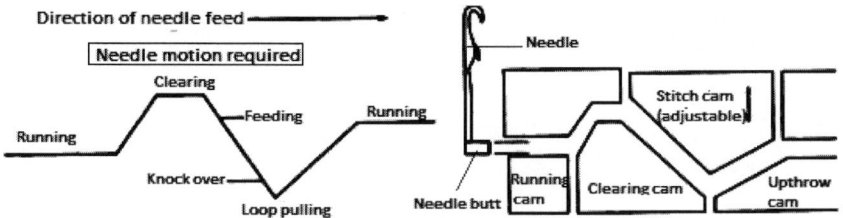

Figure 12.80. Movement of needle required (left) for stitch formation and cams arrangement (right)

whose combined effect is required. Usually four main types of knitting cams are used.

1. Raising cam: The raising cam causes the needle to be lifted to either tuck, clearing loop transfer or needle transfer depending upon machine design.

2. Stitch cam: The stitch cam controls the depth to which the needle descends thus controlling the amount of yarn drawn in to the needle loop. It is also known as knock-over cam.

3. Up throw or counter cam: The up throw or counter cam takes the needles back to the rest position and allows the formed loops to relax.

4. Guard cam: The guard cam is often placed on the butts and to prevent needles from falling out of track.

Three important functions of cam are (a) to produce motion to needles, (b) to drive the needles and (c) formation of loops.

There are two types of cams: (1) engineering cam which is circular type and (2) knitting cam which have angular motion. The engineering cams are four types: single active, cam and counter cam, box cam, counter cam/pot cam. Knitting cams are also four types: stitch cam, raising cam, up throw cam and guard cam.

Following is the movement of a needle in a stitch formation. Once a loop formation is completed, the needle is in rest position. It is prevented from rising as the needle rises, by holding-down sinkers or web holders that move forward between the needles to hold down the sinker loops. The hook will be in closed position (Fig. 12.81a). Next movement is latch opening. From the rest position, the butt of the needle moves to the clearing cam track whereby the needle rises. At this position, the previous loop opens the latch as the old loop is held down by the sinkers (Fig. 12.81b). The sinkers hold the fabric already formed while the needles rise for the next stitch formation cycle. The sinkers also support the fabric when the previous course is knocked over. Sinkers are driven by special cams whose shape depends on the type of the sinker itself.

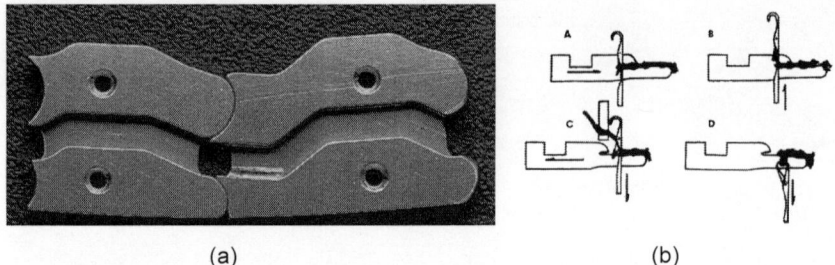

(a)   (b)

Figure 12.81. (a) Sinker. (b) Position of needles and web holding sinkers.

Apart from holding the fabric down, it has other functions like loop formation and knocking over. On bearded needle, weft knitting machines of the straight bar frame and sinker wheel type (as on Lee's hand frame), the main purpose of a sinker is to sink or kink the newly laid yarn into a loop (see Fig. 12.81b) as its forward edge or catch (C) advances between the two adjacent needles.

The needle moves further up to the top of the cam when the old loop comes completely out of the hook and moves to the latch spoon on to the stem. At this point, the feeder guide plate acts as a guard to prevent the latch from closing the empty hook (c). Further the needle butt enters the stitch cam and needle hook starts to descend, the old loop stays in position which lifts the latch at the same time the hook catches the new thread which is fed through a hole in the feeder guide. As the needle goes further down, the old loop completely closes the latch with the new inside. Old loop slips over

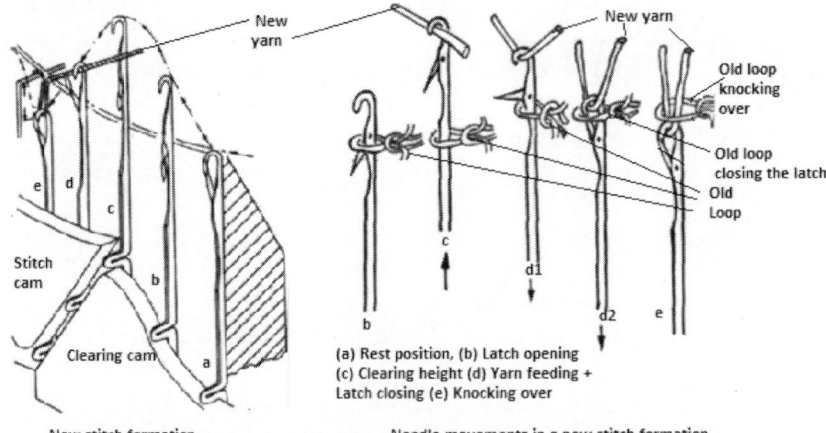

Figure 12.82. Stitch formation and needle movements

the needle when the needle comes to the lowest position and the new thread forming an new loop (d). Fig. 12.82 gives the details of the stitch formation, needle movement and cam positions. Lastly, the needle moves the bottom most position when the old loops slips out from the needle over the hook, while the hook is with the new yarn ready for the next loop formation.

In a simple machine, the cams are screwed to the cam frame and command a single movement of the needle and the selection possibilities will be very restricted. In fact, in this feed system, the needles must knit or remain idle and can probably knit only one design (this is the typical situation of jersey knitting machines). To modify the pattern, it is necessary to change the cam. These technical limits have been overcome by increasing the number of needle butts and the corresponding cam tracks necessary to drive the needle. Now machine manufacturers are able to offer modern single-bed machines with up to five selection tracks.

When we move from single bed to double bed machines, there are two sets of needles, and the cams for each beds/needles, the actions are little more complicated. One set of needles are fitted in the grooves of a cylinder bed while the other set of needles are positioned at 90° with respect to the cylinder (or cylinder needles) on a special circular plate called dial. Each bed has its own cams and is fastened to two cam frames, one around the cylinder and the other above the dial. In these machines other than extra cam tracks, there are two design possibilities: (a) revolving cam frames and stationary needle beds (for continuous fabric manufacturing machines) and (b) stationary cam frames and revolving needle beds (for cloth manufacturing machines). Today manufacturers offer rib knitting machines, i.e., machines with the needles of

 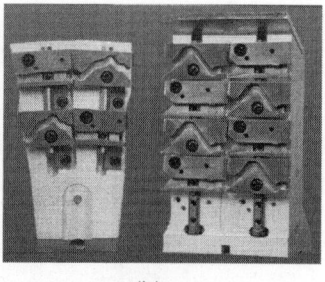

(a)   (b)

Figure 12.83. (a) Knitting head of a double bed machine. (b) Cams for double-bed machine (two tracks for the dial and four tracks for the cylinder)

the dial staggered with respect to the needles of the cylinder, featuring up to four tracks on the cylinder and two tracks on the dial (Fig. 12.83).

Interlock is another design which can be done on double-bed machines. The long and short needles of an interlock knitting machine are placed opposite to one another on dial and cylinder; the needles work alternatively in two consecutive feed systems. Another important feature of the double bed-machines is, as the fabric formed by the rising needles of one needle bed is held by the needles of the opposite needle bed during the stitch formation cycle, that there is no need of sinkers.

In an effort to increase the production manufacturers increased the number of feed systems either by extending the range of diameters available or by reducing the size of feed systems. Nowadays 60-inch diameter and 3–4 feed systems per inch for single bed machines and 2–2.4 feed systems per inch for rib knitting machines are quite common. Many other methods like adjusting the cams with an external screw, curvilinear cams sliding inside closed tracks for higher speeds, etc., are also introduced for increasing production.

With the introduction of modern machines, there are many possibilities to increase the production and design possibilities even up to jacquard designs based on these basic principles. In circular knitting machines, as explained earlier, it can be understood that each needle is fed with a separate yarn and the knitting takes place as spiral band of the thickness of the number of loops made by the total needles used. As the machine makes one revolution, each system knits one course. These courses are superimposed to form an endless band that winds around spirally to make the tube of cloth, as shown in Fig. 12.84.

Imagine that a tube of cloth is knitted on a multifeed machine and one of the systems is threaded with black yarn. The bottom of Fig. 12.84 represents

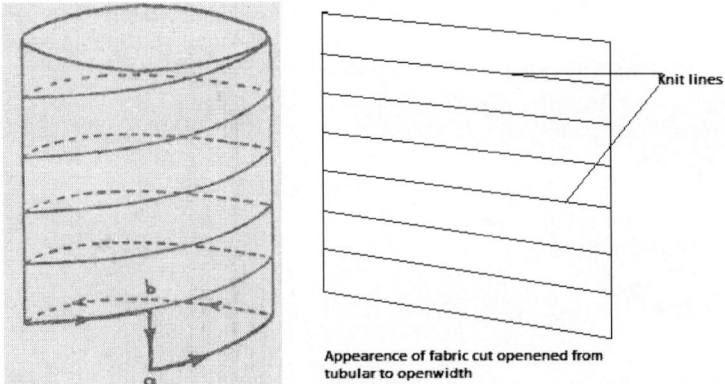

Appearence of fabric cut openened from
tubular to openwidth

Figure 12.84. Schematic representation of knitted band

a cutting line that follows the circular course a–b and then the vertical wale b–a, the black yarn thus making it possible to visualize the knitted band.

This is an inherent problem in circular knitting and if we cut open the grains of the fabric would not be horizontal, it will be in an angle, this is

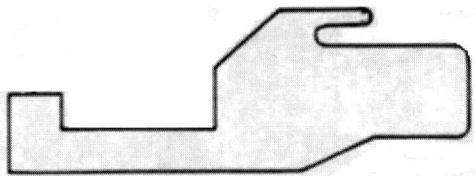

Figure 12.85. Spirality of the fabric

called the spirality of the fabric which can give problem in garments. This is a problem a processor has to attend, which will be dealt separately in processing section (Fig. 12.85).

12.2.2   Basic knitting terms

Course: The row of loops or stitches running across the width of a fabric corresponding to the weft (filling) of a woven fabric.

In weft-knitted fabrics, a course is composed of yarn from a single supply termed a course length. A pattern row is a horizontal row of cleared loops

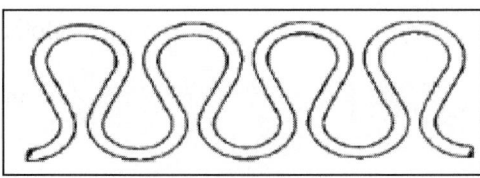

Figure 12.86. Course

produced by one bed of adjacent needles. In a plain weft-knitted fabric, this is identical to a course but in more complex fabrics, a pattern row may be

Figure 12.87. Wale

composed of two or more course lengths. In warp knitting, each loop in a course is normally composed of a separate yarn (Fig. 12.86).

Wale: In knit fabric, a column of loops running lengthwise of the fabric (Fig. 12.87).

Course count: The number of courses in a knit fabric per unit length measure. For example, courses per inch.

Wales count: The number of wales in a knit fabric per unit length measure. For example, wales per inch.

Kink of yarn: A length of yarn that has been bent into a shape appropriate for its transformation into a weft-knitted loop (Fig. 12.88).

Knitted loop: A kink of yarn that is intermeshed at its base, i.e., when intermeshed two kink of yarn is called loop (Fig. 12.88).

Knitted stitch: Stitch is a kink of yarn that is intermeshed at its base and at its top. The knitted stitch is the basic unit of intermeshing and usually consists of three or more intermeshed loops, the centre loop having been drawn through the head of the lower loop which had in turn been intermeshed through its head by the loop which appears above it (Fig. 12.88).

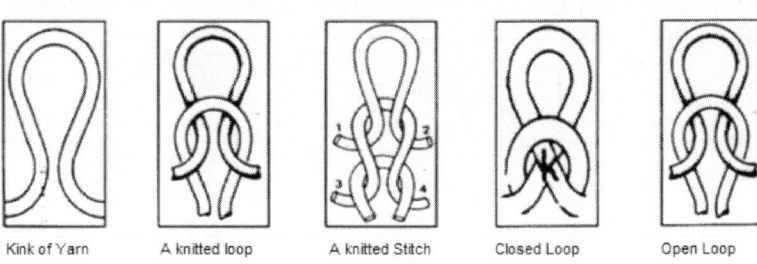

Figure 12.88. Various types of loop formation

Top arc: The upper curved portion of the knitted loop is called top arc (Fig. 12.89).

Bottom half arc: The lower curved portion that constitutes in a weft-knitted loop, half of the connection to the adjacent loop in the same course (Fig. 12.89).

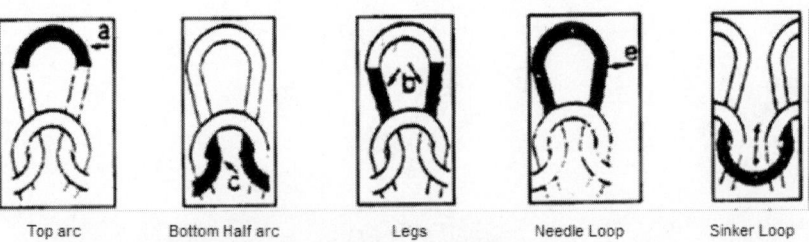

Figure 12.89. Various types of knits

Needle loop: The needle loop is the simplest unit of knitted structure. Needle loop formed by the top arc and the two legs of the weft-knitted loop (Fig. 12.89).

Legs or side limbs: The lateral parts of the knitted loop that connects the top arc to the bottom half arcs.

Face loop: A knitted loop formed on the cylinder needles on a knitted machine (Fig. 12.90)

Back loop: A knitted loop formed on the dial needles on a knitted machine (Fig. 12.90).

Sinker loop: The yarn portion that connects two adjacent needle loops belonging in the same knitted course. Bottom arc is also called a sinker loop (Fig. 12.89).

Open loop: A knitted loop of which a thread enters and leaves at the opposite sides without crossing over itself (Fig. 12.88).

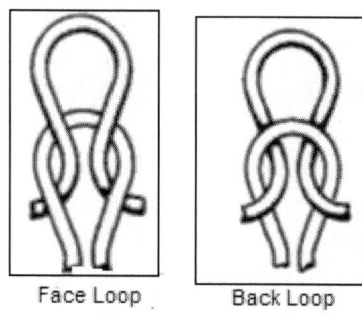

Face Loop   Back Loop

Figure 12.90. Face and back loops

Closed loop: A knitted loop of which a thread enters and leaves at the opposite sides with crossing over itself. It is made by a special needle (Fig. 12.88).

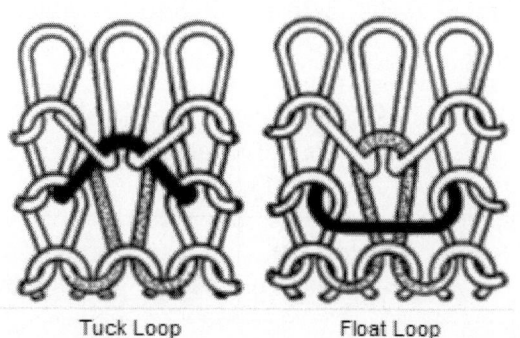

Tuck Loop   Float Loop

Figure 12.91. Tuck and float loops

Tuck loop: A knitted stitch when a needle receives a new yarn without loosing its old loop. A tuck loop makes the fabric wider, thicker and also makes the fabric slightly less extensible, whereas float loop makes the fabric narrower, thinner and much less extensible (Fig. 12.91).

Float loop: A knitted stitch when a needle holds its old loop and does not receive a new yarn. It connects two loops on the same course but not in adjacent wales. Also called miss loop (Fig. 12.91).

Knitted loop structure: The properties of a knitted structure are largely determined by the interdependence of each stitch with its neighbours on either side and above and below it. Knitted loops are arranged in rows and columns roughly equivalent to the weft and warp of woven structures termed 'courses' and 'wales', respectively.

Tuck stitch: Tuck stitch is made when a needle takes a new loop without clearing the previously formed loop so that the loops are accumulated on the needles. The number of consecutive tucks on any one needle is limited by the amount of yarn that the needle can hold, the maximum being four to five loops in the usual case. Tuck stitches tend to reduce the length of the fabric and increase in width resulting in the fabric being thicker with less extension in width (Fig. 12.92).

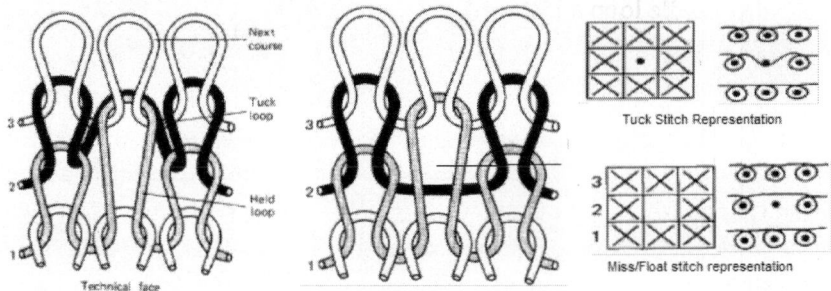

Figure 12.92. Representation of tuck and float stitches

Miss/float stitch: A miss stitch is produced by a needle holding the old loop (the needle is not raised, which effectively means the needle is missed), while the two adjacent needles are raised and cleared to produce a new knitted loop. The float will lie freely on the reverse side of the held loop and in the case of rib and interlock it will be inside the fabric. Miss stitches reduce the width of the fabric, since the wales are pulled closer together and the held loop robs yarn from adjacent loops. This tends to improve the fabric stability (Fig. 12.92).

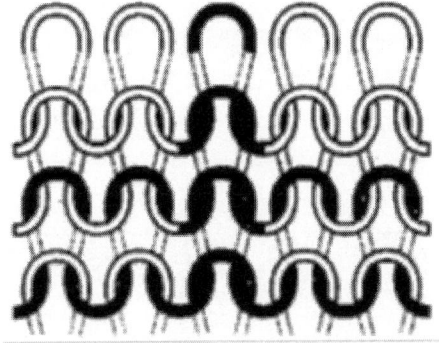

Figure 12.93. Knitted structure

Stitch: In knitting, a stitch is the loop geometry of a particular pattern repeat. It may be in the form of a knitted, a tuck or a float loop (Fig. 12.93).

Loop or stitch length: The length of yarn knitted into one stitch in a weft-knitted fabric. Stitch length is theoretically a single length of yarn which includes one needle loop and half the length of yarn (half a sinker loop) between that needle loop and the adjacent needle loops on either side of it. Generally, the larger the stitch length, the more elastic and lighter the fabric, and the poorer its cover opacity and bursting strength (Fig. 12.94).

Stitch length, l = One needle loop + Two half a sinker loop

Face Loop: Also called plain stitch or jersey stitch or flat stitch. A stitch that is so intermeshed in the fabric that its legs are situated above the top arc of the stitch formed in the same wale in the previous course. This side of the stitch shows the new loop coming through the viewer as it passes over and covers the head of the old loop. Face loop stitches tend to show the side limb

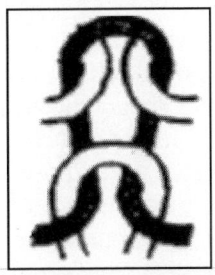

Loop/stitch Length

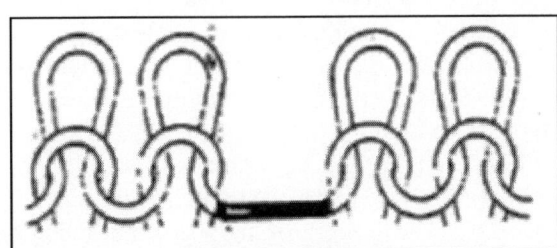

Extended Sinker Loop

Figure 12.94. Loop length and extended sinker loop

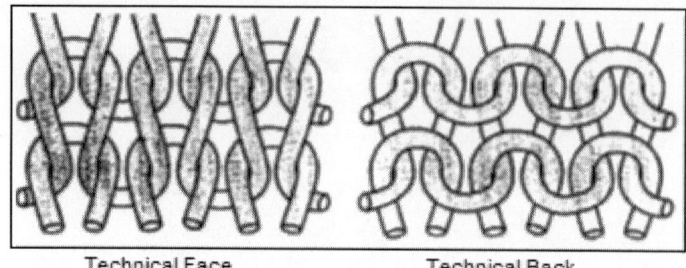

Technical Face   Technical Back

Figure 12.95. Technical left and back sides

or legs of the needle loops or overlaps as a series of interfitting 'V's. The notation of the face loop is ⊙ and by graph paper is ✗.

Technical face or right side: The under surface of the fabric on the needles will thus only show the face stitches in the form of the side limbs or legs of the loops or overlaps as a series of interfitting 'V's.

Reverse or back loop (stitch): Also called purl stitch. A stitch that is so intermeshed in the fabric that the top arc and the bottom arc are situated above the legs of the stitch formed in the same wale in the previous and in the following course. This is the opposite side of the stitch to the face loop side and shows the new loop meshing away from the viewer as it passes under the head of the old loop. Reverse stitches show the sinker loops in weft knitting and the under laps in warp knitting most prominently on the surface. The reverse loop side is the nearest to the head of the needle because the needle draws the new loops downwards through the old loops. The notation of back loop is ⍵ and by graph paper is ○.

Technical back or left side: The upper surface of the fabric on the needles will only show reverse stitches in the form of sinker loops or under laps and the heads of the loops (Fig. 12.95).

Double thread stitch: Also called double loop stitch or spliced stitch. A stitch formed from two ends of yarn.

Yield: The amount of the fabric delivered off a knitting machine in terms of its weight per unit length or area, or the number of linear units delivered per unit weight. For example, gram per metre, grams per square metre, and metre per kilogram.

Course length: The amount of yarn used in forming all the knit loops in one course of knitted fabric. Also called run-in.

Gaiting: The spacing of the needles in the cylinder or dial in relation to each other on rib and interlock machines. In rib knitting, the needles of the cylinder are between the needles of the dial. In interlock gaiting, the needle of the cylinder is directly opposed to the needles of the dial (opposed to each other).

Timing: The order the needles in the dial and cylinder go through the knitting cycle in relationship to one another. The cylinder needles that correspond to dial needles may go through the knitting cycle before or after the dial needles.

Dial height: The distance between the bottom edge of the dial section on a knitting machine at its perimeter from the corresponding upper edge of cylinder at its perimeter.

Tricot: A type of warp knitting in which spring bearded needles are normally used to make fine fabrics with usually one to three warps are used.

Raschel: A type of warp knitting in which plain and jacquard fabrics can be made. Raschel fabrics are normally coarser than other type of warp knits, but a wide range of fabrics can be made. Raschel machines may have one or two sets of needles and up to 30 guide bars.

Guide bars: A mechanism on a warp knitting machine which directs warp yarns to the knitting needles, and their movement is controlled so that patterns can be knit.

Needle bar: A flat metal plate with slots (tricks) cut into it at regular intervals into which needles slide during the knitting process.

Running length: In warp knitting, the number of inches of yarn needed to knit one rack of fabric.

Rack: A warp knitting measure of 480 m. Tricot fabric quality is judged by the number of inches per rack.

Inch quality: A measure of quality of warp knit fabric, the number of inches of fabric per rack.

Full set: A term that indicates that all guide eyes in a guide bar each have a yarn from the warp.

Part set: A term that indicates that all guide eyes in a guide bar do not have yarn from the warp.

Positive feed: When the yarn is metered off the warp beam by a metering device.

Negative feed: When the yarn is pulled off the warp beam by the knitting action of the needles during loop forming step.

Pattern wheel: A cylinder or wheel upon which a pattern chain is placed which has links of different heights so as to move the guide bars throughout its pattern.

Stitch density: The term stitch density is frequently used in knitting instead of a linear measurement of courses or wales, it is the total number of needle loops in a square area measurement such as a square inch or square centimetre. It is obtained by multiplying, for example, the number of courses and wales, per inch together. Stitch density tends to be a more accurate measurement because tension acting in one direction in the fabric may, for example, produce a low reading for the courses and a high reading for the wales, which when multiplied together cancels the effect out. Usually pattern rows and courses are, for convenience, considered to be synonymous when counting courses per unit of linear measurement.

Stitch density = Wales per inch (wpi) × Courses per inch (cpi)

Single-faced structures: Single-faced structures are produced in warp and weft knitting by the needles (arranged either in a straight line or in a circle) operating as a single set. Adjacent needles will thus have their hooks facing towards the same direction and the heads of the needles will always draw the new loops downwards through the old loops in the same direction so that intermeshing points 3 and 4 will be identical with intermeshing points 1 and 2.

Double-faced structures: Double-faced structures are produced in weft and warp knitting when two sets of independently controlled needles are employed with the hooks of one set of knitting or facing in the opposite direction to the other set. The two sets of needles thus draw their loops from the same yarn in opposite directions, so that the fabric, formed in the gap between the two sets, shows the face loops of one set on one side and the face loops of the other set on the opposite side. The two faces of the fabric are held together by the sinker loops or under laps which are inside the fabric so that the reverse stitches tend to be hidden.

Balanced structure: This is a double-faced structure which has an identical number of each type of stitch produced on each needle bed and therefore showing on each fabric surface usually in the same sequence. These structures do not normally show curling at their edges.

12.2.3   Circular knitting machines

The term 'circular' covers all those weft knitting machines whose needle beds are arranged in circular cylinders and/or dials, including latch, bearded or (very occasionally) compound needle machinery, knitting a wide range of fabric structures, garments, hosiery and other articles in a variety of diameters. Circular knitting machines are either of body size or larger, having a single cylinder or double cylinder, cylinder and dial arrangement, as is also the case with small diameter machines for hosiery. The modern circular knitting machine is a highly engineered, electronically controlled, precision knitting system capable of producing high quality fabric at very high speeds.

The main features of a circular knitting machine are:

1. The frame or body is circular according to needle bed shape supports the majority of the mechanisms of the machine

2. The yarn supply system or the creel for holding the yarn packages

3. Yarn tensioning devices

4. Yarn feed control

5. Yarn stop motion

6. Yarn feed carriers or guides

7. The knitting system, which includes the housing and driving of knitting elements and needle selection device

8. The fabric take-down mechanism

9. Start, stop and inching buttons

10. The automatic lubrication system

The yarn from the package is fed to the knitting machine needles through guides, tensioners, stop motion. The knitted fabric which is in tubular form is taken down inside the cylinder in between two guide rollers to make the fabric flat and finally rolled on the cloth roller. After accumulation into a practical roll size, it can be cut off and taken out and can be processed either in tubular or in open width form after slit opening. Normally, circular knitting also adopts the same knitting principles as the flat-bed machines. The circular machine starts to knit when the cam systems on the needle beds (cylinder and dial) move along the surface quite similar to that of the carriage on a flat-bed machine. The only difference is that the operation is continuous as cam system of the circular machine does not need to stop during knitting because there is no beginning or end of a course (see knitting action below).

12.2.3.1   Classification of circular knitting machines

Mainly there are three types of circular knitting machines.

1. Revolving cylinder latch needle machines: They produce the most popular knitted fabric form. There are two kinds of revolving cylinder latch needle machines:

   (a) Open top or sinker top or single jersey machines

   (b) Dial and cylinder machine

   Open top machines have one set of needles usually arranged in the cylinder. Except in the case of certain effect fabric machines such as pelerine, cylinder and dial machines are of either the rib or the interlock type. These machines may or may not have patterning capabilities.

2. Revolving cylinder bearded needle single-jersey fabric machine: There are two types of circular bearded needle single-jersey fabric machines.

   (a) Sinker wheel machine or French or Terrot type machine

   (b) Loop wheel frame or English type machine

They have low productivity but can produce fancy and superior quality fabrics. Both types of machines have (i) needles fixed in needle bed, (ii) revolving needle bed, (iii) ancillary elements moving yarn and loops along the needle stems and (iv) accommodates less number of needles. It produces tubular fabrics with its technical back facing outwards.

3. Circular garment length machines: They are generally of body width size or larger having a cylinder and dial arrangement or a double cylinder and they are of the small-diameter hosiery type with either a single cylinder, a cylinder and dial or double cylinders.

12.2.3.2   Features of fabric manufacturing machines

They are mainly circular machines, knitting tubular fabric in a continuous uninterrupted length of constant width. Large diameter, latch needle machines and knit fabric at high speed (also known as yard goods or piece goods' machines). The fabric is manually cut away from the machine; usually in roll form, after a convenient length has been knitted. Most fabric is knitted on circular machines, either single cylinder (single jersey) or cylinder and dial (double jersey), of the revolving needle cylinder type, because of their high speed and productive efficiency. Sinker wheel and loop wheel frames could knit high quality specialty fabrics, with bearded needles, although circular machines employing bearded needles are now obsolete. The production rates of these machines were uncompetitive. Unless used in tubular body width, the fabric tube requires splitting into open width. Now a days, machines are designed for producing open width fabric directly on the machine to avoid setting creases formed when tubular fabric made flat before rolling, especially in the case of lycra fabrics. The productivity, versatility and patterning facilities of fabric machines vary considerably. Generally cam settings and needle set-outs are not altered during the knitting of the fabric.

12.2.3.3   Features of garment length manufacturing machines

The garment-length machine has the following main features: (a) they include straight bar frames, most flats, hosiery, leg wears and glove machines, and circular garment machines including sweater strip machines. They knitting garment length sequences, which have a timing or counting device to initiate an additional garment-length programming (collectively termed 'the machine control') mechanism. This coordinates the knitting action to produce a garment-length structural repeat sequence in a wale-wise direction. The garment width may or may not vary with in the garment length. They are coarser gauge machine than fabric machines. It automatically initiates any alteration to the other facilities on the machine needed to knit a garment-length construction sequence instead of a continuous fabric. The machine control may have to initiate correctly timed changes in some or all of the following: cam settings, needle set-outs, feeders and machine speeds. It must be able to override and cancel the effect of the patterning mechanism in rib borders and be easily adjustable for different garment sizes. The fabric take-down mechanism must be more sophisticated than for continuous fabric knitting. This take-down mechanism has to adapt to varying rates of production during the knitting of the sequence and on some machines be able to assist both in the setting up on empty needles and in the take away of separate garments or pieces on completion of the sequence. Garments may be knitted to size either in tubular or in open width; in the latter case, more than one garment panel may be knitted simultaneously across the knitting bed. Large-diameter circular machines and wide Vee-bed flat machines can knit garment blanks that are latter split into two or more garment widths (blanket-width knitting). They produce knitwear, outer wear and under wear. Under wear may be knitted either in garment length or in fabric form, whereas knitwear is normally in garment-length form, which is a generic term applied to most weft-knitted outer wear garments such as pullovers, jumpers, cardigans and sweaters, usually knitted in machine gauges coarser. Jersey wear is a generic name applied to weft-knitted fabric (single jersey, double jersey). It is cut and made-up from fabric usually knitted on large circular machines (26 or 30 inches diameter), although there are larger and smaller diameter machines used. Generally, gauges are finer than E14.

12.2.3.4   *Plain or single jersey circular latch needle knitting machine*

Plain fabric is a single jersey fabric and is produced by one set of needle. Most of single-jersey fabric is produced on circular machines whose latch needle cylinder and sinker ring revolve through the stationary knitting cam systems, which together with their yarn feeders are situated at regular intervals around the circumference of the cylinder. The yarn is supplied from cones, placed either on an integral over head bobbin stand or on a free-standing creel, through tensioners, stop motions and guide eyes down to the yarn feeder guides. The fabric, in tube form, is drawn downwards from inside the needle cylinder by tension rollers and is wound on to the fabric batching roller of the winding down frame. The winding-down mechanism revolves in unison with the cylinder and fabric tube and is rack-lever operated via cam followers running on the underside of a profiled cam ring. As the sinker cam plate is mounted outside on the needle circle, the centre of the cylinder is open and the machine is referred to as an open top or sinker top machine. Compared with a rib machine, a plain machine is simpler and more economical with a potential of more feeders, higher running speeds and the possibility

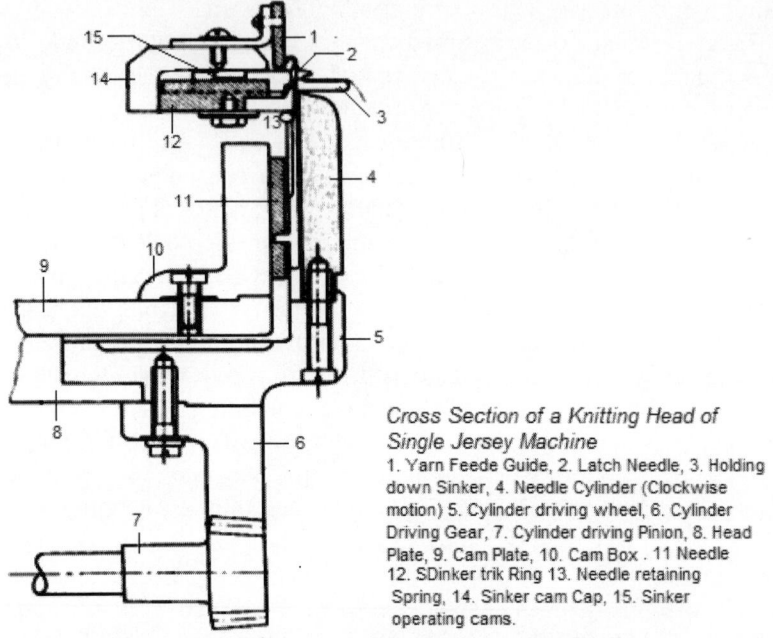

Cross Section of a Knitting Head of Single Jersey Machine
1. Yarn Feede Guide, 2. Latch Needle, 3. Holding down Sinker, 4. Needle Cylinder (Clockwise motion) 5. Cylinder driving wheel, 6. Cylinder Driving Gear, 7. Cylinder driving Pinion, 8. Head Plate, 9. Cam Plate, 10. Cam Box . 11 Needle 12. SDinker trik Ring 13. Needle retaining Spring, 14. Sinker cam Cap, 15. Sinker operating cams.

Figure 12.96. Cross section of a knitting head of single jersey machine

of knitting a wider range of yarn counts. The most popular diameter is 26 inches giving an approximate finished fabric width of 60–70 inches.

An approximate suitable count may be obtained using the formula $Ne = \dfrac{G^2}{18}$, where Ne = cotton count or English system and G = gauge in needle per inch (npi). For fine gauges, a heavier and stronger count may be necessary. Fig. 12.96 shows a cross section of the knitting head all of whose stationary parts are shaded.

The cam system consists of needle cam system and sinker cam system. Fig. 12.97 shows the arrangement and relationship between the needle cams and sinker cams as the elements pass through in a left to right direction with the letters indicating the positions of the elements at the various points in the knitting cycle. The parts of the needle cam race is shown in Fig. 12.97.

Circular knitting cam systems only allow for unidirectional knitting. Cam systems generate both the needle and the sinker movement for single jersey machines and cylinder and dial movement for double jersey machines. The needle track shows the typical three-stage needle displacement of: (1&4) the raising or clearing cam, (2&3) the lowering or stitch cam and (5&6) the guard cam that returns the needle to its entry position for the next cam system. The sinker track shows the engaged position (section 7) when the needle

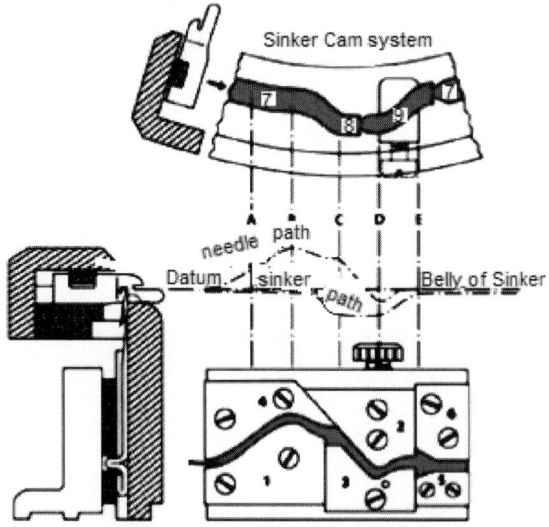

Figure 12.97. Needle cam system

is clearing. The sinker disengages in sections 8 and 9 so that knock-over can take place and reengages into section 7. The moment diagrams of the needles and sinkers are also shown in between cams.

12.2.3.5   Knitting action

Knitting involves the following actions (Fig.12.98):

Tucking in the hook or rest position: The sinker is forward, holding down the old loop while the needle rises from the rest position.

Clearing: The needle has been raised to its highest position, clearing the old loop from its latch.

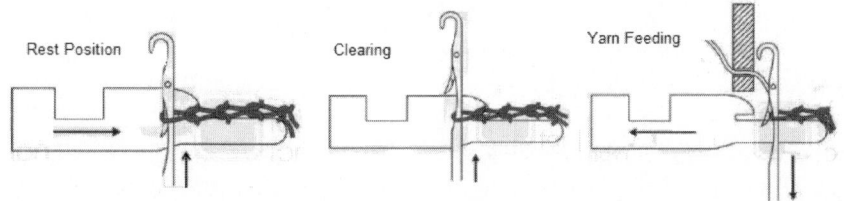

Figure 12.98. Actions involved in knitting process

Yarn feeding: The sinker is partially withdrawn allowing the feeder to present its yarn to the descending needle hook and also freeing the old loop so that it can slide up the needle stem and under the open latch spoon.

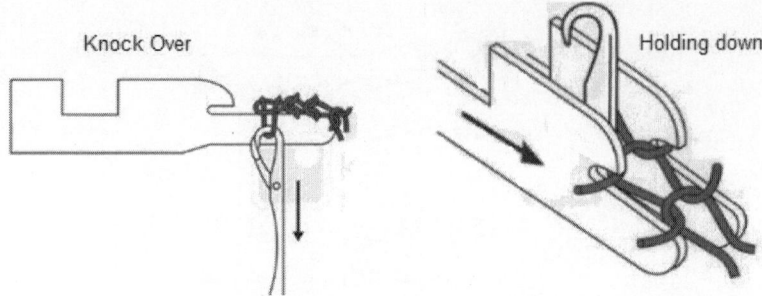

Figure 12.99. Knock-over

Knock-over: The sinker is fully withdrawn while the needle descends to knock-over its old loop on the sinker belly (Fig. 12.99).

Holding down: The sinker moves forward to hold down the new loop in its throat whilst the needle rises under the influence of the up throw cam to the rest position where the head of the open hook just protrudes above the sinker belly (Fig. 12.99).

12.2.3.6   Multisystem circular machine

Similar to a flatbed machine, multisystem circular knitting is also possible. In multisystem, several cam systems knit at the same time and each of cam system is having its own supply of yarn for its own course. So, when the machine runs, all systems move together and hence eight courses of fabric are in knitting at the same time. In other words, at the end of one revolution of the cam system, all courses of fabric are completed. Similarly, if there is more cam systems around the machine, there will be more fabric courses being produced in a single revolution of the machine, for example, say if there are 20 cam systems, 20 courses of fabric will be completed in one revolution of the cam system. The number of cam system is limited by the diameter of the machine, which is obvious. For example, a 30-inch diameter machine may have 72–90 cam systems. Since each cam system must have its own yarn supply and hence a yarn feeder, such machine can be referred as 30-inch, 90-feed machine. From above figure, further, it can be seen that whether there are eight systems or 80 systems, the space taken up by the machine will not be changed.

As against flat-bed machine the cam systems of a circular machine always operate at their maximum speed. A circular machine will have much more cam systems compared to flat-bed machines hence the production is also much higher. For example, a double system machine with 100-inch needle bed produces about 45 courses per minute and a 30-inch, 90-feed circular machine produces about 2700 courses per minute.

The knitting action is achieved by the relative motion between the cam plates, and the needle butt and the same needle action can be achieved by the cam plate is moving across the needle butt or by needle butt moving across the cam plate. Thus, there are two types of circular machines distinguished by the rotation of the machine.

1. Cam box revolving machine
2. Cylinder revolving machine

The latter machine is simpler in construction and consumes less power than cam box revolving machine since there are less moving components. As a matter of fact, most of the circular machines are cylinder revolving type. Only those machines such as the garment length machines are cam box revolving because of their complexity. Those are machines with 6–18 feeds producing complex knitting structures which cannot be accomplished if the machine is cylinder revolving.

It is obvious that circular knitting machine is naturally the choice for the volume production. Since it is ideal for volume production, there are custom built circular machines. For example, plain knit fabric done on circular machine with just one set of needles in the cylinder is available for plain knit only. All other knit structures requiring the second set of needles will be done on other machines.

12.2.3.7   Rib circular knitting machine

Another important circular knitting machine used in mass production of fabric is rib circular machines. In these machines, there are two sets of needles – one set of needles on the circumference of a vertical cylinder and a second set of needles, arranged perpendicular to the first set and mounted on a horizontal dial. On most of the circular knitting machines, the cylinder and dial rotate, whereas the cams with yarn feeder guides are stationary. Fig. 12.100 shows a cross-sectional view of the region containing the knitting elements of a rib (double knit) circular knitting machine. The set-up of the cylinder (Fig. 12.100, #3) with its knitting elements (#1–9) is the same as with plain circular knitting machines. In a horizontal (rib) dial (#10), grooves (#11) are milled in. The latch needles (#12) are housed and guided in these grooves. The dial needle (#12) obtains its motion for stitch formation through its butt (#13), which extends into a cam track (#14). This cam track (#14) is formed by the cam parts (#15 and 17), which in turn are fixed to a dial cam plate (#18). During the rotation of the cylinder and the dial, the cylinder needle (#1) is moved vertically and the dial needle (#12) is moved horizontally, corresponding to the shape of the cam track in the cylinder and dial cams. In a gauge range from 5 to 20 npi, an approximate suitable count may be G2

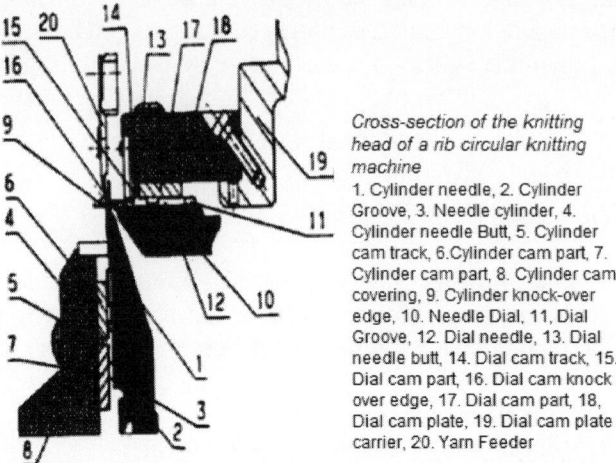

Cross-section of the knitting head of a rib circular knitting machine
1. Cylinder needle, 2. Cylinder Groove, 3. Needle cylinder, 4. Cylinder needle Butt, 5. Cylinder cam track, 6.Cylinder cam part, 7. Cylinder cam part, 8. Cylinder cam covering, 9. Cylinder knock-over edge, 10. Needle Dial, 11, Dial Groove, 12. Dial needle, 13. Dial needle butt, 14. Dial cam track, 15. Dial cam part, 16. Dial cam knock over edge, 17. Dial cam part, 18, Dial cam plate, 19. Dial cam plate carrier, 20. Yarn Feeder

Figure 12.100. Cross section of the knitting head of a rib circular knitting machine

obtained using the formula Ne = $\dfrac{G^2}{8.4}$, where Ne = cotton count or English system and G = npi.

12.2.3.8   *Fabric formation in rib knitting machines*

As explained earlier the formation of the fabric is by the knitting action of both the dial and cylinder needles in the rib machines. In this needle timing, which is the relationship between the loop-forming positions of the dial and cylinder needles measured as the distance in needles between the two stitch cam knock-over points, plays an important part. Needle timing influence the appearance, the quality and properties of the fabric produced on a rib circular knitting machine. Collective timing adjustment is achieved by moving the dial cam plate clockwise or anti-clockwise relative to the cylinder. Individual adjustment at particular feeders as required is obtained by moving or exchanging the stitch cam profile. Depending on the coordination between the cylinder and dial cams, one differentiates between synchronized timing (also known as point, jacquard or 2 × timing) and delayed timing (also referred to as rib or interlock timing) (Fig. 12.101). In synchronized timing, the cylinder and the dial needles knock-over their knitted loops at the same time. It is the term used when the two positions coincide with the yarn being pulled in an alternating manner in two directions by the needles thus creating a high tension during loop formation. Structures knitted using synchronized timing will be loose and consist of uneven stitches.

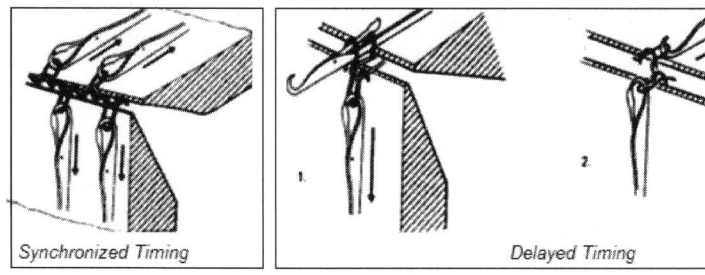

Figure 12.101. Synchronized and delayed timings

In delayed timing, the dial needles knock-over their knitted loops later than the corresponding cylinder needles. With delayed timing, the dial knock-over occurs after about four cylinder needles have drawn loops and are rising slightly to relieve the strain. The dial loops are thus composed of the extended loops drawn over the dial needle stems during cylinder knock-over, plus a little yarn robbed from the cylinder loops. The dial loops are thus larger than the cylinder loops and the fabric is tighter and has better rigidity; it is also heavier and wider, and less strain is produced on the yarn.

12.2.3.9   Knitting action in synchronized timing

In delayed timing, the dial needles knock-over their knitted loops later than the corresponding cylinder needles. With delayed timing the dial knock-over occurs after about four cylinder needles have drawn loops and are rising slightly to relieve the strain. The dial loops are thus composed of the extended loops drawn over the dial needle stems during cylinder knock-over, plus a little yarn robbed from the cylinder loops. The dial loops are

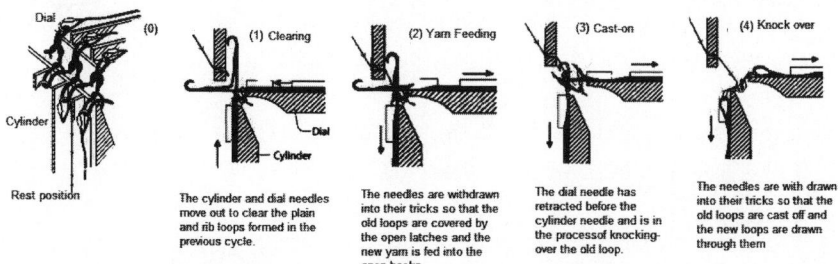

Figure 12.102. Knitting action in synchronized timing

thus larger than the cylinder loops, and the fabric is tighter and has better rigidity, it is also heavier and wider, and less strain is produced on the yarn (Fig. 12.102).

12.2.3.10   Knitting action in delayed timing

The processes involved in knitting action are shown in Fig. 12.103.

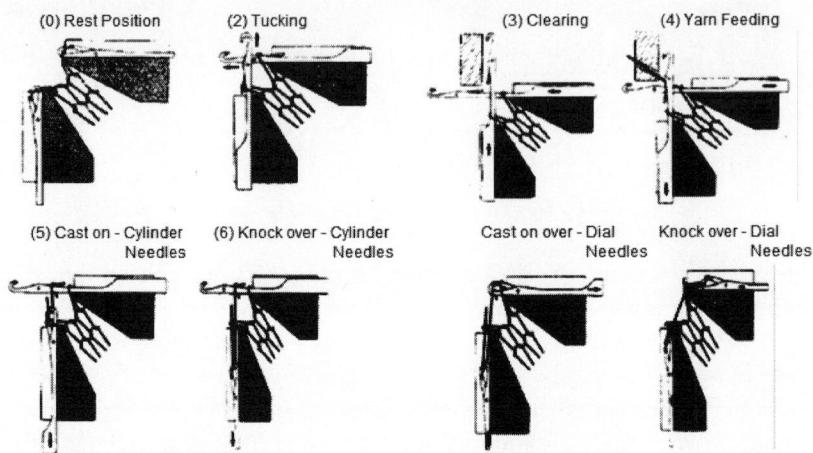

Figure 12.103. Knitting action in delayed timing

We shall not go for further details regarding knitting in this book.

12.2.4   Common weft: knitted fabrics (structures)

Single jersey: The simplest and most basic structure is the 'plain knit' which is also called 'single knit'. Plain is a knit structure family, which is produced by the needles of only one set of needle with all the loops intermeshed in the same direction (Fig. 12.104).

This structure has the maximum covering power. Plain fabric is the commonest weft-knitted fabric and is produced by widely different sorts of knitting machinery in all forms from circular fabric piece goods to fully

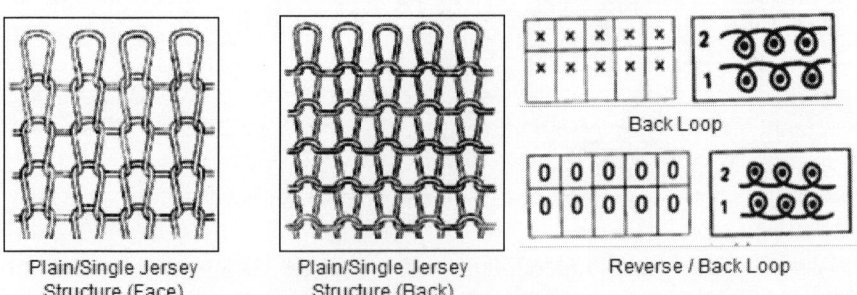

Figure 12.104. Single jersey

fashioned panels. They are used for basic T-shirt undergarments, men's vest, ladies hosiery, fully fashioned knitwear, etc.

12.2.4.1   Rib structure or rib fabric

Rib, also called 'double knit', is the second family of knit structures and widely used. Rib is knitted on machines with two sets of needles allowing them to intermesh when raised and this needle arrangement is called rib gaiting. In flat machines, this way of arrangement of two sets of needles are called V bed as from the side they look like an inverted 'V'. The simplest rib structure is called a 1 × 1 Rib. When all the needles on both beds are active in knitting, 1 × 1 rib structure is produced. When every third needle is in active (two active needles positioned in between two inactive needles), a 2 × 2 rib is produced (Fig. 12.105).

These structures are widely used for waistbands, cuffs and collars which are typical applications, together with whole garments of a fitting nature. Rib

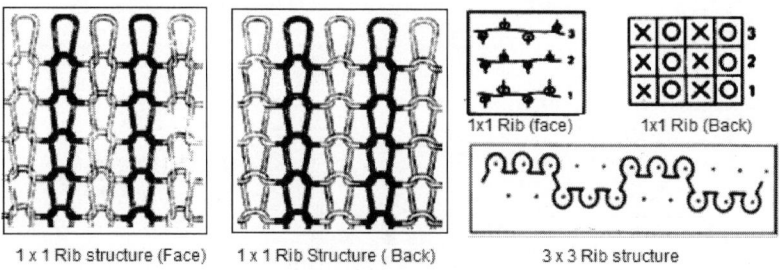

1 x 1 Rib structure (Face)   1 x 1 Rib Structure (Back)   3 x 3 Rib structure

Figure 12.105. Rib structure

fabrics are used where portions of garments are expected to cling to the shape of the human form and yet be capable of stretching when required. Cotton rib knitted fabric, bottom of the sweater, skirt belt, various types of fancy borders, underwears, sweaters, etc.

12.2.4.2   Purl knit

The third family of knit structures is the purl knit also called links/links (German), left/left, reverse/reverse, etc. This structure also needs two sets of needles for forming the loops. They are knitted on machines with special equipment (one set of needles), which are double-ended latch needles and special devices to drive them, allowing loops to be intermeshed in two directions. There are two types of purl machine: (a) flat purls, the needle beds of these machines are set on the same plane instead of being in an inverted 'V' formation and (b) circular purls, which have two superimposed cylinder one above the other so that the needles move in a vertical direction, both types of machines are capable of producing garment length or other article sequences.

12.2.4.3   Knitting process on links/link machine

There is one set of needles for both needle beds and the procedure is as follows: (a) The needle is positioned in the front needle bed in which it has just formed a face loop. (b) The needle slides through the loop towards the rear needle bed. (c) A reverse loop is produced by the other needle head through the previous face loop within the same wale. (d) The needle slides through the loop towards the front needle bed (Fig. 12.106).

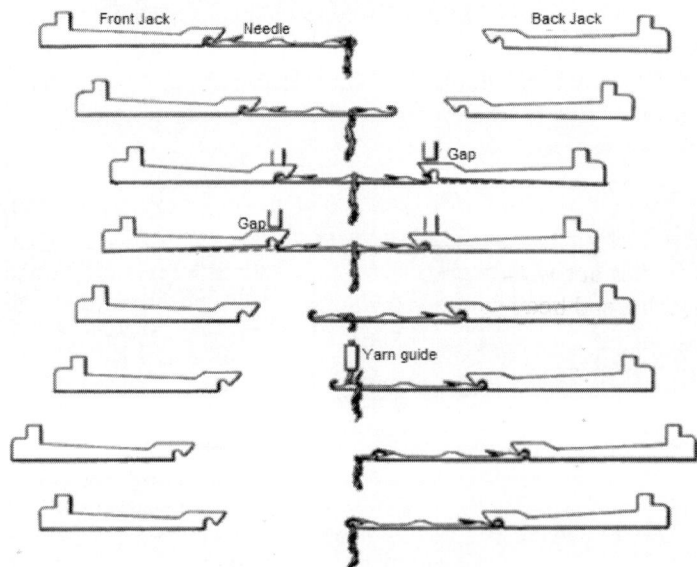

Figure 12.106. Knitting process on links/link machines

12.2.4.4   Purl fabrics are knitted on 'V' flat bed

The basic requirement of the machine is a loop transferability. The simplest 1 × 1 purl structure is produced according to the following procedure: (a) A course is knitted on the front needle bed while the rear bed remains idle. (b) All the loops are mechanically taken from the needles of the front needle bed and transferred to the empty needles of the rear bed. (c) A course of reverse loops is now knitted by the needles of the rear bed through the transferred loops. (d) All the loops are now transferred from the rear needles to the front needles, which in the next cycle knit through them (Fig. 12.107).

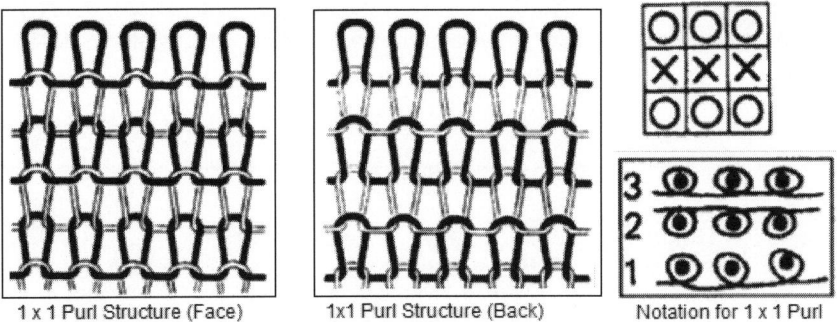

| 1 x 1 Purl Structure (Face) | 1x1 Purl Structure (Back) | Notation for 1 x 1 Purl |

Figure 12.107. Purl fabric

Not all loops have to be transferred after the completion of each course. The 2 × 2 purl structure is produced when two courses are knitted on each needle bed before the transfer operation.

When the knitting machine is sophisticated enough to handle both loop transfer and needle selection, the variety of possible purl structures is unlimited.

Purl fabrics are widely used for baby wear, children's clothing, sweater, knitwear, thick and heavy outer wear, under garments, etc.

12.2.4.5   Interlock fabric

Interlock is another common fabric used to make garments widely. Interlock is another 1 × 1 rib variant structure which is produced on specially designed machines. These machines possess two sets of needles (short and long needles) in both cylinder and dial and at least two feeders. For normal interlock, the needles in both cylinder and dial are arranged to be alternately long and short, and at the odd feeders, the long needles are selected to knit and at the

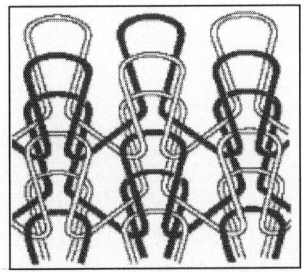
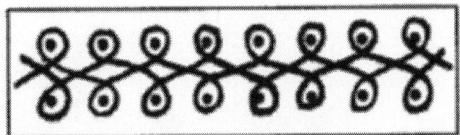

| 1 x 1 Interlock Structure | Notation for 1 x 1 Interlock structure |

Figure 12.108. Interlock fabric

even feeders, the short needles are selected to knit. At each feeder a 1 × 1 rib structures knitted at adjacent feeders interlace each other and form a double 1 × 1 rib fabric. So interlock fabric is produced by two 1 × 1 rib structure interlace to each other. Interlock has the technical face of plain fabric on both sides (Fig. 12.108).

Table 12.3. Comparison of basic weft-knitted structures

Structure	Plain	1 × 1 Rib	1 × 1 Purl	1 × 1 Interlock
Appearance	Different of face and back, V shapes on face and arc on back	Same on both sides like face of plain SJ fabric	Both face appearance is same, like the back face of plain SJ	Both face appearance same, like the face of plain SJ
Extensibility, lengthwise, widthwise, area	Moderate 10–20% High 30–50% Moderate–High	Moderate Very high 50–100% High	Very high High Very high	Moderate Moderate Moderate
Thickness and warmth compared to woven using the same yarn	Thicker and warmer	Much thicker and warmer	Very much thicker and warmer	Very much thicker and warmer
Moving	From either end	Only from end knitted last	Either end	Only from end knitted last
Curling	Tendency to curl	No tendency to curl	No tendency to curl	No tendency to curl
End uses	Ladies stockings, fine cardigans, men's and ladies shirts, dresses, base fabric for coatings, etc.	Socks, cuffs, waistbands, collars, men's outer wear, knitwear and underwear.	Children's clothings, knitwear, thick and heavy outer wear.	Underwear, shirts, suits, trouser suits, sportswear and dresses.
Physical appearance				

Now we shall look into different derivatives of this basic knits.

12.2.5   Stitch notation

There are many types of notation used in knitting field. However, the following three types of representations are widely used:

1. *Line diagram*: It is easily understandable but sometimes it is difficult to represent complex designs by this method.

2. *Symbolic notation on graph paper*: Versatile and any type of loop or knitted structure can be represented on graph paper with the help of some symbol. Each square in the graph paper is used to represent one loop. The horizontal rows are used as courses, whereas the vertical columns are used to indicate the wales in the knitted structure.
    (a) A cross mark (x) inside a square represents face loop.

    (b) A blank circle (o) inside a square represents back loop.

    (c) A dot (•) inside a square represents tuck loop.

    (d) A blank square represents float or miss loop.

3. *Schematic or diagrammatic notation*: This is done on point paper. The schematic diagram describes the movement of the yarn across the cross section of the needle (point) during loop formation. The techniques of representation of knitted loops are listed as follows and are internationally accepted and widely used in different countries.
    (a) The notation face loop is ☙ and notation of back loop is ☙.

    (b) Tuck loop face is represented as ⌣ and back as ⌢.

    (c) Float loop face is represented as ⸺ and back as • .

4. *German notation*: The symbols used in German system (DIN 62050) of graphical representation knitted loops are different and are shown below. This method uses **|** for a face loop and **∩** for a back loop. A **V** marks a front tuck and a **∧** marks a rear tuck as shown below.

Face Loop   Back Loop   Tuck Loop (Face)   Tuck Loop (Back)   Float Loop

12.2.5.1   *Derivatives of plain single jersey knit*

1. *Cross-miss knit design*: Cross miss is a miss-knit single jersey structure. So one set of needle is used to produce this structure. The repeat of the structure completes on two courses. Knitting sequence for a repeat is as follows:

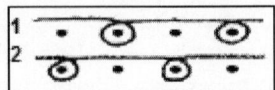

2. *Bird eye*: Bird eye is a knit-miss single jersey structure. So one set of needle is used to produce this structure. The repeat of the structure completes on four courses. Knitting sequence:

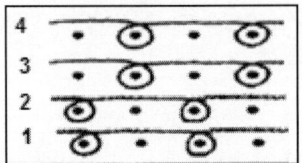

3. *Weft locknit design*: In weft locknit, one set of needle is used to produce this structure. The repeat of the structure completes on four courses. Knitting sequence:

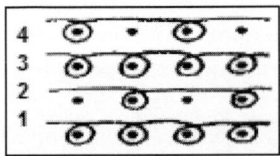

4. *Mock rib design*: Mock rib is a knit-miss single jersey structure. So one set of needle is used to produce this structure. The repeat of the structure completes on two courses. Knitting sequence for a repeat as follows:

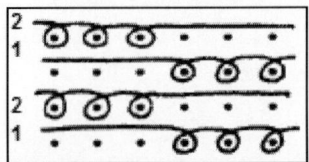

5. *Single cross-tuck design*: Single cross tuck is a knit-tuck single jersey structure. So one set of needle is used to produce this structure. The

repeat of the structure completes on two courses. Knitting sequence for a repeat as follows:

6. *Polo pique or double cross tuck*: Polo pique is a knit-tuck single jersey structure. So one set of needle is used to produce this structure. It is a very popular structure to produce T-shirts. The prominency of the design appears on the back side of the fabric. The repeat of the structure completes on four courses. Knitting sequence for a repeat:

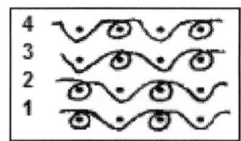

7. *Single Lacoste or Fred Perry design*: For single Lacoste is a knit-tuck structure, one set of needle is used to produce this structure. It is also a very popular structure to produce cut and sew knitwear. The prominency of the design appears on the back side of the fabric. The repeat of the structure completes on four courses. Knitting sequence:

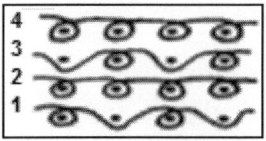

8. *Double Lacoste design*: Double Lacoste is a tuck-knit single jersey structure. So one set of needle is used to produce this structure. It is also a very popular structure to produce cut and sew knitwear. The prominency of this design near to the single Lacoste fabric. The repeat of the structure completes on six courses. Knitting sequence:

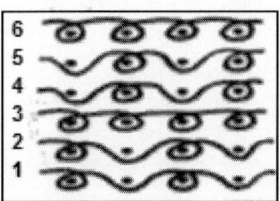

9. *Simple crepe design*: A simple crepe design may be made by tuck-knit or miss knit single jersey structure. So one set of needle is used to produce this structure. The repeat of the structure completes on four courses. Knitting sequence:

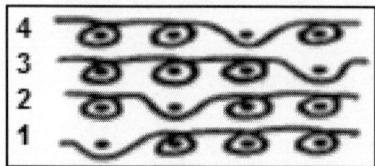

10. *Cellular blister or popcorn design*: It is a tuck-knit plain structure using one set of needle. The prominency of the design appears on the backside of the fabric. The repeat of the structure completes on eight courses. Knitting sequence:

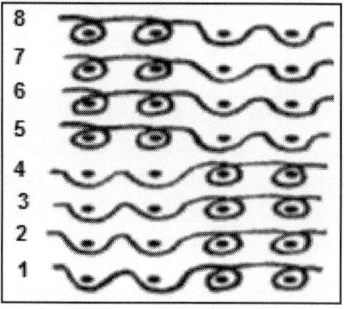

11. *Twill effects*: This can be made by tuck-knit or miss-knit or knit-tuck-miss single jersey basis. So one set of needle is used to produce this structure. The prominency of the design appears on the backside of the fabric. The main feature of this structure is that the diagonal line (twill line) appears on the fabric surface like as woven twill fabric. The repeat of the structure completes on several courses. The following figures show the knitting sequence for a repeat:

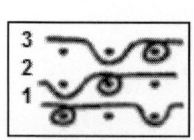 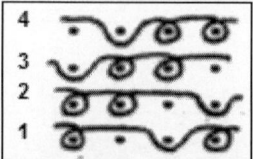 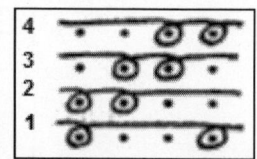

12.2.6   Derivatives based on rib structure (double jersey)

1. *Double pique*: Double pique is a double jersey fabric made on a rib basis, using a selection of knitted, loops and floats. The two most important sequences are known as Swiss double pique and French double pique, respectively, and the knitting sequences for each are shown in the following figure. Double pique is also known as wevenit, rodier and overnit.

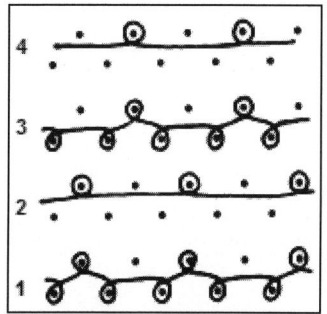

Swiss Double Pique   French Double Pique

2. *Half cardigan rib or royal rib*: It is a rib-based structure in which a great number of tuck stitches are added to make the fabric heavy, wide and soft. From the below notation diagram, it is clear that two knitting sequences are required to produce one repeat of this type of fabric. A special effect is produced when one half of the cardigan repeat is substituted for a regular 1 × 1 rib structure. The new fabric is called a 'half cardigan' and is produced. One side of the fabric, in this case the reverse side, is produced with tuck stitches and therefore looks like a 'cardigan'.

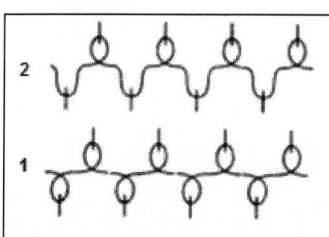
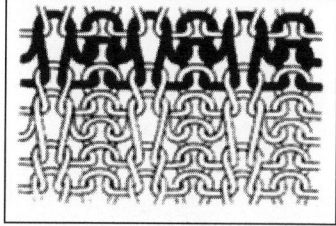

3. *Full cardigan rib or polka rib*: It is another variation of the 1 × 1 rib structure. In this case, even more tuck stitches are introduced which makes the fabric wider, heavier, bulkier and less flexible then the half cardigan or the usual 1 × 1 rib. Contrary to the previous example the

full cardigan is symmetric on both sides. Two yarns are inserted into the fabric to complete one full course, i.e., loops on the one needle bed and loops on the other needle bed.

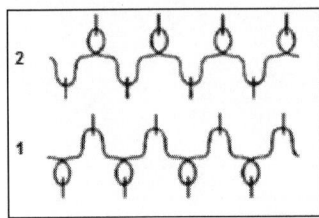 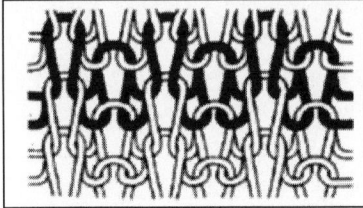

4. *Half milano rib*: A weft-knitted rib-based fabric, consisting of one row of 1 × 1 rib and one row of plain, knitting made on either set of needles. The appearance and characteristics of the fabric are related to the ratio of the course lengths of first (1) and second (2). The knitting sequence for a repeat is as follows:

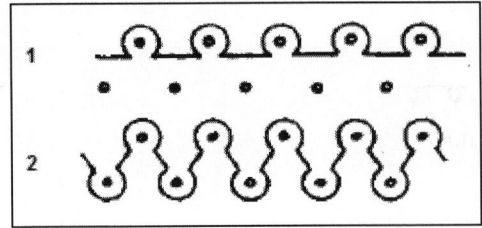

5. *Milano*: Each complete repeat of the structure consists of three components knitted in the sequences shown to give one row of 1 × 1 rib and one row of plain tubular knitting, the two component parts of tubular knitting usually being similar. The appearance and characteristics of the fabrics are related to the ratio of the course lengths of two rows.

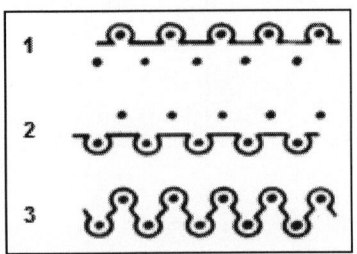

6. *Roma rib*: A weft-knitted rib-based structure. Each complete repeat of the structure consist of two components knitted in the sequences

shown to give one row of 1 × 1 rib and one row of plain knitting. The knitting sequence for a repeat is as follows:

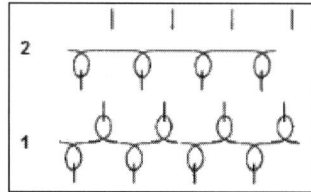

7. *Lacoste pique*: The Lacoste pique is produced by using a selection of knitted loops and tuck loops on single jersey machines. It can also be produced on rib-based machine, but it should be remembered that for the production of this fabric, only one bed is active and other bed is inactive. In the following figure, the front bed is active and back bed is inactive. The knitting sequence:

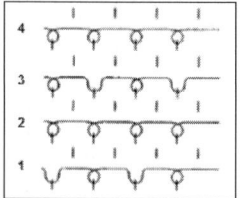 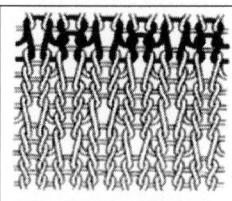

8. *Gaberdine or 2 × 2 twill design*: It is a double blister fabric of a four-needle width repeat, with the dial needles all knitting the backing at every third (ground) feed. The following is the constriction notation:

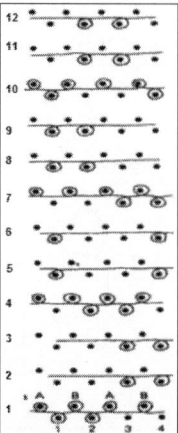

9. *Poplin*: It is a single blister design with a two-needle width repeat.

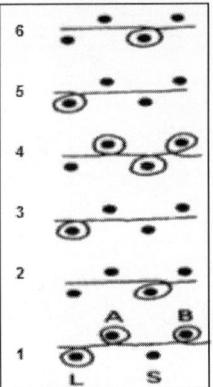

12.2.7   Derivatives of interlock structure

Most interlock variation structures are made on modified interlock machines with a six or eight feeder repeat sequences as only alternate needles in one bed are in action in a course.

1. *Single pique or cross-tuck interlock structure*: It was one of the first to be produced, by placing tuck cams in the dial at every third feeder. The tuck stitches throw the fabric out approximately 15% wider than normal interlock to a satisfactory finished width of over 60 inches, they break up the surface uniformity and help to mask feeder stripiness but they also increase fabric weight. Single pique is a tuck-knit interlock structure. So interlock needle gating system is used to produce this structure. Long and short needles in dial and cylinder, long needles facing short needles and vice-versa.

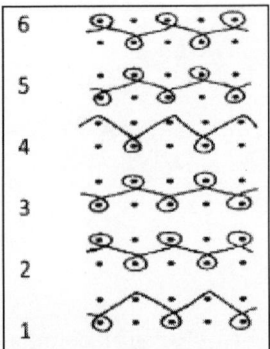

2. *Texi pique structure*: Texi pique is a tuck-knit interlock structure. So interlock needle gating system is used to produce this structure. Long and short needles in dial and cylinder, long needles facing short needles and vice-versa. The repeat of the structure completes on six feeders. They are wider and bulkier and show the same pique effect on both sides of the fabric.

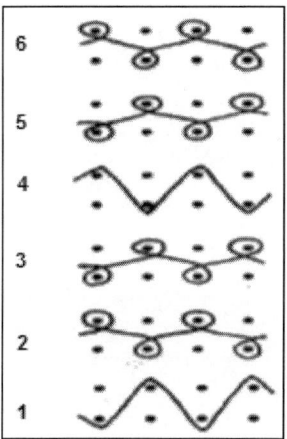

3. *Cross-miss structure*: It is the knit miss equivalent of single pique but it is narrower and lighter in weight. Cross-miss is a miss-knit interlock structure. So interlock needle gating system is used to produce this structure. Long and short needles in dial and cylinder, long needles facing short needles and vice-versa. The repeat of the structure completes on six feeders.

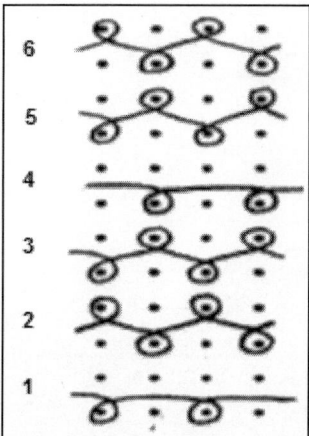

4. *Piquette*: Knitted almost same way as above structure using miss-knit interlock base. The repeat is as follows:

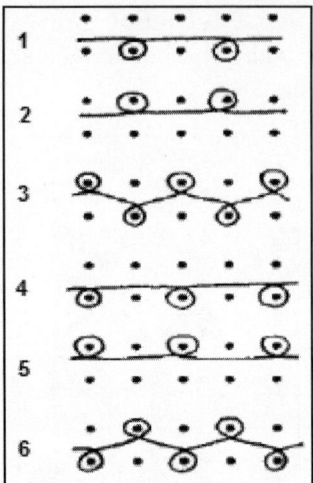

There are many other structures, such as pin tuck structure, bourre-let structure, jersey cord structure, common roma structure, Punto-di-Roma structure, cortina structure, evermonte structure, etc., that are interlock derivatives but not very common hence not described here.

A.1   Fibre identification by various methods

There are many methods available. Given below are some of the methods in practice. Some identification test may be repeated in another method (ensure the strength of the chemicals used).

A.2   Staining with Neocarmine W

A small sample of material was first prewetted with alcohol and then immersed in the Neocarmine solution for 3–5 min. After removal, it was rinsed for 2 min in running water and dried. The test can be basically done on white or pale dyed material only. In case of polyester, Neocarmine MS will only stain polyester and no other fibre (Table A.1).

Table A.1. Fibre identification after staining with Neocarmine W

Fibre	Microscopic characteristics	Stain	Chemical reactions
Wool	Overlapping outer scales with internal lengthwise lines. Coarse fibres may have internal dark blotches or rope-like marks.	Yellow	Dissolved in 2% potassium hydroxide solution, boil for 15 min
Recovered wool	Scales more or less pronounced, fibrillated brush-like fibre ends	Yellow-brown	–
Coarse hairs	Broad, often opaque and form a very broad striated cylinder	Yellow	–
Camel hair	Colour is yellow-grey to brown	–	–
Wool or lint fibre	Scaly, unmarked lengthwise stripes	–	–

Bristle	Dark brown to black, generally contain blotches or continuous channelled marking with grainy content (natural dyes)	–	–
Cashmere	Colour is white, grey or brown. The outwear layer consist of fine well-defined scales	Yellow	–
Wool or lint fibre	Display coarse scales and split fibres	–	–
Bristle	Rope-like marking with fine grained content	–	–
Mohair	Coarse stripes, wide, regular split in the fibres, the scales are thin and often narrow to a point, fine hairs unmarked	–	
Angora	Scales as with wool, but more pointed, groups of markings arranged like a string of pearls; coarse hairs have several rows of markings	–	–
Raw silk	Single fibres are stuck together with gum (sericin)	Olive green	Soluble in formic acid zinc chloride solution, warm or boil for 1 min in concentric nitric acid
Degummed silk	Structureless, smooth, transluscent irregular lengthwise	Dull yellow	
Tussah silk (wild silk)	Individual fibres broad, band like with fine lengthwise stripes and creases, slightly twisted	Green	Dissolved by boiling a few minutes in concentric nitric acid
Raw cotton	Thin ribbon like with lengthwise twists in either direction	Light blue (red spots)	Swells before dissolving in cuprammonium hydroxide

Mercerized cotton	More cylindrical than ribbon-like structure with very few twists	Deep blue (red spots)	Dissolves without swelling in cuprammonium hydroxide
Viscose	Longitudinal stripes, varying in intensity	Red to reddish purple	Swells before dissolving in cuprammoinum hydroxide
Cupro	Structureless smooth translucent	Deep blue	–
Cellulose diacetate	1–3 longitudinal stripes, otherwise structureless	Yellow (red or green spots)	Soluble in acetone, glacial acetic acid, not readily soluble in dichloromethane
Cellulose triacetate	–	Yellow (green spots)	Soluble in cold dichloromethane and glacial acetic acid, practically insoluble in acetone
Kasein	Partially smooth, partially striped longitudinally, usually with grained deposits	Yellow-orange	
Polyamide	Smooth structureless	Green(yellow spots)	Soluble in cold 855 formic acid, turbid or precipitation on addition of water
Polyacrylonitrile	Weak longitudinally striped slight twist	Only slightly stained	Soluble in cold nitric acid and boiling DMF
Polyester	Smooth and structureless		Soluble in molten phenol and boiling o-dichlorobenzene
Polyvinyl Chloride	Smooth, fine to thick, some very thick sections	Soiled	Soluble in boiling dichloromethane and cold DMF

A.3   Fibre identification by dry distillation

A fibre sample is heated in a dry test tube. The gases released are tested with a moistened litmus paper.

Acid vapours turn red litmus to blue

Alkaline vapours turn blue litmus to red

Acidic vapours are released by natural fibres, recovered cellulosic fibres and acetate fibres.

Alkaline vapours are released protein fibres, i.e., wool, silk and polyamide poly acrylonitrile.

Note: Polyester fibres in this test shows initially natural and then acidic.

A.4   Fibre identification by solvent separation

Treat with the mentioned solvent (Table A.2):

- Cold: 30 min agitated at room temperature
- Hot: 5 min at the boiling temperature of the solvent being used

1. If fibre is not soluble, a new sample is tried with the next solvent.
2. If a blend is used, the sample must be dried between each stage and used for the next relevant solvent.
3. To ensure the evaluation of the solubility test, a control reaction is carried out. Especially in case of blends.
4. After the solvent test approximately twice, the volume of a 'non-solvent' (water, methanol) is added and observed.

Table A.2 Fibre identification and solvent separation method

	Fibre to be dissolved in	Control reaction with
1	Dichloromethane: (a) cold: triacetate, (b) hot: after chlorinated PVC	Methanol: precipitation
2	Acetone (a) 21/2 acetate	With water turbidity
3	Tetra hydrofuran: (a) PVC not after chlorinated, (b) PVDC polyvinyldene chloride	Methanol: precipitation
4	Dimethyl formamide – PA6, PUE, PAC	Water precipitation
5	Formic acid (98%) – PA 6.6 (silk decompose)	Water: precipitation
6	O-Dichloro benzene – PETP, PBTP, polyester	Cool-off: turbidity
7	Xylene – polyethylene, polypropylene	Cool-off: turbidity

8	Caustic soda 5% – wool, silk	Cool-off with hydrochloric acid: turbidity
9	Formic acid/zinc chloride* – 10 min 700C – Viscose, modal	Turbidity with water
10	Sulphuric acid 98% cold: cotton, flax	

* 20% by weight of zinc chloride (anhydrous) dissolved in 80% by weight of concentric formic acid

A.5   Fibre identification by chemicals

Rapid fibre identification of man-made fibres can be done by solubility tests given in Table A.3.

Table A.3 Fibre identification by chemicals

1	Cuoxam (cuprammonium, Schweitzer's reagent)	Soluble: stir-cellulose
2	Glacial acetic acid – Cold	Soluble: CA, CT
3	Glacial acetic acid – boiling	Soluble: PA 6, PA 6.6, PA 11
4	6N hydrochloric acid	Soluble: Vinyl (PVA+)
5	Nitric acid concentrate – cold	Soluble: PAN
6	Sulphuric acid concentrate – cold	Soluble: PES
7	Dimethyl formamide	Soluble: PVC

A.6   Fibre identification by solubility

Pure man-made fibres can be identified by their differing solubility in organic solvents.

Man-made fibres unblended:

(1) Acetone cold 10 min:
      Not dissolved: go for (2)
      Dissolved – acetone/water, 80:20 cold 5 Min
            (a) Soluble: CA
            (b) Not soluble: add chloroform cold 5 min
                Soluble: CT
                Not soluble: MAG
(2) Treat with glacial acetic acid, boiling:
      Not dissolved: go to (3)
      Dissolved – treat with cyclohexanone, boiling
            (a) Soluble: PA 11

(b) Not soluble: dimethyl formamide/formic acid 85%
(75:25), boiling
Soluble: PA 6
Not soluble: PA 6.6
(3) Dimethyl formamide:
Not soluble: go to (4)
Soluble: PVC
(4) Dimethyl formamide, boiling:
Not soluble: go to (5)
Soluble: PAN
(5) Nitrobenzene, boiling
Soluble: PES

A.7   Dry distillation method of identification

The dry distillation method explained earlier is repeated, and the escaped vapours are tested with moistened pH paper (in case of wool and animal hairs and casein [but not silk] may be tested with moistened lead acetate paper which turns black). The odour of the emitted gases is also checked. Table A.4 summarizes the pH indication and the odour of the emitted gas.

Table A.4 pH identification and the odour of the emitted gas

pH	Fibre	Odour
1	CLF	Pungent, HCl smell
2 to 3	CA	Vinegar smell
3 to 4	PES	Sweety aromatic, slightly pungent
4 to 5	PVAL	Burnt sugar
5 to 6	Cellulose fibres	Burning paper smell
5 to 6	PE	Burning paraffin
6 to 7	PP	Burning paraffin
9 to 10	Protein fibres	Burning hair
10 to 11	PA	Burning hair (weak), not pungent
10 to 11	EL	Like mouse droppings
10 to 11	PAN	Sweetly aromatic

A.8   Grouping by solubility and again identification by other solvents

A.8.1   Group 1

Soluble in 0.25% sodium carbonate: calcium alginate
Soluble in trypsin 40°C
Slowly soluble in 1% sodium hydroxide at boil
Soluble in 75% w/w sulphuric acid

A.8.2   Group 2

Fibres soluble within 15 min in calcium chloride: 90% formic acid in the ratio 1:10 – diacetate, triacetate, nylon 6, nylon 66, vinylal, bombyx silk

Diacetate: Soluble in 70% v/v acetone in 1 min. Soluble in glacial acetic acid, m-cresol, concentric nitric acid, 60% w/w H_2SO_4, tetra hydrofuran, dioxan at 80°C, formdimethylamide at boil

Triacetate: Soluble in glacial acetic acid in 4 min, chloroform, commercial xylene, m-cresol, concentric nitric acid, 60% w/w H_2SO_4, dioxan at 80°C, formdimethylamide at boil

Nylon 6: 4.4 N hydrochloric acid in 2 min, 35% w/w sulphuric acid, 5 N HCl at65°C, m-cresol, concentric nitric acid, 60% w/w H_2SO_4, formdimethylamide at boil

Nylon 66: 35% w/w sulphuric acid, 5 N HCl at 65°C, m-cresol, concentric nitric acid, 60% w/w H_2SO_4

Vinylal: 35% w/w sulphuric acid in 2 min, 5 N HCl at 65°C, m-cresol, concentric nitric acid, 60% w/w H_2SO_4

Bombyx silk: 5 N HCl at 65°C in 3 min, concentric nitric acid, 60% w/w H_2SO_4, Cupra

A.8.3   Group 3

Fibres soluble within 5 min in N sodium hypochlorite + 0.5 sodium hydroxide: regenerated protein, wool, tussah silk

Regenerated protein: Dissolves in trypsin at 40°C, within 15 min

Tussah silk: Dissolves in trypsin at 40°C, within 18 min; 60% w/w H_2SO_4, cuprammonium hydroxide

A.8.4   Group 4

Fibres soluble within 8 min in 1,4 butyrolacetone: copolymer of vinyl, chlorinated vinyl acetate, poly(vinyl chloride), modacrylic

Chlorinated vinyl acetate: Dissolves in chloroform within 30 s, soluble in xylene, soluble in dioxan 80°C, formdimethylamide at boil

Modacrylic: Dissolves in commercial xylene in 5 min, formdimethylamide at boil, chlorinated poly(vinyl chloride) – soluble in commercial xylene, soluble in dioxan 80°C, formdimethylamide at boil

A.8.5   Group 5

Fibres soluble within 10 min in concentric sulphuric acid: nylon 11, acrylic, regenerated – cellulose bleached, cotton, acetylated cotton, polyester

Nylon 11: Soluble in 3 min in m-cresol, 60% w/w sulphuric acid at 60°C it gels, Soluble in formdimethylamide at boil.

Acrylic: Soluble in 3 min in concentric nitric acid, in 75% sulphuric acid, soluble in formdimethylamide at boil

Regenerated cellulose, bleached cotton: Soluble in 60% w/w sulphuric acid in 7 min, soluble in 75% w/w sulphuric acid

Acetylated cotton: Soluble in 15 min in cuprammonium hydroxide, soluble in 75% w/w sulphuric acid

Polyester: Soluble in 75% w/w sulphuric acid in 20 min

A.8.6   Group 6

Fibres not soluble in all the above solvents: polyvinyl chloride, copolymer of vinylidene chloride–vinylchloride, polyolefin, polyvinyl chloride, fluorocarbon

Polyvinyl chloride: Soluble in 1 min in tetrahydrofuran, soluble in dioxan 80°C, commercial xylene at boil, formdimethylamide at boil

Copolymer of vinylidene chloride–vinylchloride: Soluble in dioxan at 80°C, commercial xylene at boil, formdimethylamide at boil

Fluorocarbon: Formdimethylamide at boil in 30 s

Note: All wherever soluble means soluble at room temperature unless otherwise mentioned.

A.9   Fibre identification by Shirlastains

The use of Shirlastains A, B, C, D and E in identifying fibres is briefly given below with fibre identified and the colours produced.

A.9.1   Shirlastain A

This stain is developed for the identification of natural and regenerated fibres. The pure fabric or yarn (i.e., uncontaminated by size, fillings, resins and dyes) should be thoroughly wetted out and immersed in a cold solution of Shirlastain A. The sample is thoroughly washed off in cold water and compared with the multifibre test strip. Table A.5 gives the colour produced by various fibres which can be identified by Shirlastain A.

Table A.5 Fibres identified by Shirlastain A

Fibre	Colour	Fibre	Colour
Natural fibres		Degummed Silk	Brownish Orange
Raw cotton	Pale purple	Tussah	Chestnut Brown
Scoured cotton	Lilac	*Regenerated fibres*	
Mercerized cotton	Purple	Cellulose rayon	Pink (Cold) Purple (Boil)
Acetylated cotton	Off-white	Cupra. rayon	Blue
Boiled linen	Dark purplish grey	Cellulose acetate	Bright greenish Yellow
Fully boiled linen	Violet blue	Cellulose triacetate	Unstained (Cold)Bt. Yellow (Boil)
Raw hemp	Dark purplish grey (bright)	Fibrolene	Orange Yellow (Cold) Black (Boil)
Fully boiled hemp	Reddish violet blue	*Synthetic fibres*	
Raw ramie	Lavender	Polyamide	Pale Dull Yellow
Boiled ramie	Deep lavender	Polyester	Unstained
Raw jute	Golden brown	Acrylic	Unstained
Boiled jute	Bronze purple	POP	Unstained
Wool	Golden yellow (cold), copper brown (boil)	Polyester	Unstained
Chlorinated wool	Orange (cold), black (boil)		
Gum silk	Dark brown		

A.9.2   Shirlastain C

This stain gives further discrimination between the various cellulosic fibres. The test is conducted as for Shirlastain A.

Table A.6 Differentiation of various cellulosic fibres using Shirlastain C.

Fibre	Colour
Raw cotton	Mauve to reddish brown
Scoured cotton	Off-white to greyish pink
Bleached cotton	Pink
Grey cotton	Pink
Unretted flax fibre	Dark green to reddish brown
Retted flax fibre	Light greenish yellow to dull grey-green
Unbleached flax yarn	Grey
Boild flax yarn	Light grey, or green, flecked pink
Flax canvas	Dull grey tinged pink
Unbleached hemp	Dark reddish or greenish grey
Bleached hemp	Pale pinkish grey or dull pink
Raw jute fibre	Dark or brownish green
Raw jute yarn	Purplish grey
Boild jute yarn	Very dark green
Raw ramie	Slight greyish purple
Boild ramie	Yellowish green

A.9.3   Shirlastain D

This is a powder and after dissolving in water, it is used for discriminating cotton from regenerated cellulose.

A.9.4   Shirlastain E

This is also a powder. The stain is made by dissolving 0.1 g of the powder in 50 ml of dilute acid (5 ml of 2N sulphuric acid and 45 ml distilled water). In use, the stain is brought to boil, stirring all the time and the pure sample is immersed in the boiling liquor for 2 min. After washing of in warm water the sample should be compared with multifibre strip (Table A.7).

Table A.7 Fibre differentiation using Shirlastain E

Fibre	Colour
Cotton	Dull pink
Viscose rayon	Dull pink
Wool	Dark green
Silk	Dark green
Secondary acetate	Orange

Triacetate	Yellowish green
Polyamide 66	Light brown
Polyamide 66	Dark brown
Regular polyester	Cream
Acrilan 1656	Grey
Courtelle	Dull yellow
Orlon 42	Red

It is better to remove any finishing or additions on the fibre, fabric before testing. Since synthetic resin finishing and others may affect the analysis considerably or render it completely inaccurate.

Table A.8 Fibre identification of fibres using alkali and acids

Test à Fibre	Soda ash 40% sol.	Caustic soda 25% sol.	Sodium hypo chloride	Hydrochloric acid 40%	Nitric acid 15%	Nitric acid 70%	Sulphuric acid 15%	Sulphuric acid 70%	Burning in flame
Cotton	Swells	Swells, Shines	Whitened	Turns yellowish	Opens up	Looses strength Dissolves slowly	Dissolves on heating	Dissolves quickly	Burns continuously leaving grey ash of burning paper smell
Jute	-do-	-do-	-do-	–	–	-do-	-do-	Dissolves	-do-
Coir	–	–	Colour turns pale	–	–	–	Dissolves on prolonged heating	Dissolves slowly	-do-, black ash
Viscose	Swells	Swells and slowly dissolves	Gets weakened	Turns yellow	Dissolves on heating	Dissolves	Dissolves	Dissolves quickly	Burns continuously leaving grey ash of burning paper smell
Silk	Looses strength	Dissolves	Dissolves slowly	–	–	Dissolves partially	–	Dissolves	Self extinguishing, leaves crushable, black beads
Wool	-do-	-do-	Dissolves	–	–	Dissolves slowly	–	Dissolves slowly	Self extinguishing, leaves crushable, black beads, Fish
Polyester	–	–		–	–	Dissolves slowly on prolonged treatment	Transparent hard beads	Dissolves slowly	Burns, stops leaving semi
Acrylic	–	–	–	–	–	Looses strength, dissolves slowly	–	Turns yellowish brown	Dissolves, turns yellowish brown
Nylon	–	–	–	–	–	Looses strength	–	Dissolves slowly	Burns, stops, out of flame, leaving dark hard beads

Table A.9 Methods of identifying synthetic fibres – solvent method

Solvent test pathway	Fibre
Acetone: soluble. Acetone/water 80:20 5 min, cold — **Soluble**	Secondary acetate
Acetone/water cold — Insoluble; glacial acetic acid, boiling — **Soluble**	Triacetate
Chloroform, 5 min, cold — Soluble; Cyclohexanon boiling — **Soluble**	Dyenel
Chloroform, 5 min, cold — Soluble; Cyclohexanon boiling — **Insoluble**	Polyamide 11
Chloroform, 5 min, cold — Insoluble; Dimethyl formamide/formic acid 85% 75:25 boiling — **Soluble**	Polyamide 6
Chloroform, 5 min, cold — Insoluble; Dimethyl formamide/formic acid 85% 75:25 boiling — **Insoluble**	Polyamide 66
Tetrahydrofuran, cold — **Soluble**	Polyvinyl chloride
Tetrahydrofuran, cold — Insoluble; Dimethyl formamide, hot — **Soluble**	Acrylic
Dimethyl formamide, hot — Insoluble; Xylene, hot — **Soluble**; Nitrobenzene, boiling — **Insoluble**	Polypropylene
Dimethyl formamide, hot — Insoluble; Xylene, hot — Insoluble; Nitrobenzene, boiling — **Soluble**	Polyester

Table A.10 Guide for dissolving the second component in acrylic (and other) blends

Solvents	Treatment temp. °C	Time (min)	Polyacrylonitrile	Spun rayon	Cotton	Wool	Polyamide 6	Polyamide 66	Polyester	Cellulose acetate	Cellulose triacetate
Acetone	57	2								L	
Gl. acetic acid	20–25	2								L	(L)
Formic acid 85%	20–25	2					L	L		L	(L)
Phosphoric acid 85%	20–25	45		(L)	(L)		L	L		(L)	(L)
Sulphuric acid 67%	20–25	45		L	L		L	L		L	L
Glacial acetic acid: H2SO4:H2O (5:3:1)	50–55	20		L	L		L	L		L	L
Glacial acetic acid	114–116	2					L	L		L	L
Phosphoric acid 85%	110–120	2		L	L		L	L		L	L
Sulphuric acid 67%	110–130	1		L	L		L	L		L	L
Formic acid 85%	105–107						L	L		L	L
Caustic soda 8%	100–102	1				L					
Calc. Thiocyanate Sat. Soln.	95	5					(L)	(L)		(L)	(L)

Note: Blank – not soluble; (L) – slightly soluble; L – soluble.

Table A.11 Methods of identifying fibres – solvent method II

Acetone, cold 10 min						
Soluble						
Dilute solution with water	Insoluble	Caustic soda 5% boil 3 min		Insoluble		
Clouding	Soluble	Cool solution, acidify with HCl acid	Rinse sample with water, glacial acetic acid, boil 1 min			
	Clouding	Soluble	Cool solution and dilute with water	Tetra hydrofuran, cold		
		Insoluble				
	Lead acetate test	Positive				
		Negative	Clouding	Soluble	Dilute with water	
				Insoluble	Dimethyl formamide, boil for 2 min	

Table continued...

Table A.11 Continue

	Clouding	Soluble	Insoluble				
		Cool, dilute with water and add 2 drops of NaOH, 36 °Be	Xylene boil				
			Flocculation	Soluble	Insoluble		
				Cool solution	Nitrobenzene, boil, 2 min		
				Flocculation	Soluble	Insoluble	
					Cool Solution		
					Flocculation		
Secondary acetate, post chlorinated PVC	Wool	Silk	Polyamide	Polyvinyl chloride	Acrylic	Polypropylene	Polyester	Cellulosic

Table A.12 Fibre identification by action of acid and alkali

	Soda ash 45% Soln.	Caustic soda 26% Soln.	Sodium hypochlorite	Hydroch loric acid 40%	Nitric acid 16%	Nitric acid 70%	Sulphuric acid 16%	Sulphuric acid 70%	Remarks
Cotton	Swells	Swells and shines	Whitened	Turns yellowish	Opens up	Weakens and dissolves slowly	Dissolves on heating	Dissolves quickly	Resistance to alkalis
Jute	-do-	-do-	-do-	-do-	-do-	-do-	-do-	Dissolves	Rough handle
Coir			Colour turns pale				Dissolves on prolonged heating	Dissolves slowly	
Viscose	Swells	Swells, slowly dissolves	Gets weakened	Turns Yellow	Dissolves on heating	Dissolves	Dissolves	Dissolves quickly	Soft filaments good lustre
Silk	Looses strength	Dissolves	Dissolves slowly				Dissolves slowly	Dissolves	Lustrous filaments
Wool	-do-	-do-	Dissolves			Dissolves slowly		Dissolves slowly	Rough crimpy fibres
Polyester						Dissolves slowly on long treatment	Transparent hard beads	Dissolves slowly	Resistance to chemicals
Acrylic						Weakens, dissolves slowly		Turns yellowish brown	Translucent, more voluminous
Nylon						Looses strength		Dissolves slowly	Strong more elastic

Table A.13 Identification of fibres by solubility in acids and solvents

	Acetic acid	Acetone	Sodium hypochlorite (5%)	Hydrochloric acid (20%)	Formic acid (85%)	1.4 Dioxane	m-xylene	Cyclohexanone	Dimethyl formamide	Sulph. acid (59.5%)	Sulph. acid (70%)	m-cresol
Temp (°C)	20	20	20	20	20	101	139	156	90	20	38	139
Time (min)	5	5	20	10	5	5	5	5	10	20	20	5
Cotton	Insoluble	Insoluble	Insoluble	Insoluble	Insoluble	Insoluble	Insoluble	Insoluble	Insoluble	Insoluble	Soluble	Insoluble
linen	Insoluble	Insoluble	Insoluble	Insoluble	Insoluble	Insoluble	Insoluble	Insoluble	Insoluble	Insoluble	Soluble	Insoluble
Wool	Insoluble	Insoluble	Soluble	Insoluble	Insoluble	Insoluble	Insoluble	Insoluble	Insoluble	Insoluble	Insoluble	Insoluble
Silk	Insoluble	Insoluble	Insoluble	Insoluble	Insoluble	Soluble	Soluble	Soluble	Soluble	Insoluble	Insoluble	Soluble
Polyester	Insoluble	Insoluble	Insoluble	Insoluble	Insoluble	Insoluble	Insoluble	Insoluble	Insoluble	Soluble	Soluble	Insoluble
Rayon	Insoluble	Insoluble	Insoluble	Insoluble	Insoluble	Insoluble	Insoluble	Insoluble	Insoluble	Soluble	Soluble	Soluble
Acetate	Soluble	Soluble	Insoluble	Insoluble	Soluble	Soluble	Insoluble	Soluble	Soluble	Soluble	Insoluble	Insoluble
Acrylic	Insoluble	Insoluble	Insoluble	Insoluble	Insoluble	Insoluble	Insoluble	Insoluble	Soluble	Insoluble	Insoluble	Insoluble
Anidex	Insoluble	Insoluble	Insoluble	Insoluble	Insoluble	Insoluble	Insoluble	Insoluble	Insoluble	Insoluble	Insoluble	Insoluble
Aramid	Insoluble	Insoluble	Insoluble	Insoluble	Insoluble	Insoluble	Insoluble	Insoluble	Insoluble	Insoluble	Insoluble	Soluble
Nylon	Insoluble	Insoluble	Insoluble	Soluble	Soluble	Insoluble	Insoluble	Insoluble	Insoluble	Soluble	Soluble	Soluble
Nytril	Insoluble	Insoluble	Insoluble	Insoluble	Insoluble	Insoluble	Insoluble	Soluble	Insoluble	Insoluble	Insoluble	Insoluble
Olefin	Insoluble	Insoluble	Insoluble	Insoluble	Insoluble	Insoluble	Soluble	Soluble	Insoluble	Insoluble	Insoluble	Insoluble
Spandex	Insoluble	Insoluble	Insoluble	Insoluble	Insoluble	Insoluble	Insoluble	Insoluble	Soluble	Insoluble	Insoluble	Insoluble
Teflon	Insoluble	Insoluble	Insoluble	Insoluble	Insoluble	Insoluble	Insoluble	Insoluble	Insoluble	Insoluble	Insoluble	Insoluble
Vinyon	Insoluble	Soluble	Insoluble	Insoluble	Insoluble	Soluble	Soluble	Soluble	Soluble	Insoluble	Insoluble	Soluble

A.10   Quick identification of common fibres by burning test

Cellulose acetate (CA) burn with an acidic odour and melts at the same time. A hard ball remains. The thread can be broken easily. Soluble in acetone and glacial acetic acid.

Cotton (Co) burn readily and leave behind little powdery ashes. Odour of burning paper when flame is extinguished and left to smoulder.

Viscose (CV) has similar properties like cotton, except for the poor wet strength.

Polyamide (PA) (nylon) melts immediately with a 'celery' like smell when burn. Leave a glossy hard white to brown ball. Soluble in formic acid.

Polyester (PES) is difficult to burn and does so with black smoke and a sharp aromatic odour. A hard ball remains (dark in colour). Take time to burn/ melt than acrylic fibres.

Polyacrylonitrile (PAC) (acrylic) burns very quickly and with a sooty smoke. Melts and drip. Burning smell is very sweet and the residue is brittle. Soluble in concentrated H_2SO_4.

Wool (WO) burns poorly with the odour of burning hair and the flame dies when source is removed. The burnt section is thicker and brittle, so it can be powdered between fingers (Fig. A.1).

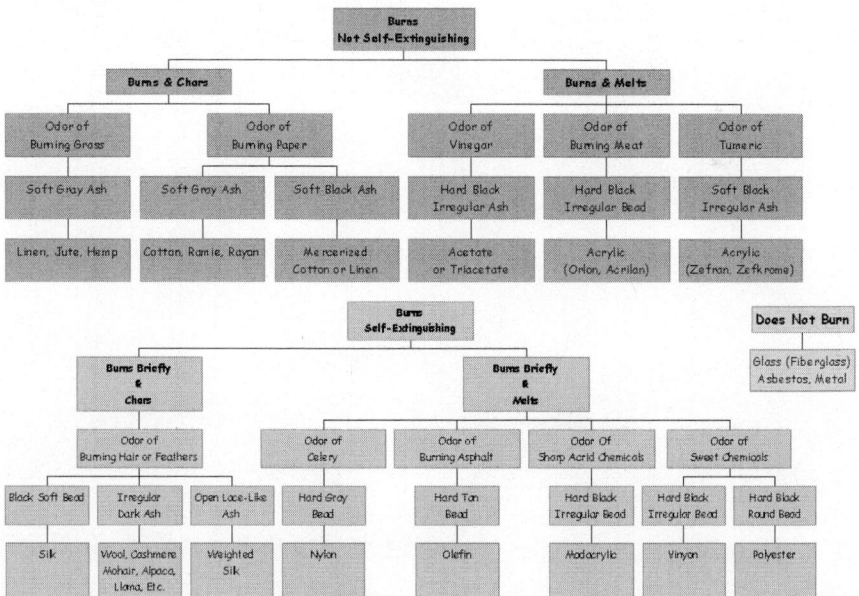

Figure A.1. Fibre burn chart

References

1. Andreoli, C. Reference Book of Textile Technologies – Manmade Fibre.

2. Basu, A. (Ed.). 2015. Advances in Silk Science and Technology. Woodhead Publishing Limited, Cambridge, UK.

3. Bille, H., Feinauer, A., Gebert, K., Kirner, U., Leube, H., Richter, P., Huttiqer, W., Schbnpflug, E., Thurner, K., Weigold, T.S., Wolf, H., Wurz, A. 1968. Symposium on the Dyeing and Finishing of Polyester Fibres and their Blends.

4. Bralla, J.G. 2007. Handbook of Manufacturing Processes: How Products, Components and Materials are Made. Industrial Press.

5. Bunsell, A.R. (Ed.). 2009. Handbook of Tensile Properties of Textile and Technical Fibres. Woodhead Publishing Limited, Cambridge, UK.

6. Cashmilon Technical information. 2003. Cashmilon in General, Asahi Kasei.

7. Chawla, K.K. 1998. Fibrous Materials. The Press Syndicate, University of Cambridge, UK.

8. Cheremisinoff, N.P. (Ed.). 1998. Advanced Polymer Processing Operations. Noyes Publications, Westwood, NJ, USA.

9. Cook, J.G. 2001a. Handbook of Textile Fibres. Vol. I. Natural Fibres. Woodhead Publishing Limited, Cambridge, UK.

10. Cook, J.G. 2001b. Handbook of Textile Fibres. Vol. II. Man-Made Fibres. Woodhead Publishing Limited, Cambridge, UK.

11. Cristina, P. 2011. Polyurethane Elastomers From Morphology to Mechanical Aspects. Springer-Verlag/Wien, New York, USA.

12. Deopura, B.L., Alagirusamy, R., Joshi, M., Gupta, B. (Eds.). 2008. Polyesters and Polyamides. Woodhead Publishing Limited, Cambridge, UK.

13. Eichhorn, S.J., Hearle, J.W.S., Jaffe, M., Kikutani, T. (Eds.). 2009. Handbook of Textile Fibre Structure – Vol. 1: Fundamentals and Manufactured Polymer Fibres. Woodhead Publishing Limited, Cambridge, UK.Franck, R.R. (Ed.). 2001. Silk, Mohair, Cashmere and Other Luxury Fibres. 2001. Woodhead Publishing Limited, Cambridge, UK.

14. Franck, R.R. (Ed.). 2005. Bast and Other Plant Fibres. Woodhead Publishing Limited, Cambridge, UK.

15. Freti, F. 2004. Fondazione. ACIMIT, Milano, Italy.

16. Gandhi, K.L. (Ed.). 2012. Woven Textiles Principles, Developments and Applications. Woodhead Publishing Limited, Cambridge, UK.

17. Gokarneshan, N. 2004. Fabric – Structure Design. New Age International (P) Ltd., Publishers, New Delhi, India.

18. Gordon, S., Hsieh, Y.L. (Eds.). 2006. Cotton: Science and Technology. Woodhead Publishing Limited, Cambridge, UK.

19. Hongu, T. 2001. New Fibers. Woodhead Publishing Limited, Cambridge, UK.

20. Hongu, T., Phillips, G.O. 2001. New Fibers. Woodhead Publishing Limited, Cambridge, UK.

21. Hongu, T., Phillips, G.O., Takigami, M. 2005. New Millennium Fibers. Woodhead Publishing Limited, Cambridge, UK.

22. Hongu, T., Phillips, G.O., Takigami, M. 2005. New Millennium Fibres. Woodhead Publishing Limited in association with The Textile Institute, Cambridge, UK.

23. Jinlian Hu. 2011. The Hong Kong Polytechnic University, Hong Kong, Adaptive and Functional polymers, Textiles and their Applications. Imperial College Press, London, UK.

24. Johnson, N.A.G., Russell, I.M. (Eds.). 2008. Advances in Wool Technology. 2008. Woodhead Publishing Limited, Cambridge, UK.

25. Kaplan, N.S. 2009. A Practical Guide to Fibre Science. Abhishek Publications, Chandigarh, India.

26. Karger-Kocsis, J. (Ed.). 1999. Polypropylene An A-Z Reference. Institute for Composite Materials Ltd. University of Kaiserslau tern Germany and Technical University of Budapest Hungary, Kluwer Academic Publishers, Dordrecht, The Netherlands.

27. Kothari, V.R. 2011a. Application of Contemporary Fibres in Apparels. Coconut Shell Fibre, Apparel Views, July.

28. Kothari, V.R. 2011b. Application of Contemporary Fibres in Apparels. Coconut Shell Fibre, Apparel Views, September.

29. Kothari, V.R. 2011c. Application of unconventional fibres in apparels. Indian Textile Journal.

30. Kothari, V.R. 2011d. New Fibres for Home Textiles, Home Textile Views, April–June.

31. Kozłowski, R.M. Woodhead Publishing Limited, Cambridge, UK.

32. Kozłowski, R.M. (Ed.). 2012a. Handbook of Natural Fibres. Vol. 1: Types, Properties and Factors Affecting Breeding and Cultivation. Woodhead Publishing Limited, Cambridge, UK.

33. Kozłowski, R.M. (Ed.). 2012b. Handbook of Natural Fibres. Vol. 2: Processing and Applications.

34. Lewin, M. (Ed.). 2007. Handbook of Fiber Chemistry. CRC Press, Taylor & Francis Group, Boca Raton, FL, USA.

35. Lord, P.R., Mohamed, M.H. 1992. Weaving. Conversion of Yarn to Fabric. Merrow Publishing Co. Ltd., Shildon, Co., Durham, UK.

36. Lyocell Technical literature Ix4009E, Ciba Specialty Chemicals Inc., November 1999.

37. Mark, J.E. (Ed.). 1999. Polymer Data Handbook. 1999. Oxford University Press.

38. Melliand Textilberichte, 49, pp. 1053–1093.

39. Morton, W.E., Hearle, J.W.S. 2008. Physical Properties of Textile Fibres. Woodhead Publishing Limited, Cambridge, UK.

40. Murugesh Babu, K. 2013. Silk Processing, Properties and Applications. Woodhead Publishing Limited, Cambridge, UK.

41. Mussig, J. (Ed.). 2010. Industrial Applications of Natural Fibres. Department of Biomimetics, Hochschule Bremen – University of Applied Sciences, Bremen, Germany, John Wiley and Sons.

42. Needles, H.L. 1986. Textile Fibres, Dyes, Fineshes, and Processes. A Concise Guide. Noyes Publications, Park Ridge, NJ, USA.

43. Nemr, A.E. 2011. Textile: Types, Uses and Production Methods. Nova Science Publishers, New York, USA.

44. Nicholson, J.W. 1997. The Chemistry of Polymers. Second Ed. Royal Society of Chemistry, Cambridge, UK.

45. Polyester Spandex Fibre Technology Type S-5 – Technical Information Bulletin, Globe Manufacturing Corporation. 9903.

46. Prof. Dr. rer. nat. Hans-Karl Rouette, 2000. Encyclopedia of Textile Finishing, Springer.Richard Furter, R., Uster, 2004. Application Report – Physical properties of spun yarns.

47. Rouette, H.K. 2000. Encyclopedia of Textile Finishing. Springer.

48. Sadov, F., Adov, M., Korchagi, A., Matetsky, A. 1973. Chemical Technology of Fibrous Materials. MIR Publications, Moscow.

49. Sharma, A. 2007. Processing and Coloration of Bamboo fibres. RSWM Ltd, Rajasthan, India.

50. Simpson, W.S., Crawshaw, G.H. 2002. Wool. Science and Technology. Woodhead Publishing Limited, Cambridge, UK.

51. Spandex Fiber Technology – Polyether Type S-85, Technical information Bulletin, Globe Manufacturing Corporation. 9903.

52. Tao, X. (Ed.). 2001. Smart Fibres, Fabrics and Clothing. Woodhead Publishing Limited, Cambridge, UK.

53. Technical Bulletin – Cotton Spun Yarns for Knit and Woven Fabrics. 2003. ISP 1006, Cotton Inc, Cary, NC, USA.

54. The Dyeing of Polyester Fibres. 1964. Imperial Chemical Industries Limited, Dyestuff Division.

55. Ugbolue, S.C.O. (Ed.). 2009. Polyolefin Fibres Industrial and Medical Applications. Woodhead Publishing Limited, Cambridge, UK.

56. Vassiliadis, S. 2011. Advances in Modern Woven Fabrics Technology. InTech, Rijeka, Croatia.

57. Wakelyn, P.J., Bertoniere, N.R., French, A.D., Thibodeaux, D.P., Triplett, B.A., Marie-Alice, R., Goynes, Jr., W.R., Edwards, J.V., Hunter, L., McAlister, D.D., Gamble, G.R. 2007. Cotton Fiber Chemistry and Technology, CRC Press, Taylor & Francis Group, New York, USA. Encyclopedia of Polymer Science and Technology. John Wiley & Sons, Inc.

58. Woodings, C. (Ed.). 2001. Regenerated Cellulose Fibres. Woodhead Publishing Limited, Cambridge, UK.

59. Yehia El Mogahzy, Understanding the Fiber-to-Yarn Conversion System Part II: Yarn Characteristics (elmogye@auburn.edu).